HUMAN GENETICS

2nd edition

HUMAN GENETICS

2nd edition

Anne Gardner and Teresa Davies

Bristol Genetics Laboratory, Southmead Hospital, Bristol, UK

Viva Books

New Delhi | Mumbai | Chennai | Kolkata
Bangalore | Hyderabad | Kochi | Guwahati

First Indian Edition 2010
Reprinted 2012

VIVA BOOKS PRIVATE LIMITED

- 4737/23 Ansari Road, Daryaganj, New Delhi 110 002
  E-mail: vivadelhi@vivagroupindia.net, Tel. 42242200
- Plot No. 76, Service Industries, Shirvane, Sector 1, Nerul, Navi Mumbai 400 706
  E-mail: vivamumbai@vivagroupindia.net, Tel. 27721273, 27721274
- Jamals Fazal Chambers, 26 Greams Road, Chennai 600 006
  E-mail: vivachennai@vivagroupindia.net, Tel. 28290304, 28294241
- B-103, Jindal Towers, 21/1A/3 Darga Road, Kolkata 700 017
  E-mail: vivakolkata@vivagroupindia.net, Tel. 22836381, 22816713
- 7, Sovereign Park Apartments, 56-58, K. R. Road, Basavanagudi, Bangalore 560 004
  E-mail: vivabangalore@vivagroupindia.net, Tel. 26607409, 26607410
- 101-102 Mughal Marc Aptt., 3-4-637 to 641, Narayanguda, Hyderabad 500 029
  E-mail: vivahyderabad@vivagroupindia.net, Tel. 27564481, 27564482
- First Floor, Beevi Towers, SRM Road, Kaloor, Kochi 682 018, Kerala
  E-mail: vivakochi@vivagroupindia.net, Tel: 0484-2403055, 2403056
- 232, GNB Road, Beside UCO Bank, Silpukhuri, Guwahati 781 003
  E-mail: vivaguwahati@vivagroupindia.net, Tel: 0361-2666386

Published by arrangement with

Scion Publishing Limited
Bloxham Mill, Barford Road,
Bloxham OX15 4FF
UK

ISBN : 978-81-309-1364-3

Published by Vinod Vasishtha for Viva Books Private Limited, 4737/23, Ansari Road, Daryaganj New Delhi - 110 002. Printed and bound by Raj Press, R-3, Inderpuri, New Delhi - 110 012.

Contents

Preface

We were pleased to be given the opportunity to write this new edition and update each chapter, but we were surprised at the extent of revision needed in certain areas.

Genetics has always been a constantly changing discipline, illustrated by both scientific discoveries and technical innovation. We now know that humans have considerably fewer genes than was previously thought, and genomic imprinting has been subsumed into the expanding area of epigenetics. Perhaps the greatest change in genetic analysis has been the introduction of automated techniques, leading to high-throughput screening and sequencing in molecular genetics, and the use of microarrays in molecular cytogenetics.

As in the first edition, the content should be relevant to both first year undergraduates and newcomers to the medical profession. This book is intended as an introduction to human genetics, to supplement other basic genetics textbooks available elsewhere. We hope that this new edition will encourage students to explore areas of interest to them in more depth – and we feel sure that genetics will prove as fascinating to our readers as it always has to those working in the field.

Anne Gardner and Teresa Davies
May 2009

Acknowledgements

We would like to thank the staff at the Bristol Genetics Laboratory for supplying the various images used in this book, Martin Davies for providing some diagrams and also Rod Howell for his previous contribution to the first edition.

Abbreviations

ADA	adenosine deaminase
AFP	alpha-fetoprotein
ALL	acute lymphoblastic leukaemia
AML	acute myeloid leukaemia
APKD	adult polycystic kidney disease
APL	acute promyelocytic leukaemia
ARMS	amplification refractory mutation system
ARPKD	autosomal recessive polycystic kidney disease
AS	Angelman syndrome
ATLL	adult T-cell leukaemia–lymphoma
ATP	adenosine triphosphate
BAC	bacterial artificial chromosome
BCC	basal cell carcinoma
BMT	bone marrow transplantation
BSA	bovine serum albumin
CDK	cyclin dependent kinase
CEP	chromosome enumeration probe
CF	cystic fibrosis
CFTR	cystic fibrosis transmembrane conductance regulator
CGH	comparative genomic hybridization
CHT	congenital hypothyroidism
CML	chronic myeloid leukaemia
CNV	copy number variation
CPM	confined placental mosaicism
CSCE	capillary conformation sensitive electrophoresis
CVS	chorionic villus sampling
DECIPHER	Database of Chromosomal Imbalance and Phenotype in Humans using Ensembl Resources
DGGE	denaturing gradient gel electrophoresis
dNTP	deoxynucleoside triphosphate
DSS	dosage-sensitive sex reversal
ECARUCA	European Cytogeneticists Association Register of Unbalanced Chromosome Aberrations
EPR	enzymatic photoreactivation
ESAC	extra structurally abnormal chromosome
Fab	fragment antigen binding
FAPC	familial adenomatous polyposis coli
FBS	fetal blood sampling
Fc	fragment crystallizable
FISH	fluorescence *in situ* hybridization

FITC	fluorescein isothiocyanate
FSHD	facioscapulohumeral dystrophy
GIFT	gametes intra-Fallopian transfer
HA	heteroduplex analysis
HCG	human chorionic gonadotrophin
HD	Huntington disease
HFEA	Human Fertilisation and Embryology Authority
HGVS	Human Genome Variation Society
HNPCC	hereditary non-polyposis colon cancer
HSR	homogeneously staining region
Ig	immunoglobulin
IS	International Scale
ISCN	International System for Human Cytogenetic Nomenclature
ITD	internal tandem duplication
IUGR	interuterine growth retardation
IVF	*in vitro* fertilization
LINES	long interspersed nuclear elements
LMP	last menstrual period
LOH	loss of heterozygosity
MCADD	medium chain acyl-CoA dehydrogenase deficiency
MDS	myelodysplastic syndromes
MEN	multiple endocrine neoplasia
M-FISH	multicolour FISH
MHC	major histocompatibility complex
MLPA	multiple ligation-dependent probe amplification
MMR	mismatch repair
MOM	multiples of the median
MPD	myelodysplastic/myeloproliferative disease
MRC	Medical Research Council
MRD	minimal residual disease
mRNA	messenger RNA
MSAFP	maternal serum alpha-fetoprotein
MUD	matched unrelated donor
NBCC	naevoid basal cell carcinoma
NER	nucleotide excision repair
NIPD	non-invasive prenatal diagnosis
NOR	nucleolar organizer
NTD	neural tube defect
OLA	oligonucleotide ligation assay
PAC	P1 phage-derived artificial chromosome
PAPPA	pregnancy-associated plasma protein A
PAR	pseudo-autosomal region
PCR	polymerase chain reaction
PEO	pentaethylene oxide
PIGD	pre-implantation genetic diagnosis
PKU	phenylketonuria
PWS	Prader–Willi syndrome
qPCR	real time (quantitative) PCR

QF-PCR	quantitative fluorescence PCR
RFLP	restriction fragment length polymorphism
rRNA	ribosomal RNA
RSV	Rous sarcoma virus
RT-PCR	reverse transcriptase PCR
SCD	sickle cell disease
SCE	sister chromatid exchange
SCID	severe combined immunodeficiency disease
SDS	sodium dodecyl sulphate
SINES	short interspersed nuclear elements
SNP	single nucleotide polymorphism
SSCP	single-stranded conformational polymorphism
STR	short tandem repeat
SVAS	supravalvular aortic stenosis
TCR	T-cell receptor
TF	transcription factor
TGF	transforming growth factor
TNF	tumour necrosis factor
TS	tumour suppressor
TSH	thyroid stimulating hormone
UPD	uniparental disomy
UTR	untranslated region
VCFS	velo-cardio-facial syndrome
VNTR	variable number of tandem repeats
XIST	X-inactivation-specific transcript
XP	xeroderma pigmentosum
YAC	yeast artificial chromosome

CHAPTER 1

Organization of DNA

Learning objectives

After studying this chapter you should confidently be able to:

- **Outline the levels of secondary structure of DNA leading to the chromosome**
  DNA is the hereditary information stored in the cell nucleus. In association with histone proteins, it is condensed into 46 chromosomes.
- **Describe the internal structure of the DNA molecule**
  DNA is an anti-parallel double helix, comprising an outer sugar–phosphate backbone and four internal bases held together by hydrogen bonding in A:T and G:C base pairs. This ensures that one DNA strand is complementary to another, and can therefore be replicated accurately.
- **Outline the principles of DNA replication**
  Each pre-existing (**conserved**) DNA strand serves as a pattern or **template** for the new complementary strand.
- **List the phases of the cell cycle**
  At interphase the cell cycle is divided into four phases called G_0, G_1, G_2 and S. Mitosis (M) is the stage at which cell division occurs.
- **Describe the stages of mitosis and meiosis and explain the resultant numbers of chromosomes**
  There are two kinds of cell division; mitosis is somatic cell division and results in 46 chromosomes per cell, meiosis occurs in germ cells and results in 23 chromosomes per cell.
- **Define exons and introns with respect to a typical gene, and describe the organization of genomic and repetitive DNA**
  Genes comprise conserved coding exons and polymorphic non-coding introns. Repetitive DNA sequences are also present in large quantities.
- **Describe a normal karyotype**
  The 46 human chromosomes are arranged pictorially in pairs in descending size order, with the sex chromosomes at the end. Banding enables each pair to be differentiated from any other pair.

1.1 CHROMOSOME STRUCTURE AND THE DNA MOLECULE

Our bodies consist of around nine billion cells. Most of our cells contain a nucleus with its nuclear membrane, surrounded by cytoplasm. Within the cytoplasm are found various structures known as **cell organelles**.

The nucleus contains the hereditary information stored in the form of **deoxyribonucleic acid (DNA)**. This is known as **nuclear DNA**.

Although the structure and functions of the cytoplasmic organelles are too many to list individually, the mitochondria (singular, mitochondrion) are of interest as they have their own DNA, known as **mitochondrial DNA**.

Eukaryotes, which include such diverse organisms as yeasts and humans, are characterized by DNA contained within a nuclear membrane, and the presence of cell organelles.

Prokaryotes, for example bacteria, have no clear division between nucleus and cytoplasm and may only have one simple 'chromosome' called a nucleoid (see *Box 1.1*).

Box 1.1 Prokaryotes and eukaryotes

Prokaryotes have no nuclear membrane, they have no introns in their genes, and their gene controlling systems are organized to respond to rapid changes in the environment; DNA and RNA are found together. Eukaryotes have more complex control systems. Eukaryotic RNA is transported from the nucleus to the cytoplasm.

When a cell is not dividing, it is in a state known as **interphase**. At this time, the DNA is in a very decondensed state. From the point at which a cell begins to divide, however, the DNA begins to condense into the highly coiled shapes known as **chromosomes**.

The nuclei of body (**somatic**) cells have 46 chromosomes comprising 22 pairs of autosomes and one pair of sex chromosomes, whereas in eggs or sperm (the **germ** cells) there are 23 chromosomes. When fertilization occurs the correct number of chromosomes is restored.

If a chromosome is gradually unravelled, it can be seen that the DNA is supercoiled into a solenoid-like form, which in turn comprises thinner fibres. As these are unwound it can be seen that the DNA is wrapped around protein structures called **nucleosomes** (see *Figure 1.1* and *Box 1.2*). If the histone proteins of the nucleosome are removed, the basic form of DNA known as

Box 1.2 Nucleosomes

A nucleosome comprises eight protein molecules called **histones**. These are basic proteins comprising four pairs called H2A, H2B, H3 and H4 respectively. In its partially folded state, DNA is bound to a fifth histone H1, thought to participate in the control of gene expression by tighter binding of the DNA to prevent transcription.

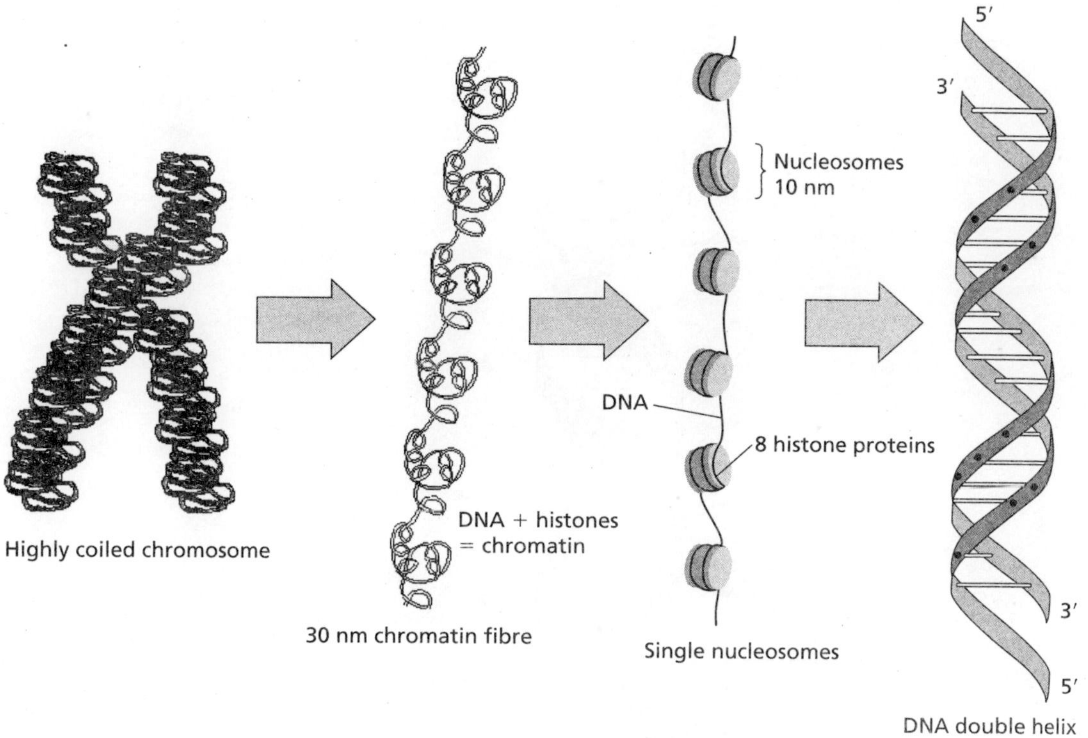

Figure 1.1
Uncoiling a chromosome to reveal the DNA double helix.

the **double helix** is revealed. This comprises two separate strands wound around each other, such that when viewed from above the helix normally appears to spiral in a clockwise or right-handed direction (see *Figure 1.2*). A human diploid cell nucleus (i.e. 46 chromosomes) contains about 2 metres of DNA helix.

The main spiral of the double helix is known as the **sugar–phosphate backbone.** The phosphate is attached to the outer part of a five-sided sugar 'ring' – the deoxyribose molecule. Attached to the inner part of the sugar ring is one of four different nitrogenous bases. The bases are arranged towards the centre of the two single helices, and lie almost flat, in a stacked formation (*Figure 1.3*).

Two **bases** (each comprising two organic rings), adenine (**A**) and guanine (**G**), are called **purines**. The bases cytosine (**C**) and thymine (**T**) (each comprising one organic ring) are called **pyrimidines** (see *Figure 1.4*); sometimes a cytosine may be modified by the addition of a methyl (CH_3) group attached to the fifth atom of the carbon ring, and this cytosine is then said to be methylated (see *Box 1.3* for difference between DNA and RNA).

Organic rings may contain carbon, nitrogen and/or oxygen atoms. The carbons are consecutively numbered in a clockwise direction around the ring

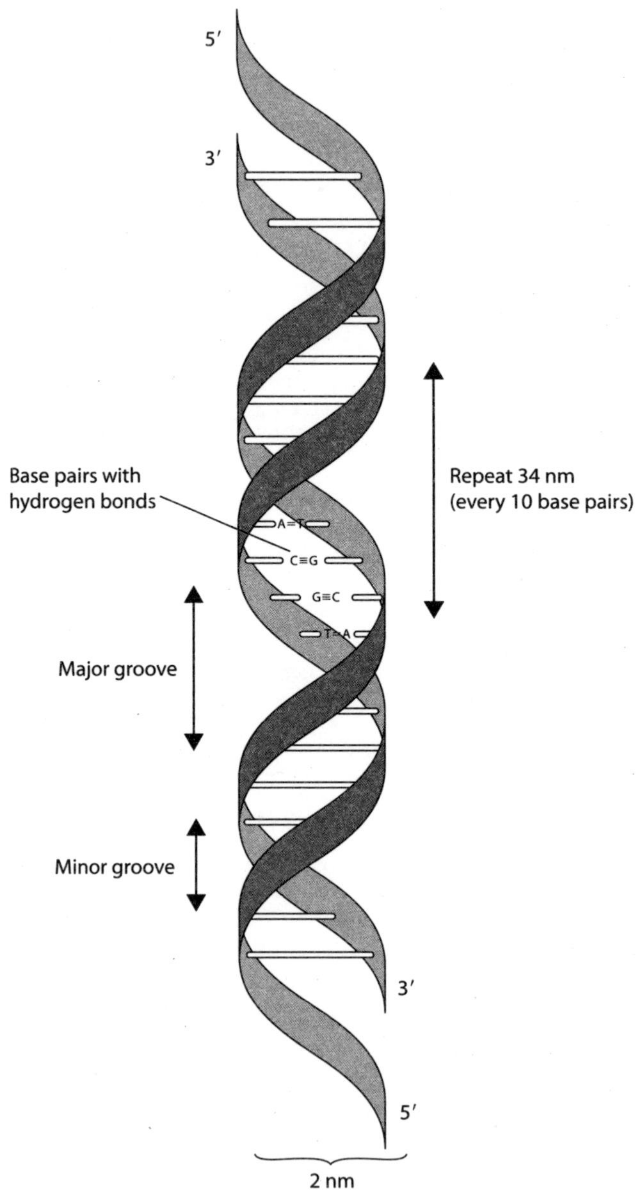

Figure 1.2
The DNA double helix.

of the pentose sugar. It is then possible to describe how various molecules may be attached to each other at a particular carbon position on a ring (see *Figure 1.3*). In *Figure 1.3*, the phosphate group can be seen attached to the 5′ (five **prime**) carbon of the deoxyribose, whereas the base is attached to the 1′ carbon (see *Box 1.4* for details of terminology).

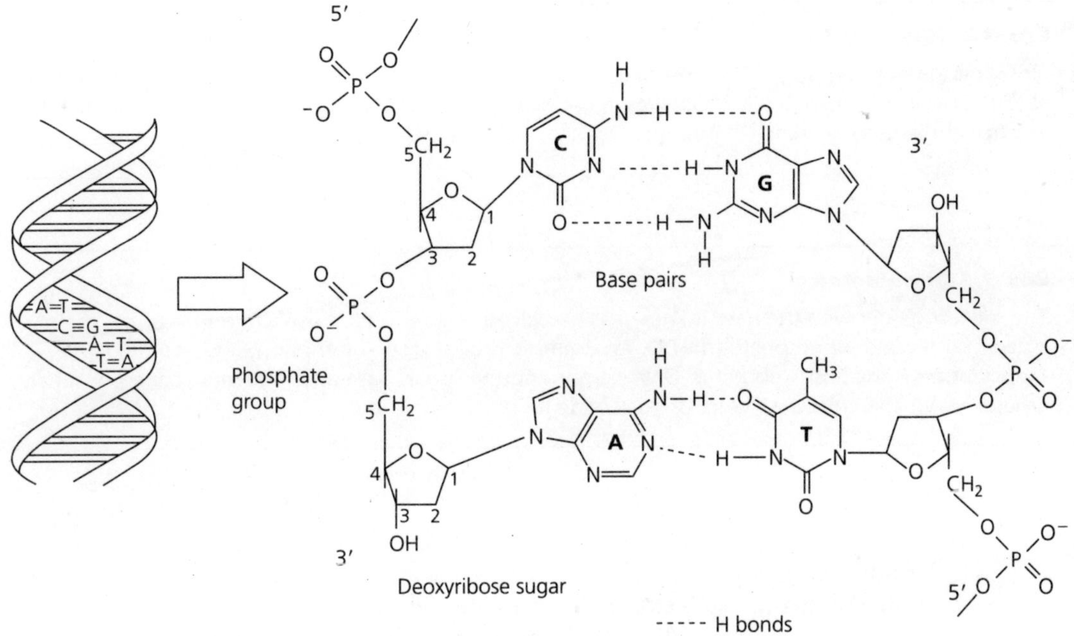

Figure 1.3
Detailed structure of the DNA helix.

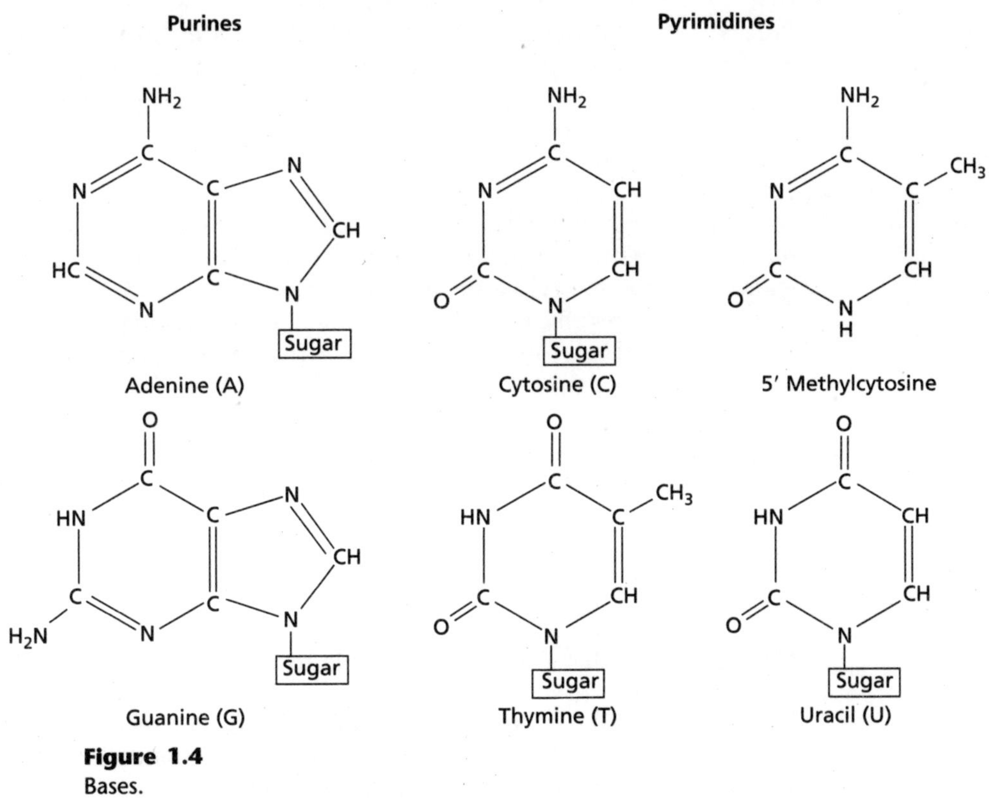

Figure 1.4
Bases.

Box 1.3 RNA and DNA

The sugar ring of messenger RNA (mRNA) is a ribose molecule with a hydroxyl (OH) group attached at the 2′ carbon, instead of a hydrogen as in the deoxyribose of DNA, mRNA uses the base **uracil (U)**, in place of the base thymine (T) found in DNA.

Box 1.4 Terminology

The sugar together with the base is known as a nucleoside; the sugar, base and triphosphate group is a deoxynucleoside triphosphate (**dNTP**). An example is deoxycytidine triphosphate, written dCTP. A **nucleotide** is the basic subunit of DNA comprising the sugar, base and interconnecting phosphate group. DNA is therefore a polymer of nucleotide units.

The 5′ end of one chain of the double helix ends in a 3′ OH group at the other end. The other chain lies **anti-parallel** to the first, as it has the 3′ OH group at the top, next to the 5′ of the first chain. This helix will end in a 5′ phosphate. This polarity is important in DNA replication, and in the action of certain modifying enzymes used in DNA techniques.

The double helix is held together with hydrogen bonds formed between pairs of bases. The base A always pairs with T, using two hydrogen bonds, whereas C pairs with G more strongly, as there are three hydrogen bonds. This is known as **complementarity**. The pairing of an A with a T, or a C with a G creates a **base pair**, written as 'bp' (note that 1000 bp = 1 kilobase (kb), and 1000 kb = 1 Megabase (Mb)).

If two DNA strands are separated (denatured), which can be achieved by heat or an alkaline pH, the hydrogen bonds are broken, but the strands may reassociate if the temperature is lowered or the pH is returned to neutral. This phenomenon is sometimes referred to as **reannealing**, or if it is part of a molecular biology technique, **hybridization**.

Because each strand of a DNA helix is **complementary** to the other, any faulty or missing bases may be replaced, as the opposing base (or sequence of bases) will provide the guidelines for repair.

1.2 DNA REPLICATION

DNA replication is possible because the sequence of bases on one strand of the double helix is complementary to that on the other.

Principle of replication

The double helix has to unwind for replication to take place. Each pre-existing (**conserved**) strand serves as a pattern or **template** for the new complementary strand. Once copying has taken place there will then be four strands: two old strands and two new strands. DNA replication is therefore said to be **semi-conservative**.

Eukaryotic replication

In the eukaryotic cell all the DNA is copied during the **S** (**synthesis**) phase of a continuing **cell cycle** (discussed on the following pages).

Eukaryotic DNA synthesis has more than one start point for replication (see *Box 1.5* for comparison with prokaryotic replication). These replicating units are called **replicons.** At the localized site of a replicon, a helicase enzyme unwinds the DNA helix, and then replication occurs in both directions at a **replication fork** (*Figure 1.5*). Replication proceeds continuously in the 5′ to 3′ direction, with nucleotides attached to the 3′ end. This **continuously formed DNA strand** is called the **leading strand**.

> **Box 1.5 Prokaryotic replication**
>
> A prokaryote such as a bacterium has one circular chromosome with a single point of origin for the start of replication, which proceeds in two directions until there are two rings.

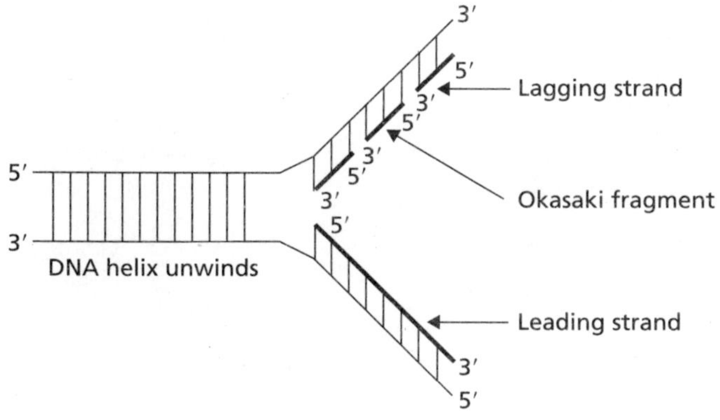

Figure 1.5
DNA replication.

In order that the DNA can be copied from the 3′ to 5′ direction, **short pieces** of around 1000 bases called **Okasaki fragments** are synthesized in the 5′ to 3′ direction; these are then joined together by a ligase (see *Box 1.6* for details on other enzymes involved in DNA manipulation). This DNA sequence is called the **lagging strand.**

DNA synthesis is finally completed when all the individual replicons meet.

Box 1.6 Some enzymes involved in the manipulation of DNA

- A **polymerase** copies DNA or RNA by incorporation of the appropriate complementary base (actually a deoxynucleoside triphosphate – a dNTP); some DNA polymerases can also proof-read by spotting mistakes and removing the wrongly synthesized bases
- An **exonuclease** removes dNTPs
- An **endonuclease** nicks a strand of DNA
- A **ligase** joins together (ligates) two pieces of double-stranded DNA at their sugar–phosphate (phosphodiester) bonds

1.3 THE CELL CYCLE

A somatic cell either exists in a resting stage, a growing stage, or a state of cell division, when it divides into two daughter cells. The progression around the cell cycle is controlled by molecules called **cyclins**, which direct other associated molecules called cyclin dependent kinases (**CDKs**). These enzymes add phosphate groups (this is known as **phosphorylation**) on to certain gene products to regulate their activity.

The process of cell division is called **mitosis**; at all other stages the cell is in **interphase**.

Interphase is divided into several other phases (*Figure 1.6*), as follows.

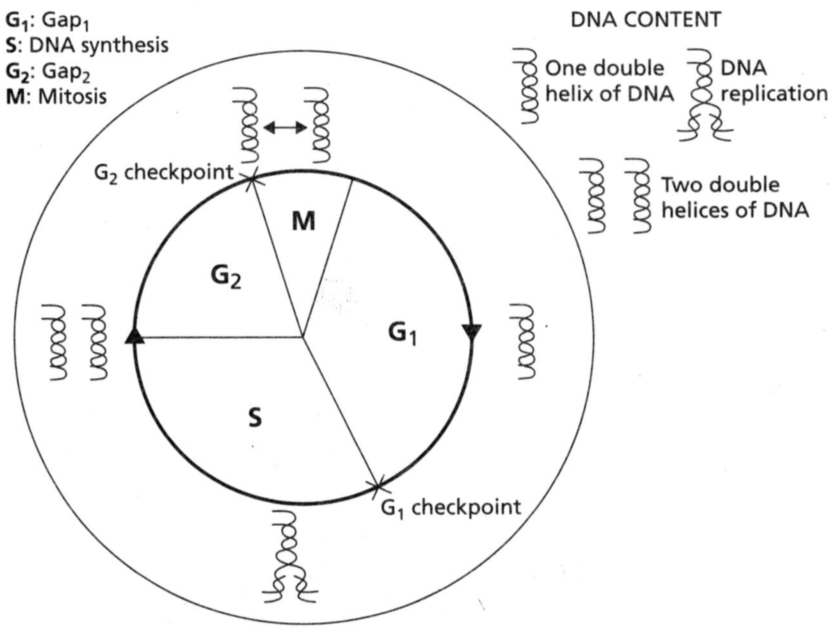

Figure 1.6
The cell cycle.

- G_0 is a resting phase from which cells move into G_1.
- G_1 (Gap_1) is a growing phase which, in actively growing cells, lasts on average around 16–24 hours, during which the cell makes various molecules necessary for the following stages. G_1 has an extremely variable timespan, depending on the growth and division requirements of the particular cell type. Rapidly growing cells such as embryonic cells may omit G_1 altogether.
- **S phase** is when DNA synthesis occurs. This lasts on average 8–10 hours.
- G_2 (Gap_2) lasts for about 4 hours during which there is protein synthesis and membrane assembly. At the end of this stage the cell is committed to mitosis, and the DNA starts to condense into chromosomes.

There are two important points in the cell cycle that commit the cell to go on to the next stage.

- The first is the **G_1 to S checkpoint**, when the cell checks that there is no DNA damage before replication, and that all the correct genes for replication are turned on.
- The second is the **G_2 to M checkpoint**, when the cell checks that all the correct apparatus for chromosome segregation has been synthesized and is present.

Mitosis is the stage at which cell division occurs. Mitosis occurs in all embryonic tissues and continues at a lower rate in most adult tissues, other than in end cells, e.g. neurones, and is vital for both tissue formation and maintenance. Mitosis lasts around 1 hour, and is itself split into several phases; this process is discussed in more detail below.

1.4 MITOSIS AND MEIOSIS

Mitosis

Mitosis (*Figure 1.7*) is the stage at which somatic cell division occurs to produce two identical daughter cells. As a result of DNA replication during S phase of the cell cycle, each chromosome in a cell entering mitosis will consist of two identical strands known as **chromatids.** Mitosis occupies only about 1 hour of the whole cell cycle. It is a continuous process but is usually described as being divided into four stages, prophase, metaphase, anaphase and telophase.

- **Prophase** begins when the chromosomes become visible following condensation. Each chromatid consists of a pair of long thin parallel strands or sister chromatids which are held together at the **centromere.** The nuclear membrane disappears and the **nucleolus** (the site of ribosomal RNA synthesis) becomes undetectable as its component parts disappear. The **centriole** divides and its two products migrate towards opposite poles of the cell.
- **Metaphase** begins when the chromosomes have reached their maximal contraction. They move to the equatorial plate of the cell and the spindle

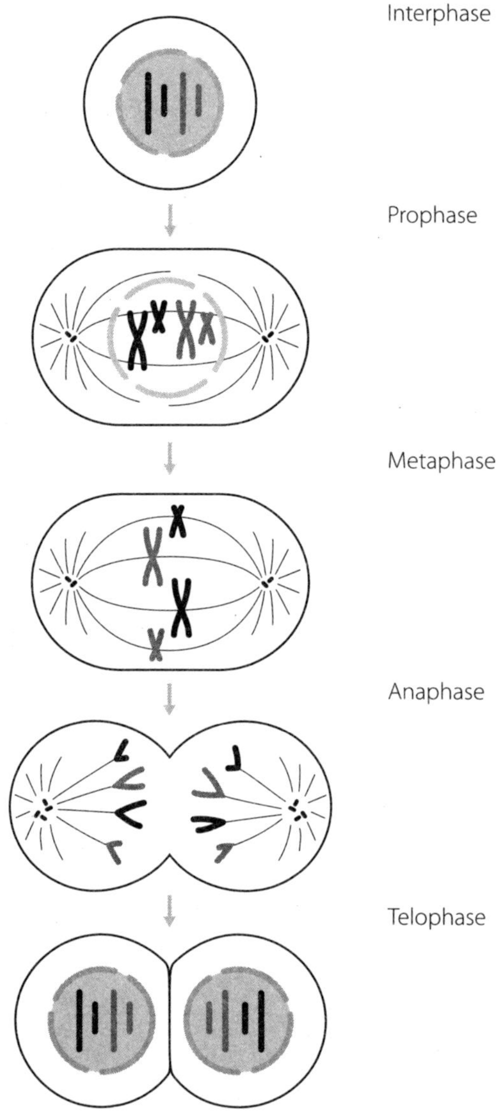

Figure 1.7
Stages of mitosis.

forms. The spindle consists of microtubules, formed by the centrioles and consisting of the protein **tubulin**. They connect the centrioles to the **kinetochore** of the centromere for each chromosome. The two chromatids of each chromosome begin to separate until they are only connected at the centromeric region.

- **Anaphase** begins when the centromeres divide and the paired chromatids separate, each to become a daughter chromosome. They move centromere first to the poles of the cell.

- **Telophase** starts when the daughter chromosomes reach each pole of the cell. The cytoplasm divides, the cell plate forms and the chromosomes start to unwind. The nuclear membrane reforms at this stage.

The two cells can now enter the cell cycle again at G_1 or enter the resting phase G_0. It is during the later stages of prophase and during metaphase that the chromosomes can be most readily examined for abnormalities by cytogeneticists.

Meiosis

Meiosis (*Figure 1.8*) is the specialized form of cell division that is used to produce gametes (also called **gametogenesis**), and it occurs in the gonads. The somatic diploid chromosomal complement ($2n$) is halved to the haploid number (n) of a mature gamete. This is done in such a way as to ensure that each gamete contains one member of each pair of chromosomes. This reduction is achieved by meiotic cell division. The chromosomes can assort independently during meiosis and it is this that leads to genetic variability (see *Box 1.7*).

Fusion of the egg and sperm restores the diploid number. See *Boxes 1.8* and *1.9* for further details on spermatogenesis and oogenesis, respectively.

Box 1.7 The generation of genetic variation/diversity

Since the chromosomes assort independently during meiosis, this results in 2^{23} or 8 388 608 different possible combinations of chromosomes in the gamete from each parent. Hence there are 2^{46} possible combinations in the zygote. There is still further scope for variation provided by crossing over during meiosis. This can result in combinations of genes on a chromosome different from that on the chromosomes in the parent. If there is on average only one crossover per chromosome (there are likely to be more than this) and a 10% paternal/maternal allele difference, then the number of possible zygotes exceeds 6×10^{43}. This number is greater than the number of human beings who have so far existed!

Box 1.8 Spermatogenesis

Spermatogenesis occurs in the seminiferous tubules of the male from the time of sexual maturity onwards. At the periphery of the tubule are **spermatogonia** of which some are self-renewing stem cells and others are already committed to sperm formation. The **primary spermatocyte** is derived from a committed spermatogonium. The primary spermatocyte undergoes the first meiotic division to produce two **secondary spermatocytes**, each with 23 chromosomes. These cells undergo the second meiotic division, each forming two **spermatids**. The spermatids mature without further division into **sperm**.

The production of mature sperm from a committed spermatogonium takes about 60–65 days. Normal semen contains 50–100 million sperm per ml. Sperm production continues (albeit at a reduced rate) into old age. The numerous replications that occur during sperm production are thought to increase the chance for mutation (see *Chapter 3*). It has been shown that the risk for several single gene mutations is increased in the offspring of older men.

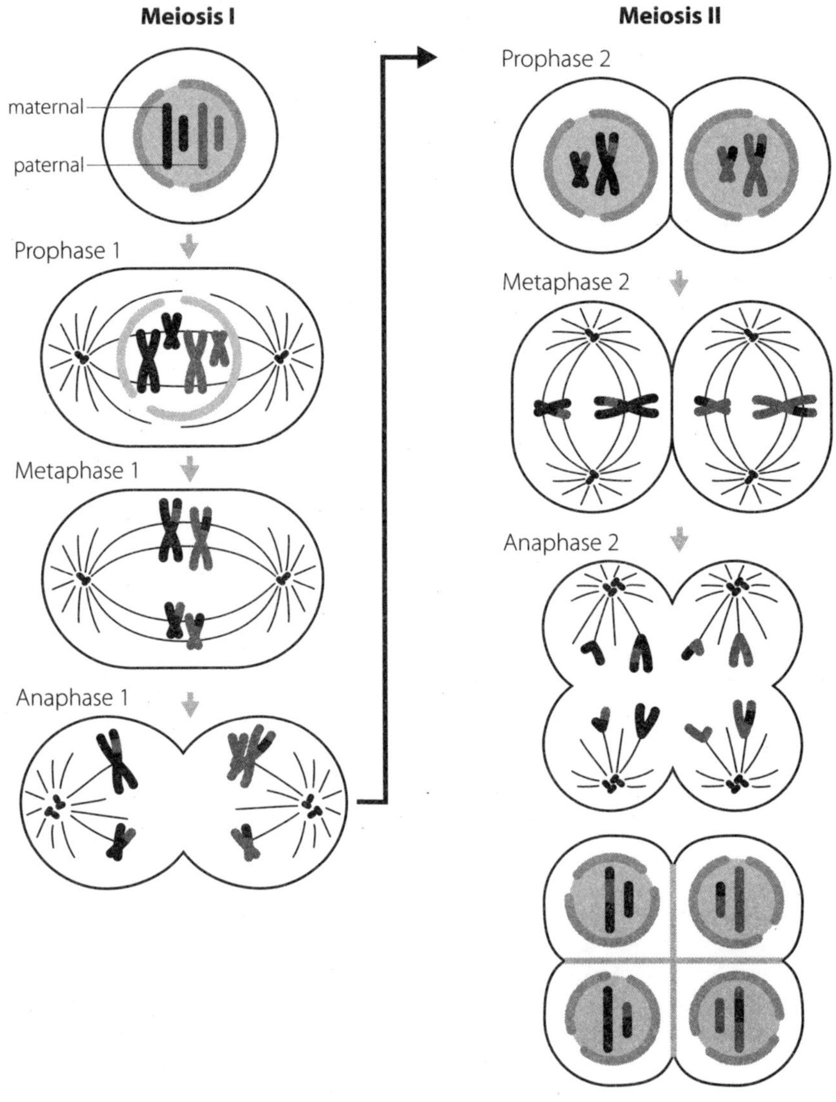

Figure 1.8
Stages of meiosis.

Meiosis consists of one DNA replication but two successive divisions: the first and second meiotic divisions.

First meiotic division (MI)

This consists of the stages prophase, metaphase and anaphase. Chromosomes at this stage consist of two chromatids. Homologous chromosomes pair, side by side, with the exception of the X and Y chromosomes in male meiosis in which only the ends pair. Exchange of homologous segments occurs between chromatids from each of the pair of homologous chromosomes – this is

Box 1.9 Oogenesis

In contrast to spermatogenesis, the process of oogenesis in the female is largely complete at birth. **Oogonia** are derived from the primordial germ cells. By about the third month of fetal life, the oogonia have become **primary oocytes** and some of these have already entered the prophase of first meiosis. The primary ooctye remains in a phase of maturation arrest in prophase I (unlike spermatocytes) until sexual maturity is reached. This stage is known as **dictyotene.** As each individual follicle matures and releases its oocyte into the Fallopian tube, the first meiotic division is completed. Hence completion of the first meiotic division in the female may take over 40 years.

The first meiotic division results in an unequal division of the cytoplasm: the **secondary oocyte** receives the great majority of the cytoplasm and the daughter cell, known as the **first polar body,** consisting largely of just a nucleus.

Meiosis II then commences during which fertilization can occur. This second meiotic division results in the formation of a **second polar body**. It is thought that the long interval between the onset of meiosis and its completion in the female contributes to the increased risk of failure of homologous chromosomes to separate during meiosis (non-disjunction) in the older mother. This non-disjunction is associated with an increased risk of trisomy for chromosome 21 (Down syndrome) and for other chromosomes (see *Chapter 5*).

known as **crossing over or recombination.** During prophase 1 in the male, the X and Y chromosomes pair in the pseudo-autosomal region, PAR (see *Chapter 2*) and there is a single obligatory crossover.

Prophase of the first meiotic division is complex and five stages can be recognized as follows.

- **Leptotene** – starts with the first appearance of the chromosomes. At this stage each chromosome consists of a pair of thread-like sister chromatids.
- **Zygotene** – during zygotene (synapsis), homologous chromosomes pair. This pairing starts at the telomeres and proceeds towards the centromeres to form **bivalents**. These are closely bound together by the **synaptonemal complex**. In the male, the paired X and Y chromosomes form a sex bivalent. It is condensed early in pachytene as the **sex vesicle**.
- **Pachytene** – this is the main stage of chromosomal thickening or condensation. Each chromosome is now seen to consist of two chromatids, hence each bivalent is a **tetrad** of four strands. Crossing over occurs during which homologous regions of DNA are exchanged between chromatids.
- **Diplotene** – during diplotene, the bivalents start to separate. Although the two chromosomes of each bivalent separate, the centromere of each remains intact. The two chromatids of each chromosome remain together. During the longitudinal separation the two members of each bivalent can be seen to be in contact at several places, called **chiasmata** (singular chiasma, see *Chapter 4*). These mark the location of crossovers, where the chromatids of homologous chromosomes have exchanged material in late pachytene. On average, there are about 52 chiasmata per cell in the human male and more in the female. At diplotene in males, the

sex bivalent opens out and the X and Y chromosomes can be seen attached to one another.

- **Diakenesis** – this is the final stage of prophase, during which the chromosomes coil more tightly.

Metaphase begins when the nuclear membrane disappears and the chromosomes move to the equatorial plane. At anaphase, the two members of each bivalent disjoin, one going to each pole. These bivalents are assorted independently to each pole. The cytoplasm divides and each cell now has 23 chromosomes, each of which is a pair of chromatids, differing from one another only as a result of crossing over.

Second meiotic division (MII)

The second meiotic division follows the first without an interphase. It resembles mitosis and includes the stages of prophase, metaphase, anaphase and telophase, with the centromeres now dividing and sister chromatids passing to opposite poles.

Meiosis has three important consequences:

- gametes contain only one representative of each homologous pair of chromosomes
- there is random assortment of paternal and maternal homologues
- crossing over ensures uniqueness by further increasing genetic variation (see *Box 1.7*)

1.5 GENES

The nuclear genome

The nuclear genome comprises the DNA, arranged in genes contained within chromosomes, in the cell nucleus. There are thought to be 20–25 000 genes in the human genome.

Eukaryotic genes comprise **exons** and **introns** (*Figure 1.9*). The exons are the important coding sequences; any change in this code (a **mutation**) may lead to an abnormal protein being produced and it is the production of these

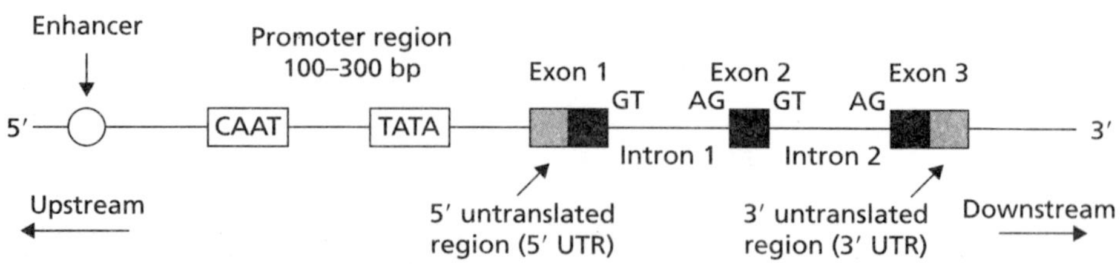

Figure 1.9
Structure of a sense strand of a eukaryotic gene.

abnormal proteins that forms the fundamental basis of many genetic diseases. Exons are consecutively numbered from the 5′ end of the gene as exon 1, exon 2 and so on. In between genes, and indeed in between exons of genes, are the intervening sequences or introns. It is interesting to note that the sex determining region (*SRY*) gene and the human mitochondrial genome do not have any introns. The stretches of DNA in introns appear to contain no useful coded information. Although this DNA may originally have coded for genes, these no longer function, so a mutation in an intron would then be unlikely to cause a genetic disorder.

A particular fragment of DNA is not necessarily identical in different people. Benign differences are known as **polymorphisms.** As we will see in *Chapter 7* on molecular genetics, these polymorphisms can be exploited by molecular biologists to study genetic diseases in families because they are inherited in a Mendelian fashion.

It has been suggested that 80–90% of the human genome comprises non-coding stretches of DNA, and that 40–50% of this is repetitive in nature (there are varying degrees of this repetitiveness, as shown in *Table 1.1*). Repetitive human DNA is also called satellite DNA.

Gene size in humans may range from around 100 bp (for the tyrosine transfer RNA) to 250 kb (for the moderately large gene for cystic fibrosis), up to 2.4 Mb for one of the largest human genes known – the gene coding for the protein dystrophin which is dysfunctional in Duchenne muscular dystrophy. Although this gene is 2.4 Mb long, it is still too small to be seen at the level of resolution of a light microscope.

The direction to the left of the 5′ end of a gene is said to be **upstream**, whereas the direction to the right of the 3′ end of a gene is **downstream** (*Figure 1.9*).

The mitochondrial genome

Mitochondria have their own circular genome comprising mitochondrial DNA (mtDNA). The genome is around 16.5 kb in size. Two circular strands known as the H and L strands code for 13 proteins involved in energy production and 22 **transfer RNA (tRNA)** molecules, together with two ribosomal RNAs (rRNA).

Some proteins are produced by a combination of nuclear genes and mitochondrial genes. The nuclear products are transported through the cytoplasm to the mitochondria where they associate with mitochondrial gene products to form a functional protein. An example is the cytochrome c oxidase complex, which comprises three subunits coded by mitochondrial genes and ten subunits coded by nuclear genes.

1.6 CHROMOSOMES AND CHROMATIN

The structural organization of DNA does not stop at the DNA level. Within the chromosome are found specialized categories of chromatin.

Table 1.1 Repetitiveness at the DNA level

Name of DNA sequences	Comments
Unique	Single copy DNA; there is about 10% of coding DNA in the whole human genome.
Microsatellite	Also known as short tandem repeats (STRs) of 1–4 bp, e.g. CA, TA, tetranucleotides; found on all chromosomes.
Minisatellite	Variable number of tandem repeats (VNTRs) of 9–64 bp; these were the original 'fingerprinting' probes and are found on all chromosomes, especially near telomeres. Telomeres, 6 bp, TTAGGG, up to several kb; found at the ends of all chromosomes.
Interspersed	SINES (short interspersed nuclear elements). An example is the human Alu repeat of about 280 bp; up to 1 million copies, mostly in the euchromatic (G light) areas of chromosomes.
Interspersed	LINES (long interspersed nuclear elements). The L-1 or Kpn repeat is 1.4–6.1 kb; up to 100 000 copies, mostly in the heterochromatic (G dark) areas of chromosomes.
Satellite	Alphoid (α) repeat of 171 bp; up to several Mb in length. Found in centromeric heterochromatin of all chromosomes.
Satellite	Beta (β) repeat of 68 bp; 100 kb up to several Mb in length. Found in centromeric heterochromatin of chromosomes 1, 9, 13, 14, 15, 21, 22, Y.
Satellite 1	Repeat of 25–48 bp; A–T rich. Found in the centromeric heterochromatin of most chromosomes, and other heterochromatic regions.
Satellite 2	Repeat of 5 bp; probably found on all chromosomes.
Satellite 3	Repeat of 5 bp; probably found on all chromosomes.
Triplet repeats	Repeats of 3 bp, usually CGG, CAG, CCG or CTG. Tens of copies in the normal person can expand to 1000s in an affected individual. Unique in that these dynamic mutations may expand from generation to generation and actually cause certain genetic syndromes. See also *Chapters 3* and *7*.

Heterochromatin and euchromatin

Heterochromatin comprises highly condensed chromatin fibres. There are two types of heterochromatin:

- **constitutive heterochromatin** is composed of repetitive DNA sequences containing **no transcriptionally active genes**. It replicates late in the cell cycle at the end of the S phase. Its function is unknown, but it may have a role in conserving and protecting vital genes from recombination. In human chromosomes, most of the heterochromatin is located close to the centromeres.

- **facultative heterochromatin** can be transcriptionally active or inactive, and is also late replicating. It is found in the selectively inactivated X chromosome in females.

Euchromatin contains chromosome fibres which are less densely packed than heterochromatin. Euchromatin is generally transcriptionally active, but may contain some regions of transcriptionally inactive DNA.

Nucleolar organizers (NORs), the genes controlling ribosomal RNA synthesis, occur in repeated blocks on the short arms of chromosomes 13, 14, 15, 21 and 22. These genes have to be active until mitosis and therefore remain uncoiled, while the remainder of the mitotic chromosome is condensed. The NORs are visible as secondary constrictions.

1.7 CHROMOSOMES AND NOMENCLATURE

Until the early 1970s, chromosomes were only identifiable by shape and size, as they were just stained and observed by light microscopy as solid blocks of

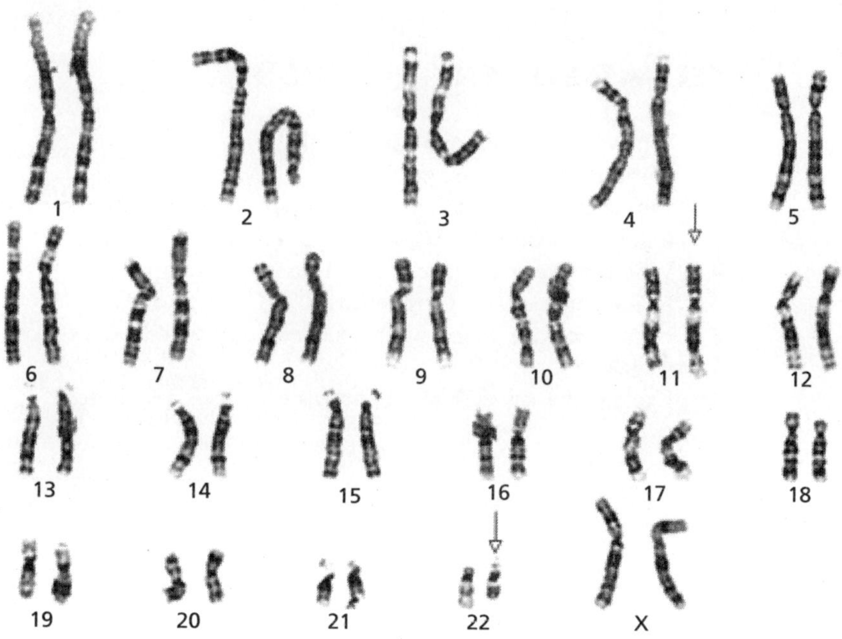

Figure 1.10
A G-banded female karyotype with 46 chromosomes including a balanced reciprocal translocation between the long arm of one chromosome 11 and the long arm of one chromosome 22. This translocation has been described in a number of families. Abnormal segregation can result in offspring with an abnormal karyotype which causes miscarriage or birth of children with an abnormal phenotype (see *Chapter 5*). The ISCN nomenclature is 46,XX,t(11;22)(q23;q11.2).

colour. From 1973, the technique of G-banding ensured that chromosomes could be differentiated from one another by their individual band patterns.

For the purposes of analysis, the 22 autosomes and the pair of sex chromosomes are arranged in a **karyotype** according to their size and group (*Figure 1.10*). By convention, autosomes are numbered in descending order of size, from 1 to 22. The short arm is designated 'p' and the long arm 'q'.

The International System for Human Cytogenetic Nomenclature, or ISCN, divides the banding pattern of each chromosome arm into regions and bands so that every band can be uniquely identified by a simple shorthand method, so that, for example 1p35 refers to chromosome 1, short arm, region 3, band 5. The numerically lowest region and band numbers are closest (or **proximal**) to the centromere; the highest numbers are at the tips, or **distal** to the centromere.

The karyotype, nomenclature and chromosome analysis are discussed further in *Chapter 5*.

SUGGESTED FURTHER READING

Strachan, T. and Read, AP. (2004) *Human Molecular Genetics,* 3rd Edition. Oxford: Garland Science.

SELF-ASSESSMENT QUESTIONS

1. Where is DNA located in the cell?
2. Outline the structure of DNA.
3. Define complementarity; write down the complementary bases to the following sequence: AGGTTCGGAT.
4. Why are exons of genes more important than introns?
5. Name three classes of repetitive DNA and give examples.
6. Name the four stages of the cell cycle. How long does a cell spend in each stage?
7. Explain the key differences between meiosis and mitosis.

CHAPTER 2

How normal genes work

Learning objectives

After studying this chapter you should confidently be able to:

- **Describe briefly the mode of action of structural and controlling genes**
  Genes can code for either protein subunits or functional RNAs. The proteins may be structural or may control other genes. The RNAs tend to control by acting in a positional (spatial) manner.
- **Outline the processes of transcription and translation**
  One strand of the double-stranded DNA comprising a gene is transcribed to produce a primary RNA transcript, which is spliced to form messenger RNA. This is then translated into protein with the help of ribosomes and tRNAs.
- **Explain the basic concept of imprinting and uniparental disomy**
  Some gene expression is dependent on the parent of origin. Because some genes only express on certain chromosomes from one parent, abnormalities may arise due to uniparental disomy, where both chromosomes containing inactive genes are inherited from the same parent.
- **Give examples of different levels of gene control**
  Although housekeeping genes are always switched on, most genes are switched on or off as required, using various levels of control. This ranges from general control such as gene promoters, methylation, gene dosage, and chromatin structure, to specific gene control such as silencers and enhancers.
- **Describe the three basic methods of DNA repair**
  The three basic types of DNA repair are enzymatic photoreactivation, excision repair and mismatch repair.
- **Outline the importance of somatic recombination and allelic exclusion in immunogenetics**
  The products involved in immune response produced by B and T cells are derived from the maternal or paternal genome by allelic exclusion. The variety of antibodies (immunoglobulins) produced by B cells could not exist in such large numbers if it were not due to somatic recombination, which uses a small number of genes that then recombine to produce a diversity not otherwise feasible.

- **Name and describe the function of three major classes of developmental genes**
  Early embryonic cells are totipotent. Developmental genes are usually controlled in cascades with appropriate feedback loops. Three such cascade pathways involve growth factors, signalling genes and homeobox genes.
- **Explain the importance of the *SRY* gene with respect to sexual development**
  A normal male karyotype is 46,XY. The male sex determining region gene (*SRY*) is found on the Y chromosome. *SRY* is thought to upregulate *SOX9*, which activates the *FGF9* gene. This ensures that the gonad becomes male.

2.1 GENE EXPRESSION

When Watson and Crick elucidated the structure of DNA in 1953, they realized that complementary base pairing could provide a model for both DNA replication and the inheritance of genetic material. During the following years, the steps by which the DNA sequences of genes were 'read' and turned into a string of amino acids comprising a polypeptide (a subunit of protein), were also determined. The 'central dogma' of molecular biology is:

$$\textbf{DNA} \rightarrow \textbf{RNA} \rightarrow \textbf{Protein}$$

DNA has to be **transcribed** into an intermediate molecule, RNA, which is then **translated** into amino acids, the correct sequences of which give rise to an active protein. In eukaryotes, DNA is found in the cell nucleus. However, RNA, amino acids and hence proteins are generally found in the cytoplasm.

It was also quickly established that particular genes coded for certain proteins which made physical parts of the human body, such as keratin found in hair and nails, or the coloured pigment found in eyes or skin. These are known as **structural genes.**

The production of proteins from a gene is called **gene expression.** Some genes may not code for an obvious end product – their proteins or RNA control other genes by switching them on and off. These are **controlling genes.** Gene control may occur directly or indirectly via a **signal transduction pathway.** Signals from outside the cell are directed into the cytoplasm and also into the nucleus. This chapter will look at some mechanisms by which controlling genes influence gene expression.

2.2 TRANSCRIPTION

The double-stranded DNA must unwind from its helix to be copied by an enzyme called RNA polymerase II. The antisense DNA strand (the one that runs in the 3′–5′ direction) is copied such that the resulting RNA reads 5′–3′ (by convention, diagrams of transcription always display the 5′ end of RNA on the left, which implies that it was copied from the antisense DNA strand). The resulting RNA will then have exactly the same sequence as the DNA 5′–3′ sense strand, except that the thymines have been replaced by uracils (*Figure 2.1*).

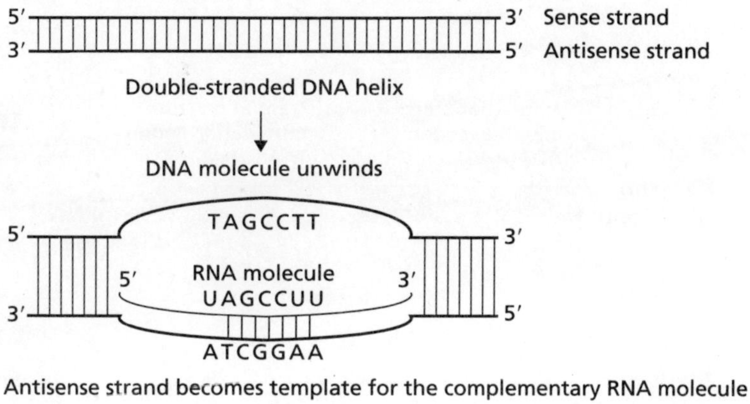

Figure 2.1
Principles of transcription from DNA to RNA.

In eukaryotes (see *Box 2.1* for a comparison with prokaryote transcription), the newly transcribed RNA (sometimes called the **primary transcript RNA**) will include sequences for introns as well as exons. As this primary RNA passes from the nucleus into the cytoplasm, the introns loop out, and are carefully excised. The splicing out of introns is so precise that a nuclear intron will always start with GT and end with AG. The exons are then **spliced** together (*Figure 2.2*). Prokaryotic RNA does not need splicing as there are no introns. Human RNA containing no introns is called **messenger RNA (mRNA)**.

Box 2.1 Transcriptional differences in eukaryotes

There are other transcriptional differences between prokaryotes and eukaryotes; eukaryotic mRNA undergoes two further steps in transcription:

- 7-methylguanosine is added to the 5′ end of the mRNA – this is known as **capping**
- a long series of adenine bases (As) is added to the 3′ end – a **poly A tail**

2.3 TRANSLATION

This is the process whereby the mRNA is translated into protein. The cytoplasm contains structures called ribosomes, which themselves are made up of protein subunits derived from **ribosomal RNA (rRNA)**. rRNA is mainly produced in the **NORs** of the nucleolus (within the nucleus). NORs are found at the satellites of acrocentric chromosomes.

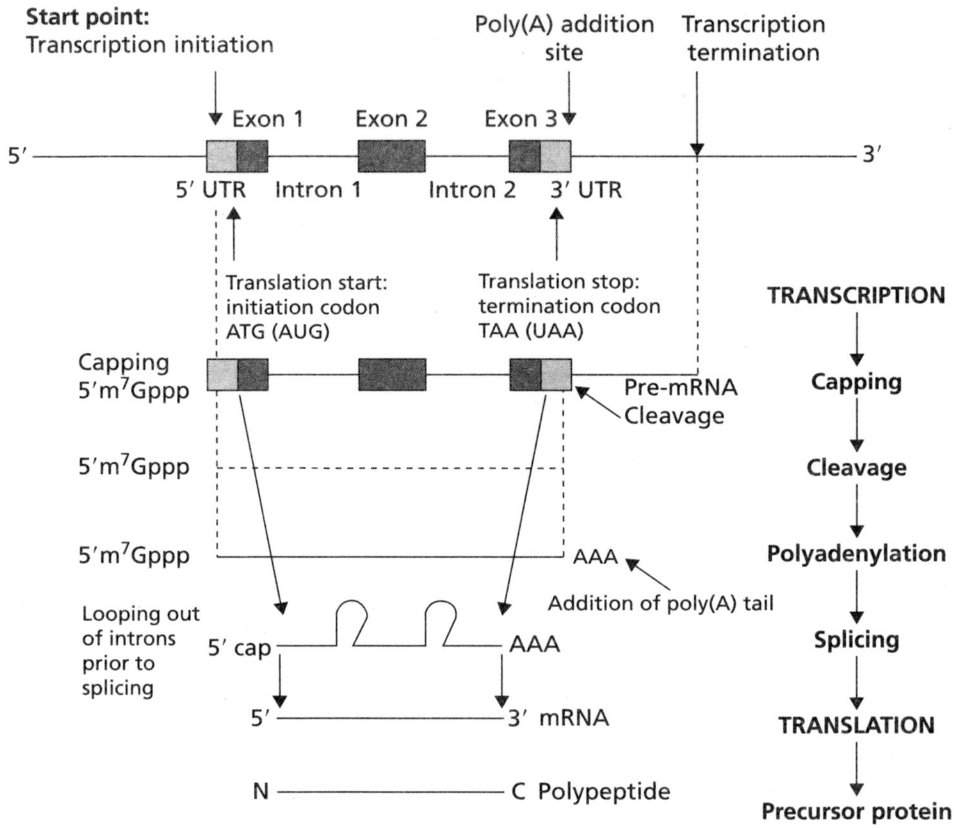

Figure 2.2
Transcription and translation.

The mRNA passes through the cytoplasm and proceeds to the ribosomes (*Figure 2.3*) where the sequence is read in groups of three bases (**triplets**) called **codons.** Apart from three 'stop codons' (see below), each triplet corresponds to a particular amino acid. As there are four bases (**A, C, G and T**) there are 64 (4×4×4) possible **codes**, but not all are used, as there are only 20 amino acids (see *Table 2.1*). As several triplets can code for the same amino

Box 2.2 Mitochondrial codons

Mitochondria do not use all the same triplet codons as nuclear DNA.

Codon	Nuclear genome	Mitochondrial genome
UGA	STOP	Trp
AUA	Ile	Met
AGA	Arg	STOP

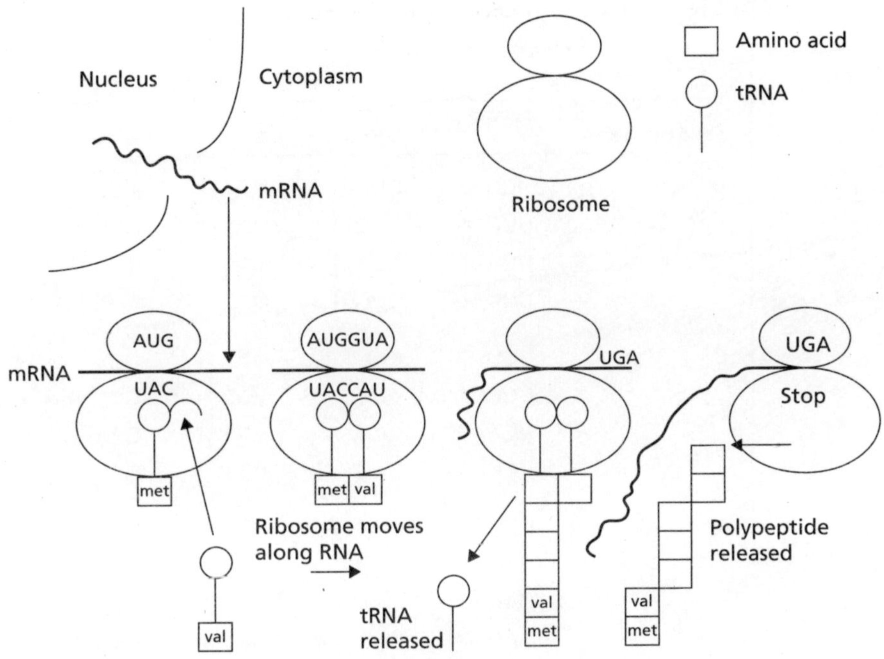

Figure 2.3
Translation of mRNA by ribosomes and tRNA molecules.

acid, this is a **degenerate** or redundant code. However, the **initiation codon** is almost always AUG (which uniquely codes for methionine) and is recognized as such by the ribosome (see *Box 2.2*).

As the triplet codes are read, the correct amino acid has to be placed in sequence almost like beads on a string. This is done by molecules known as **transfer RNAs** (tRNAs), whose appearance in some organisms resembles a cloverleaf. At the 'stem' end is attached a particular amino acid, for example, valine. On the opposing 'leaf' end is the **anticodon** for that amino acid. The mRNA codon (GUA) will be complementary to the tRNA anticodon sequence (CAU), thus the correct amino acid (V, valine) will be delivered to the growing amino acid chain (see *Figure 2.3*). Many polypeptide chains can be synthesized simultaneously in the cytoplasm by the attachment of several ribosomes to the same RNA molecule. Three stop codons (UAA, UAG or UGA) signal chain termination, and the polypeptide chain is then released from the ribosome.

2.4 CONTROL OF GENE EXPRESSION

Most genes are not **constitutively expressed**, i.e. they are not turned on all the time and in all cells. Usually genes have to be periodically switched on or off, and are regulated to produce their proteins appropriately. An example is

Table 2.1 The degenerate genetic code

	First base				
Second base	**U**	**C**	**A**	**G**	**Third base**
U	UUU F	CUU L	AUU I	GUU V	U
	UUC F	CUC L	AUC I	GUC V	
	UUA L	CUA L	AUA I	GUA V	
	UUG L	CUG L	AUG M	GUG V	
C	UCU S	CCU P	ACU T	GCU A	C
	UCC S	CCC P	ACC T	GCC A	
	UCA S	CCA P	ACA T	GCA A	
	UCG S	CCG P	ACG T	GCG A	
A	UAU Y	CAU H	AAU N	GAU D	A
	UAC Y	CAC H	AAC N	GAC D	
	UAA X	CAA Q	AAA K	GAA E	
	UAG X	CAG Q	AAG K	GAG E	
G	UGU C	CGU R	AGU S	GGU G	G
	UGC C	CGC R	AGC S	GGC G	
	UGA X	CGA R	AGA R	GGA G	
	UGG W	CGG R	AGG R	GGG G	

Key to amino acids (the three letter code is recommended by the Human Genome Variation Society):

A alanine (Ala)
C cysteine (Cys)
D aspartic acid (Asp)
E glutamic acid (Glu)
F phenylalanine (Phe)
G glycine (Gly)
H histidine (His)
I isoleucine (Ile)
K lysine (Lys)
L leucine (Leu)
M methionine (Met)
N asparagine (Asn)
P proline (Pro)
Q glutamine (Gln)
R arginine (Arg)
S serine (Ser)
T threonine (Thr)
V valine (Val)
W tryptophan (Trp)
Y tyrosine (Tyr)
X stop codon

the production of insulin in the pancreatic cells as a response to eating sugar, i.e. expression in the correct place at the correct time. Those that are turned on all of the time are obviously very important and are constantly required for the correct day-to-day functioning of the organism's cells; these genes are known as **housekeeping genes.**

Eukaryotic gene expression is usually regulated during transcription or translation, either at the molecular level or on a larger scale (see *Table 2.2* for summary).

Table 2.2 Levels of eukaryotic genetic control

Stage	Level	Examples
Transcriptional	Molecular: general	Promoters, transcription factors
	Molecular: specific	Enhancers, silencers
	Large scale: general	Methylation
		Chromatin structure
Post-transcriptional		Alternate splicing
Translational	Molecular	Translational response elements
Other		Dosage
		X-inactivation
		Genomic imprinting

Transcriptional control

The basis of molecular control

A single gene usually codes for one polypeptide (or protein) and in eukaryotes such a gene sequence is also called a **cistron**. In bacteria, one gene can control several structural genes, which may be expressed sequentially in response to (for example) a nutrient appearing in the environment, when appropriate digestive enzymes may be produced. Jacob and Monod called this arrangement an **operon**.

Promoters and transcription factors

As genes comprise both exons and introns, then the logical start position for transcription should be at the beginning (5′) of exon 1, terminating at the 3′ end of the last exon. In fact, these areas are flanked by pieces of untranslated DNA; the actual initiation and termination areas are shown in *Figure 2.2.* The untranslated region at the beginning of a gene is called the 5′ UTR (untranslated region) and at the end of the gene it is called the 3′ UTR.

Just before the initiation site (at about 25 bases 5′ of the site) a consistent run of bases, TATAA, is often found; this is known as a **TATA box**. Some genes may have another box known as the **CAAT box** with the sequence CCAAT. This is equivalent to an 'on' switch, and is known as the **promoter** region (*Box 2.3*). It is possible for a gene to have more than one promoter

Box 2.3 Consensus sequences and motifs

When DNA sequences are conserved because they have an important function, they are called **consensus sequences**. An example would be the TATA box. Sometimes amino acids are conserved in the same way, as their three-dimensional structure may be essential to the function of the protein. All these groups are known as **motifs**.

which may produce different versions of the same protein in different body tissues (muscle dystrophin may be a different length to brain dystrophin, for example). Sometimes gene expression may be a response to an external factor rather than a promoter, such that, for example, the presence of a hormone can induce transcription. Genes that respond in this way contain **response elements.**

In order for transcription to begin, RNA polymerase II (itself a small protein) has to 'log on' to the beginning of the DNA molecule to transcribe the sequence into RNA. For this purpose, small proteins known as **transcription factors** are employed to assist the RNA polymerase II. Transcription factors have different regions which take on specific conformations which fit their target DNA and the RNA polymerase II like a piece of a jigsaw. They have odd sounding names like the zinc finger motif, the leucine zipper and the helix-loop-helix.

A specific transcription factor (TFIID) binds to the promoter region first, then other transcription factors guide the RNA polymerase to its correct position which enables transcription to be turned on at a basic level. This is the transcription unit (*Figure 2.4*).

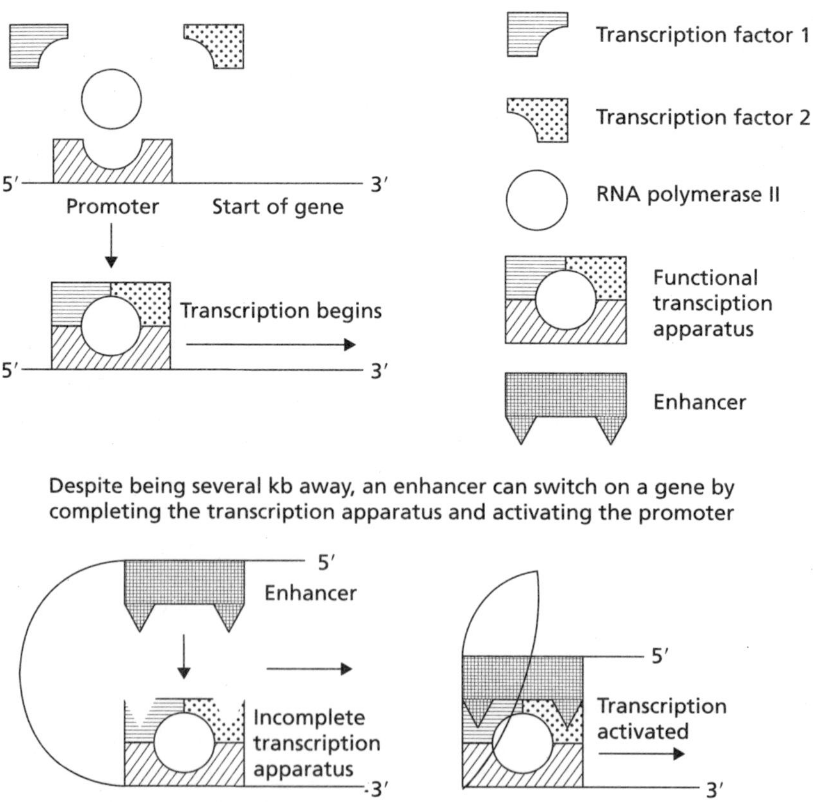

Figure 2.4
Molecules involved in transcription.

Specific transcriptional control

Sometimes formation of the transcription unit does not reach maximum efficiency; certain sequences called **enhancers** help achieve this (*Figure 2.4*) by optimizing transcription of a gene. Enhancers are interesting because they may lie several kilobases away from their target gene and its promoter. A mutation in an enhancer may cause a clinical disorder due to a lack of a particular protein, yet the actual gene seems undamaged.

The opposite of enhancers are **silencers** (or repressors) which switch genes off.

Methylation

The DNA base cytosine can be methylated by the addition of a methyl group (CH_3) (see *Figure 1.4*). Methylation therefore tends to be found in CG-rich sequences. Methylation is often seen in vertebrate DNA and generally follows the rule that active genes (i.e. those that are transcribing) have less or no methylation. Inactive (or repressed) genes show more methylation with no gene expression, as transcription factors may not recognize the DNA sequences.

However, it does not follow that all gene expression is regulated by methylation, as the non-vertebrate fruit fly *Drosophila* does not have methylated DNA at all.

At the 5′ ends of actively transcribing genes are found short fragments of DNA called **CpG islands.** In these CpG islands, the 5′ **c**ytosine is connected by a **p**hosphodiester bond to a 3′ **g**uanine. This way of writing bases therefore represents the DNA sequence read 'horizontally' rather than 'vertically' as a C:G base pair. Although not every gene has these CpG islands, they are thought to represent the beginning of **housekeeping** (or other widely expressed) genes.

CpG islands can be detected using the methylation-sensitive restriction enzyme *Hpa*II which cannot cut methylated DNA. The resulting small pieces of cleaved DNA must therefore be unmethylated, and are sometimes called **HTF** islands (*Hpa* tiny fragments).

Acetylation

Histones have chemically modifiable N-terminal tail domains. Acetylation of certain histones (e.g. H3 and H4 of the *IGF2R* gene) is associated with increased gene transcription, and thus modulates gene expression. It is thought that, together with DNA methylation, acetylation may be a mechanism for initiation, maintenance and transmission of genomic imprinting.

Chromatin structure

Chromatin structure comprises DNA and its associated histones together with other basic and acidic proteins. The primary coils of DNA are organized into higher levels of very condensed structures (see *Figure 1.1*).

The nature of the chromatin structure itself is therefore important in gene expression; an open chromatin structure free of methylation will generally indicate that those genes can become transcriptionally active. There will be room for the correct transcription factors and appropriate enzymes to bind properly.

Areas of condensed chromatin (heterochromatin) indicate suppression or inactivation of genes. There will be certain areas of constitutive heterochromatin (e.g. the centromeres) or telomeric chromatin (see *Box 2.4*) which are inactive a great deal of the time, as these domains have an important structural function: to hold the chromosome together properly. Any active genes which are inadvertently moved near to these inactive areas run the risk of being inactivated themselves. See *Chapter 3* for abnormal **position effects.**

Box 2.4 Telomere structure and function

Telomeres cap the ends of eukaryotic chromosomes and maintain their structural stability. The human telomere consists of a stretch of tandemly repeated DNA sequences comprising six nucleotides (TTAGGG), together with a telomere-binding protein which protects the terminal DNA. If telomeres are lost, the end of the DNA helix would become unstable, and may fuse with the ends of other broken chromosomes, or be subject to degradation.

It becomes apparent that genes retain specific functions partly because of the three-dimensional chromatin position in which they are naturally found, while other factors such as methylation also play an important role. It is possible, however, for whole chromosomes (or areas of chromosomes) to become inactivated, resulting in essentially haploid areas of the human genome. This is not abnormal; functional haploidy probably arose during evolutionary development to balance and protect the human genome (see also sections on dosage and imprinting below).

Post-transcriptional regulation

After transcription of the primary RNA, the point at which the introns are spliced out gives another opportunity for differential gene expression. It is possible to splice out certain exons as well as introns; this produces different mRNAs and is known as **alternative splicing.**

Translational control

The operon arrangement of the bacterial genome enables a rapid response to its environment. Translational control in humans fulfils a similar function, as examples usually arise when a rapid response is required. In a similar manner to transcriptional response elements, the presence of certain substances may elicit a translational response. When a particular molecule required for the production of haemoglobin (haem co-factor) is present, the rate of translation of globin RNA in the reticulocytes increases.

Because different cell types have different mRNAs in their cytoplasm, translational control must be rare; if it were not, we would see all possible mRNAs in every cell type. Unmodified mRNA is unstable and as it passes into the cytoplasm it is quickly degraded by RNAses. One way translational control is expressed is by protection of mRNA, in order that efficiency of translation of particular mRNAs is maximized. Protection may be achieved using long 5′ and 3′ untranslated regions, by the formation of secondary structures such as stems and loops, or by generation of hundreds of copies of important mRNAs which are protected from degradation by the ribosomes themselves.

When an egg is fertilized by a sperm, there is a large increase in the rate of protein synthesis, yet this does *not* require the production of new mRNAs after fertilization. This means that there must be pre-existing stable mRNAs in the mother's unfertilized egg which are only translated after fertilization.

Other methods of gene control

Dosage

The 46 chromosomes in a human diploid nucleus comprise 23 chromosomes from each parent. Previously it was assumed that they have equal importance in a cell. This cannot be true, however, because females have two X chromosomes, while males only inherit one X from their mother (together with a Y from their father). Without **dosage compensation** females would have twice the amount of gene product as males with only one X. One of the X chromosomes is therefore **inactivated** (see *Box 2.5*).

Box 2.5 Haploinsufficiency

Autosomal genes are present in pairs – one on each homologous chromosome. If the level of protein required for the cell to function normally requires that both genes should be active, they will exhibit **bi-allelic expression.**

In some circumstances (due, for example, to the deletion of one of a pair of alleles) only 50% of the protein will be produced from the remaining intact gene. A haploid level of protein may not be enough for the cell to behave appropriately, and it therefore displays **haploinsufficiency** (see *Chapter 3*).

X chromosome inactivation. One of the two X chromosomes in female cells is **randomly inactivated** at the early blastula stage in embryonic development. In a female nucleus the inactive X can be seen as a distinctive densely staining structure called the Barr body (Barr described its composition as sex chromatin – it is actually highly condensed heterochromatin). Females are therefore a mosaic (mixture) of cells containing maternally *or* paternally inactive X chromosomes in a ratio of approximately 1:1. The parental origin of inactivation is then preserved when a particular somatic cell divides (see *Box 2.6*).

Control of X-inactivation. X-inactivation is controlled by a gene called *XIST* (pronounced exist) found in the **X-inactivation centre** (XIC) on the long

Box 2.6 The pseudo-autosomal region

There are certain small areas which are left active on the inactivated X. One of these is the pseudo-autosomal region (PAR) on the tip of the short arm at Xp22.3, where there are important sequences shared with the Y chromosome at Yp11.3. This is the area where an obligatory crossover occurs at meiosis.

It is thought that around 15% of genes remain active on the inactive X; as they are not all in the PAR (some are on the long arm), this suggests that the inactivation proceeds individually in small increments (perhaps gene by gene).

arm of the X chromosome at Xq13. Although most of the genes on the inactivated X will not be expressed, *XIST* is expressed in females, as it has to be active to initiate the inactivation of its own X chromosome. The way *XIST* operates is by generating a functional structural RNA, not the usual protein, and its mode of action is to coat the X chromosome which is to be inactivated. All extra X chromosomes (apart from the active X) are inactivated.

Although *XIST* is responsible for the initiation and spreading of X-inactivation, the XIC is not needed for the maintenance of X chromosome inactivation, as the XIC can be lost but the genes still remain silent due to methylation (see *Box 2.7*). What is certain, however, is that the single X chromosome in the male has to remain active, and to do this it is necessary that *XIST* is switched off. Another way of expressing this is to say that the *XIST* gene on the male X is **imprinted** (see below).

Box 2.7 Unusual inactivation patterns

Not all X chromosomes are inactivated in the same way as in humans. In marsupials such as kangaroos the paternal X is always inactivated, whilst in some male scale insects the whole paternal chromosome set is inactivated!

This is not comparable to dosage in humans; the paternal X or paternal genome in these animals is imprinted (see *Section 2.5*) and therefore not functionally equivalent to the female X or the female genome.

2.5 EPIGENETICS AND GENOMIC IMPRINTING

Eamonn Maher has described epigenetics as the "silence of the genes". It may be defined as all meiotically and mitotically heritable changes in gene expression which do not involve changes in DNA sequence.

Epigenetics results in a modification in gene expression which is heritable but potentially reversible. This may be one mechanism for fine control of gene expression. Epigenetic silencing of genes can be initiated and sustained by three systems:

- DNA methylation
- RNA-associated silencing
- histone modification, whereby loss of acetylation leads to transcriptional repression

While normal processes include X-inactivation (see above), tissue-specific gene expression and genomic imprinting, it also follows that mutational changes in epigenetic pathways will lead to an epigenetic pathology (see *Chapter 3*).

Genomic imprinting

It is logical that dosage compensation exists on the X chromosome, in order that both males and females only express one X allele. It might be supposed that autosomes have no requirement for a dosage mechanism, as both males and females receive 22 autosomes from each parent. However, not every chromosome is imprinted; only individual genes or small chromosome regions on particular chromosomes are inactivated.

Genomic imprinting results in one of two otherwise identical alleles being marked (imprinted) during gametogenesis such that the homologous alleles are expressed differently depending on the parent of origin. Genomic imprinting is found in X-inactivation in females, the autosomes of somatic cells, and can also be acquired. A pathological example is the retinoblastoma tumour suppressor gene (see *Chapter 8*), whereby the promoter of the remaining normal copy of the *RB1* gene is methylated, rendering it inactive.

Resetting of the imprint in germ cells

Imprinting has already occurred in somatic cells. However, when an individual reproduces, any imprint received from their parents has to be **reset**; it is wiped out, so that in the chromosomes passed on to their own children, the new imprint is from the correct parental sex (see *Figure 2.5*).

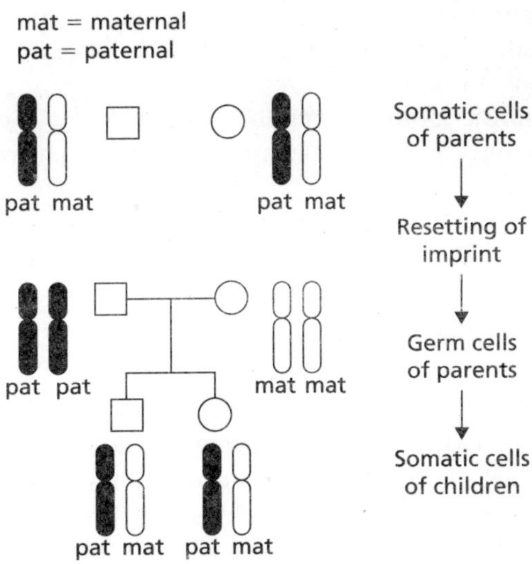

Figure 2.5
Resetting of the imprinting pattern.

The long arm of chromosome 15 at 15q11-q13 is a well-known example of human imprinting. For example, in a woman's somatic cells one chromosome 15 would be inherited from her mother (maternally imprinted); the other would be inherited from her father (paternally imprinted). In order to have normal children, her paternal chromosome 15s must be reset as maternal before they are passed on via her germ cells. Similarly, her partner's maternal chromosome 15s in his sperm will have been reset to paternal. That way the embryo will receive a correctly imprinted chromosome 15 from each parent.

Evidence for genomic imprinting

Developmental evidence. At fertilization a 23,X or 23,Y haploid sperm fertilizes an egg cell nucleus carrying a 23,X haploid set of chromosomes, which normally produces a 46,XX or 46,XY embryo with its associated placenta and membranes. This process appears to be specific to mammals. When a fertilized mammalian egg divides into a ball of cells called a blastocyst, these groups of cells differentiate into two types (lineages). The inner cell mass will give rise to the embryo, whilst the trophoblast (outer) layer becomes the placenta and membranes.

If there has been an error at meiosis in the ovum, or degradation of the female haploid set, it is possible that a sperm may enter an empty egg (which has no chromosomes of its own). If the haploid set of the sperm doubles, then we have a structure in which all the chromosomes are paternally derived (androgenetic). This structure, called a **hydatidiform mole**, consists of long strings of placental material containing many fluid-filled cysts. The cysts are derived from chorionic villi, which have no blood vessels and therefore cannot drain fluid. There is no embryo. Complete hydatidiform moles are very rare: occurring in just 1:2000 pregnancies in Europe, but they are more common in the Far East where they occur with a frequency of 1:200. A potentially lethal cancer called a choriocarcinoma may arise from the mole; this can then invade the uterine wall, but this process is detectable by monitoring hormone levels.

The paternal genome alone is therefore sufficient for the production of placental structures, but the maternal genome appears to be required for fetal development.

In an **ovarian teratoma**, an egg cell bearing a 23,X haploid set of chromosomes doubles to 46,XX, such that all the chromosomes are maternally derived (gynogenetic). The egg develops into a round tumour which, if dissected, reveals an uncoordinated mass of fetal tissues such as hair, intestines, or teeth, but no placenta. This supports the view that the maternal genome is necessary for embryonic development.

Chromosomal evidence. **Uniparental disomy (UPD)** occurs when both of a particular pair (or part of a pair) of chromosomes have been inherited from one parent. All the remaining pairs of chromosomes have been inherited equally, one from each parent. Examples of UPD include:

- **Maternal UPD 7.** If an embryo has inherited both chomosome 7s from its mother (maternal UPD 7), there is a slowing down of growth of that fetus (interuterine growth retardation: IUGR). The child is small but otherwise unaffected. This phenomenon is not seen if both chromosome 7s are paternal.

- **Paternal UPD 11.** A condition called Beckwith–Wiedemann syndrome may result in a very large infant who may have internal organs such as the liver that are larger than normal (see *Box 2.8*). A prominent feature is a large tongue. There is usually an abnormality of the short arm of chromosome 11. Beckwith–Wiedemann syndrome can arise by several different mechanisms, but often two paternal 11p15 regions are inherited together with one maternal 11, essentially resulting in trisomy 11p15. It is known that a growth gene (*IGF2*) is present in this region; this gene is active on the paternal 11 but switched off (imprinted) on the maternal 11. If more than one copy of the active paternal gene is inherited, there will be overgrowth.

Box 2.8 Critical chromosome regions and UPD

UPD of a whole chromosome is not always required – UPD of only one gene or a group of genes can result in a clinical phenotype. For example, a whole extra chromosome 11 is not needed in order to display Beckwith–Wiedemann syndrome; the critical area has been narrowed to a region at the end of the short arm at 11p15.

Evidence from imprinting syndromes. Prader–Willi syndrome (PWS) and Angelman syndrome are two phenotypically different syndromes (see *Appendix: Glossary of disorders*) whose genes map to the same locus 15q11–q13. It is now known that although the genes for Prader–Willi syndrome or Angelman syndrome lie close together, they are coded separately and differently imprinted. This can be deduced for the following reasons:

- a deletion of the paternal 15 or maternal UPD results in Prader–Willi syndrome
- a deletion of the maternal 15 or paternal UPD results in Angelman syndrome (*Figure 2.6*)

The gene preventing Prader–Willi syndrome is imprinted (switched off) on the maternal 15 and so a working paternal 15 is necessary for normal development. The gene preventing Angelman syndrome must be active on the maternal 15.

Why is imprinting required?

Evidence from hydatidiform moles and ovarian teratomas:

- Genomic imprinting may be needed at a critical stage in cell division when it is important to have only one copy of a specific regulatory gene, such as the divergence of the inner cell mass and the trophoblast layer. Maybe by inactivating (imprinting) one gene or set of genes, a suitable

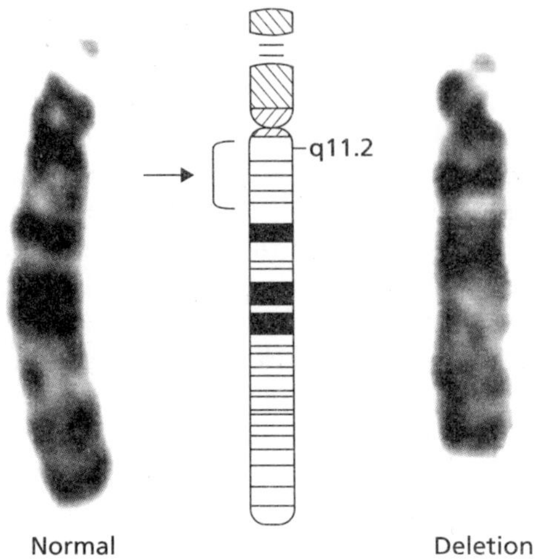

Figure 2.6
A pair of homologous chromosome 15s, with an ideogram showing the microscopically visible deletion typical of Prader–Willi and Angelman syndromes.

mechanism is created **to avoid confusion and competition between genes**. For example, the maternal genome controls the development of the embryo whilst the paternal genome controls the placental development.

- Ovarian teratomas are comparatively common in humans, so it has also been proposed that genomic imprinting **protects the female from trophoblastic invasion**, as the active paternal placental genes required for full development are not present. If genomic imprinting was not present, the trophoblast layer would also develop and a lethal choriocarcinoma would result.

Evidence from UPD 7 and Beckwith–Weidemann syndrome. Maternal UPD7 results in a small fetus, but trisomy 11p15 (which may comprise paternal UPD 11p15 and a maternal chromosome 11) results in a large infant.

It has been suggested that there is conflict between the mother and her fetus, such that the paternal genome of that fetus promotes fetal growth (the selfish gene wishing to survive). This may lower the mother's chance of a successful birth, however, so the maternal genome competes by trying to limit fetal growth (see *Box 2.9*).

2.6 DNA REPAIR

DNA mutations may arise in the following manner:

- physical damage due to external sources such as UV light or chemicals, leading to damaged bases which may be fused or cross-linked

Box 2.9 Imprinting as a short-term adaptation

Marcus Pembrey has suggested that imprinting may be useful as a short-term adaptation (2–10 generations) during times of stress or times of plenty.

It was noted that pregnant Dutch women gave birth to smaller infants during the famine of the Second World War. The post-war female infants themselves went on to have small babies; suggesting that this effect may last more than one generation. The opposite is also true. Magellan observed that a sixteenth-century Patagonian tribe was renowned for being particularly tall; this no longer remains true in the twenty-first century.

Although it appears that these observations are following acquired inheritance patterns, the explanation may be due to imprinting taking place on particular genes which have been exposed to (for example) a particular maternal diet in the first trimester of pregnancy.

Genes on the short arm of human chromosome 11 (11p15) include an insulin gene known to be capable of imprinting. It has been shown in rats that their insulin sensitivity can be decreased such that two generations of their offspring become more prone to diabetes.

The inheritance of these imprinted genes would still be passed on in a Mendelian manner.

- errors arising from mistakes in DNA replication, meiotic recombination, or faulty DNA repair, leading to incorrectly paired bases

Mutations are discussed in more detail in *Chapter 3*.

As most mutations are deleterious to the genome, surveillance at the DNA level is required. The fidelity of DNA repair is remarkable; in one year a germline cell containing 3×10^9 bp of DNA may only have 10–20 base pair changes. It is thought that DNA repair systems detect and react to abnormal conformations of the DNA helix, rather than acting in a sequence-dependent manner. There are three basic types of DNA repair.

Enzymatic photoreactivation

Enzymatic photoreactivation (EPR) is a **direct** form of repair which **reverses damage** to the DNA. For example, ultraviolet light may give rise to a fusion of two adjacent thymines, known as a thymine dimer. An enzyme called photolyase recognizes the distortion in the double helix and, when activated by light, uncouples the dimer. No DNA synthesis is required.

Excision repair

Damaged bases are recognized by enzymes which remove either one damaged base (base excision repair) or a length of 27–29 bp surrounding the damaged base (nucleotide excision repair, or NER).

In NER, the two nicks flanking the base are made by endonucleases, and a multi-subunit complex acts as an exonuclease which excises the DNA fragment containing the faulty base. The gap is repaired by a DNA polymerase (using the complementary sequence on the other strand) and is closed by a DNA ligase.

Excision repair cannot tell which is the 'correct' strand to repair; it recognizes mismatches and randomly excises the mismatched base from one strand or the other. The outcome may be favourable, or lead to a new pathway of mutation.

Mismatch repair

The most common source of mismatches arises from DNA replication and, to a lesser extent, during meiotic recombination. Mismatch repair (MMR) proteins **proof-read** the newly synthesized DNA, and detect anomalies such as small loops indicative of a mismatch. A nick is made 1–2 kb from the mismatch; this strand is degraded by exonucleases until it passes the mismatch. A DNA polymerase fills in the gaps and inserts the correct base.

MMR can recognize the new strand from the parental strand, because the parental strand is methylated at the adenine sites (in the sequence GATC), whereas the newly replicated DNA is not. This type of repair acts as an 'anti-mutator' pathway. There may be slippage and mismatching of microsatellite repeats which are found throughout the human genome, which will usually be corrected by MMR. At a particular locus these will remain the same length. If MMR is faulty, microsatellite repeats of different lengths are seen. This is the case with hereditary non-polyposis colon cancer (HNPCC), as discussed in *Chapter 8*.

2.7 IMMUNOGENETICS

The immune system

The function of cells in our immune system is to distinguish our own normal cells (self) from both external pathogens, such as bacteria and viruses, and from internal changes to normal cells, such as transformation into cancer cells (non-self).

There are two systems in the body where these fighting cells can be found; the blood system and the lymphatic system. Examples of lymphatic organs are the bone marrow which manufactures particular undifferentiated stem cells, the thymus (which is found at the base of the neck), and the lymph glands.

White blood cells (leukocytes) arise from the bone marrow and they can be divided into two cell lines:

- the **myeloid** cell line provides a very generalized natural or innate immunity to infection, which is **non-adaptive**
- the **lymphocytic** lineage provides a more specific immune response, and is part of an **adaptive** system comprising three kinds of lymphocytes: T cells, B cells and natural killer cells (or large granular lymphocytes)

Any foreign molecules, including whole bacterial cells, viruses, cellular components or proteins, can be detected by the immune system and are called **antigens**. The response of the adaptive immune system is to produce proteins called **antibodies** which recognize and bind to the antigen; these disease-causing molecules are then removed from the system. Our immune system mainly functions using two of the three types of lymphocytes.

B cells

B cells arise in the bone marrow and differentiate into plasma cells and memory cells. When a B cell meets an antigen for which it is specific, **clonal selection** occurs, whereby the B cell divides into many antibody-producing plasma cells. However, a few specialized **memory cells** are produced, which will remember the antigen should it be encountered in the future.

Antibodies are the protein product of the B cell genes; they are called **immunoglobulins** (abbreviated to Ig). As antigens can appear in the body in vast numbers of different types of molecules, it would be expected that enormous numbers of different antibodies would be needed to fit the antigens. However, there is not enough DNA to code for the millions of genes necessary to do this. Immunoglobulins therefore not only have common subunits, but also **variable regions** generated by **somatic recombination** of the DNA (i.e. the genes) during the development of the B lymphocytes. To see how this **genetic diversity** is achieved, we must look at antibody structure.

Antibody structure

Immunoglobulins comprise four chains or protein subunits. Because of the differences in weight, the two identical longer chains are called the **heavy chains** (H), the two shorter chains are **light chains** (L). They are held together with disulphide bonds in a 'Y' shape (*Figure 2.7*). The upper region that binds the antigen is called the Fab (fragment antigen binding) fragment; the lower region is the Fc (fragment crystallizable) fragment. As expected, the variable region of the light chains binds the antigen.

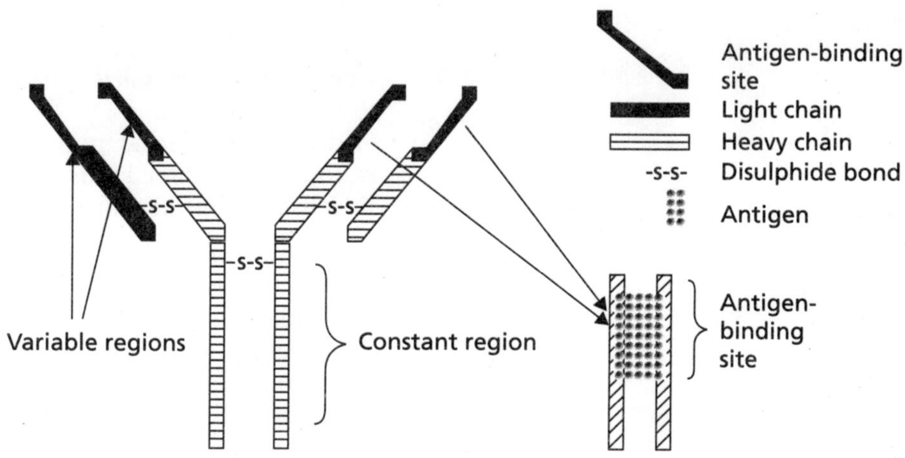

Figure 2.7
Antibody structure.

Light chains (L)

There are two types of light chains, called kappa (κ) and lambda (λ), coded by genes on chromosomes 2 and 22, respectively. However, although we produce both types of light chain, in any particular immunoglobulin molecule either two κ or two λ chains interact with the heavy chains – never one of each.

Somatic recombination of light chains. If we consider, for example, the production of a kappa light chain, we first have to look at the arrangement of genes in the germline. As an embryo, all our cells start with a common arrangement of kappa genes, coding for three different regions of the final chain. As the B lymphocyte differentiates and matures, a somatic recombination (rearrangement) occurs which is unique in that it involves post-zygotic DNA.

In the unrearranged germline, there are three segments separated by introns; the **variable** light chain region (V_L) comprising around 300 genes, a smaller **joining** segment (J_L) comprising five genes, and a single **constant** region (C). Using enzymes called recombinases, just one Vκ gene is selected, together with just one Jκ gene; these are then joined to the C gene to form a VJC unit which has a fixed antigen specificity in a particular B cell (*Figure 2.8*). If any one of the 300 separate Vκ genes can join with any one of 5 Jκ genes (and one constant), then 1500 different kappa chains are possible.

The rearrangement process in lambda light chains is similar to kappa chain synthesis and gives a similar number of rearrangement possibilities.

Heavy chains (H)

These are coded for by genes on chromosome 14. In a similar manner to light chains there is a **variable** region (V) comprising 100–300 genes, an additional **diversity** segment (D) of 12–20 genes, and 4–6 **joining** (J) genes.

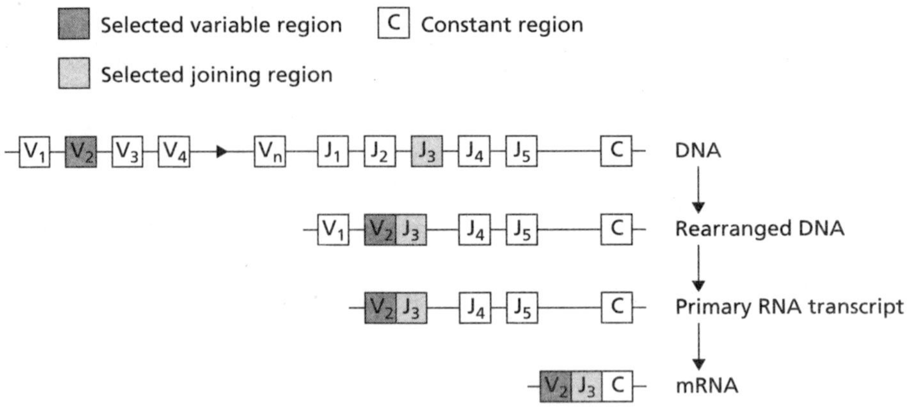

Figure 2.8
Somatic recombination of the kappa light chain genes.

The **constant** (C) region is separated by a large intron and is more complicated than the light chains in that there are five C genes in the germ cells.

Somatic recombination of heavy chains. Two rearrangements occur separately; a D gene is joined to a J gene, and then a V gene is brought next to the DJ unit giving a fixed antigen specificity. A total of at least 4800 different heavy chains are therefore possible.

Genetic diversity in B cells

It can now be seen that from a few hundred subunits of the kappa and lambda light chain genes, together with the heavy chain genes, as many as 10^{11} combinations are possible. This generates antibodies in a sufficient variety of shapes needed to bind with different antigens. IgM is the first immunoglobulin made in an immune response by the B cells themselves. It appears on the cell surface; when a matching antigen is encountered growth of the B cell is triggered.

T cells

T cells arise in the bone marrow and migrate to the thymus to mature. As they do so, they develop membrane receptors (T-cell receptors, or TCRs) which recognize particular antigens (see *Box 2.10*). Unlike the free immunoglobulins of B cells (which generate the humoral response), these receptors remain bound on the surface of the T cell (and so generate the cell-mediated response).

The structure of the TCR includes an α and β protein chain (designated Ti$\alpha\beta$, or a γ and δ protein chain (Ti$\gamma\delta$).

Box 2.10 The major histocompatibility complex

While B cells recognize antigens on their own, the TCRs need the help of other sets of proteins such as those of the **major histocompatibility complex** (MHC). The MHC region is on the short arm of chromosome 6, comprising a number of genes which are very polymorphic; this means that each gene has up to 50 alternative sequences (alleles) per locus.

Each individual therefore has a near unique combination of these genes which identify you and your cells as self. In families, closely related people have more chance of sharing these genes, which are passed down from our parents in 'blocks' or **haplotypes**. This is why organ transplants are rejected between unrelated individuals, but have more chance of succeeding within a family, as the tissue type is more likely to have a closer match of genes. Identical twins, having identical MHC genes, will match exactly. However, identical twins will have their own non-identical sets of immunoglobulins produced by differing somatic B cell recombination.

T cell diversity

T cell maturation also involves **rearrangements** of the gene segments, which will result in unique TCRs. α and γ chains have V and J segments, while β and δ chains have V, D and J segments. The β and γ chain genes are on chromosome 7, and the α and δ chain genes are on chromosome 14.

If we take the diversity of immunoglobulins (with heavy and light chains) and the two types of TCR heterodimers (αβ and γδ), the total possible number of rearrangements is enormous. Other mechanisms such as hypermutation in B cells raise the number of combinations for antigen recognition to a staggering 10^{18}.

Allelic exclusion

The components synthesized by B and T cells are either derived from the maternal chromosome or the paternal chromosome, but not both. One allele is excluded.

The chromosomal input into a B cell is from both parents, so theoretically there should be a choice of alleles (alternatives) from two parental chromosome 2s, two chromosome 22s, and two chromosome 14s. However, in each B cell only one locus from each chromosome is expressed, so the antibody comprises one heavy chain, together with either a kappa or lambda light chain.

In a particular T cell having, for example, two α and two β chain loci, only one α and one β locus are expressed, showing the allelic exclusion found in B cells.

2.8 MITOCHONDRIAL GENES

In *Chapter 1*, mitochondria were introduced as organelles found in the cell cytoplasm of eukaryotes. Each cell has several hundred mitochondria, each having an outer smooth membrane and an inner folded membrane where energy is generated for the body's metabolic processes by oxidative phosphorylation.

The mitochondrial genes are independent of the nuclear genome in that they have their own transcription and translation systems. The reason why mitochondria appear to be more independent than other cell organelles is that they may have been derived from prokaryotic bacteria which became symbiotic (lived together for mutual benefit) with eukaryotic cells early in evolution.

Oxidative phosphorylation comprises five multi-enzyme complexes. Thirteen mitochondrial genes code for some of the subunits of four of these complexes: the NADH dehydrogenase complex, the cytochrome b–c1 complex, the cytochrome c oxidase complex and the ATP synthase complex. Together with subunits encoded by the nuclear genome, these components enable molecular oxygen to be released in a usable form from cells.

The other major group of 22 mitochondrial genes code for each tRNA corresponding to a particular amino acid, e.g. ser-tRNA or leu-tRNA.

2.9 DEVELOPMENTAL GENETICS

The aim of this section is not to describe in detail the development of a fetus because that information can be obtained from a good embryology book.

Instead, in keeping with previous sections which have looked at genes and gene control, this section studies three of the many interacting cascades of genes responsible for certain aspects of developmental pathways: growth factors, signalling genes and homeobox genes.

In the early days of developmental biology, the key question with respect to a newly fertilized embryo was whether each cell (e.g. an intestinal cell or a brain cell) had a predetermined fate, or whether every cell had the potential to become any of the differentiated body cells.

An early human embryo containing only hundreds of cells can divide into two roughly equal groups, each of which develops into distinct but identical (monozygotic) twins. All the genes of the dividing embryo are therefore active at this early stage. Embryonic cells are **totipotent**, and have the potential for full growth and differentiation into any cell type (see *Box 2.11*).

Three layers of cells arise in the early embryo; the **endoderm** (= inner, giving rise to structures such as the lungs), the **mesoderm** (= middle, giving rise to the blood and genito-urinary system), and the **ectoderm** (= outer, giving rise to the spinal cord, brain and skin). But how do these cells know:

- where they are in the cell mass (**position**)?
- which end of the structure should grow (**polarity**)?
- where in that general structure should a pair of limbs or organ be situated?
- how many limbs and which way up should they be (**pattern formation**)?

As well as knowing how to differentiate and grow, cells also have to know when to slow down, stop growing or die. A fetal hand starts by looking like a

Box 2.11 Mammalian cloning

Another way that totipotency can be shown arises from the method commonly known as **cloning**. Originally demonstrated in frogs by Gurdon in the 1960s, it is now possible in mammals. Dolly the cloned sheep was born in 1997, after Ian Wilmut and his veterinary colleagues in Scotland took a mammary gland cell (which had been frozen in liquid nitrogen for 6 years) derived from the original Dolly. The nucleus was extracted by micromanipulation (using a very narrow pipette). An egg (ovum) from another sheep was taken and the nucleus removed. Dolly's nucleus was then pushed through the egg cytoplasm and given a weak electric shock to simulate fertilization.

The result was an apparently normal sheep, proving that one mammary gland cell can eradicate its mammary gland origin and 'redevelop' (dedifferentiate). This shows that all the genes must be present, and that the cell nucleus must be totipotent.

The only question remaining was how old was Dolly? Was she newborn, or was she 6 years old: the age of her original cell? Dolly died in 2003 and was found to have lung disease and arthritis. As sheep usually live until 11 or 12 years of age, this suggests she was ageing prematurely, so perhaps the age of the donor cell nucleus is not reset.

In 2007 scientists created the first cloned monkey embryos and extracted stem cells from them. This offers the hope of developing treatments in humans for a range of disorders, as well as growing replacements for damaged tissue.

Cloning is gradually being superseded by a new technique whereby a sample of skin cells is taken from an adult, and injected with genes to induce re-programming. The cells become pluripotent, resembling the stem cells found in human embryos.

mitten; the fingers are formed because cells die in the spaces in between. This is called programmed cell death (**apoptosis**), and is discussed further in *Chapter 8*.

Embryonic pattern formation

Taking the limb as an example of development, each phase can be studied with respect to the particular groups of genes involved at each stage. **Orthologs** are similar, or homologous, genes or DNA sequences between different species, e.g. *SHH* in humans, *hh* in *Drosophila*, or *SRY* in humans, *Sry* in mice. **Paralogs** are homologous genes or DNA sequences within a species, e.g. the four sets of homeobox or *HOX* clusters of human genes, such that A13 is a paralog of D13. Note that human genes are written in italicized capitals. Their gene products can also be structural RNA, and are written in non-italicized capitals. Genes from non-human species are written in lower case italic script; mouse genes have an initial italicized capital letter.

Although in the past most developmental biology has been studied through non-human sources such as the fruit fly (*Drosophila*), or vertebrates such as the mouse or chicken, it seems likely that animal models will be found to apply to humans too (except for minor differences such as wings or tails!).

Growth factors

During the initiation of the limb bud, cell division must occur by mitosis. Growth factors provide the initial stimulation for this to occur; they can also control the cell cycle during the transition from G_0 to G_1. Different kinds of cells have appropriate growth factors: fibroblast growth factors (FGF) for fibroblast cells (e.g. connective tissue), epithelial growth factors (EGF), nerve growth factors (NGF), and platelet-derived growth factor (PDGF) which is found in blood vessels and helps in clotting.

Each growth factor has a membrane-bound cellular **receptor** to which that specific growth factor binds. An example is FGFR3 (fibroblast growth factor receptor 3). Inside the cell (intracytoplasmic), the binding of a growth factor to a specific receptor causes an intracellular cascade of events culminating in chemical signals which can suppress or activate developmental genes.

Some factors can suppress growth, including the tumour necrosis factor (TNF) or the transforming growth factor (TGF).

The hedgehog signalling pathway

Two stages must occur during embryonic pattern formation. The cells must be informed of their place in a three-dimensional system and they must know their spatial orientation (see *Box 2.12*). Only then is that information interpreted to form appropriate structures, for example, specification of limb pattern such as an arm or leg.

It is believed that groups of cells have certain boundaries which are characterized by gradients (differences in concentration) of different

Box 2.12 In the context of the structure of an organism

- Anterior is the front end of an organism; in humans this is the 'head end'.
- Posterior is the hind end of an organism; in humans this is the 'tail end'.
- Ventral is the underside of an organism; in humans this is the stomach surface.
- Dorsal is the upper surface of an organism; in humans this is the back.

molecules (sometimes known by the general term **morphogens**). Cells respond to threshold concentrations of these molecules which arise from a positional signal from a specialist area of the cell: the **polarising region** (P).

Cells near the P region are exposed to high morphogen concentrations and form **posterior** digits (e.g. legs, feet, toes); cells further away become **anterior** digits (arms, hands, fingers). A human gene thought to encode such a long-range signalling molecule across a limb bud is the **sonic hedgehog gene** – *SHH* . This is one of a class of **segmental polarity genes.**

The hedgehog signalling pathway is extremely complex, but some elements are understood mostly through its *Drosophila* counterpart *hh* and its mouse counterpart *Shh.* The sonic hedgehog gene is responsible for the patterning of the ventral neural tube, the notochord, limbs, foregut and lung, controlling growth and differentiation by interacting with and indirectly inducing other banks of genes which have more specific functions. *WNT* (wingless related gene), for example, is involved in the **ventral** patterning of a limb and *DPP* (decapentaplegic) in patterning on the **dorsal** side. *HOX* genes have been extensively studied and are discussed below.

In the absence of *SHH* a gene called **patched** (*PTCH*) acts to prevent high expression and activity of a receptor called smoothened (SMO). In its normal role of growth suppressor, patched has become known as a 'gatekeeper'; a line of defence against uncontrolled skin cell proliferation resulting in tumours. This gene will be discussed further in *Chapters 3* and *8*. When *SHH* is present, it binds to *PTCH,* releasing *SMO,* which activates a series of GLI transcription factors (*Figure 2.9*).

Homeobox and paired box genes

We have seen how the limb bud is initiated, the pattern specified and the correct dorsal and ventral differentiation occurs. How do the arms and legs know where to appear along the length of the backbone?

This is the function of the **homeobox** (*HOX*) genes. There are 38 in man, divided into four groups A–D, and then numbered (e.g. 1–13) such that a particular gene may be called, for example, *HOXD13.* The *HOX* genes are actually transcribed in order (i.e. *HOXA1* to *HOXA13*) from the 5′ to 3′ direction, which corresponds to a linear time sequence from the posterior end of the embryo to the anterior end.

Originally studied in *Drosophila* larvae where they are responsible for determining the fate of each larval segment in the adult fly (e.g. antennae, wings), in humans they determine at which point along the spinal vertebrae

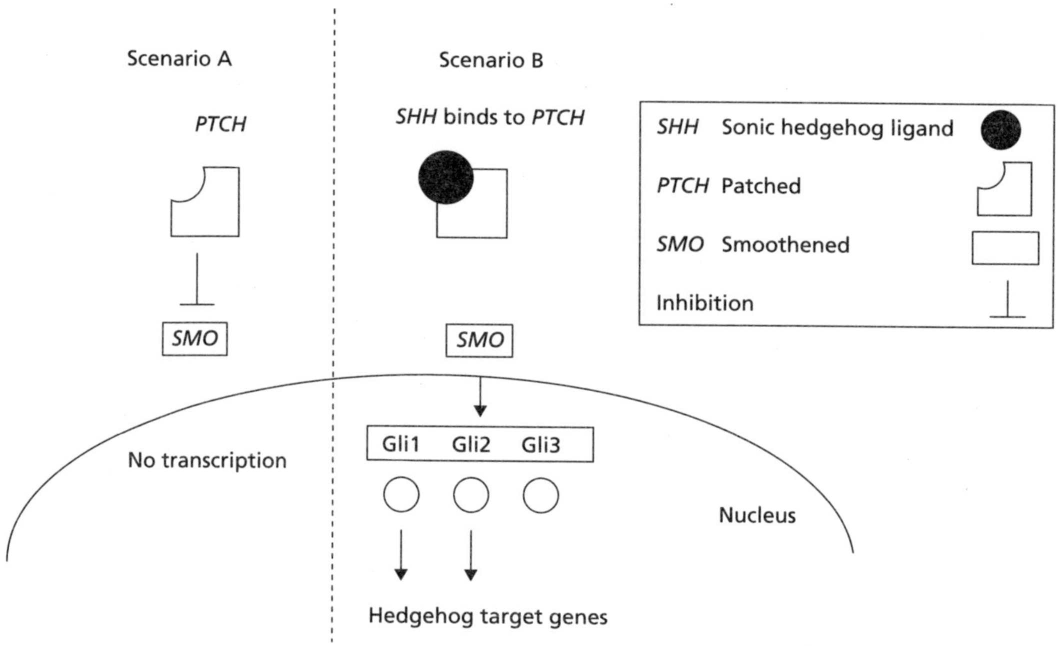

Figure 2.9
Part of the hedgehog signalling pathway.
Scenario A: in the absence of the *SHH* ligand, *PTCH* represses *SMO*, so no transcription occurs.
Scenario B: the binding of the *SHH* ligand to *PTCH* relieves *SMO* inhibition. The Gli transcription factors are activated. Gli1 and Gli2 are activators; Gli3 is a repressor.

the appropriate limbs appear (antero-posterior position **patterning**), and also ensure the correct numbers of appendages – in matching pairs.

Some functions of an associated group of genes called **paired box** (*PAX*) genes have also been identified. *PAX-3* controls the growth and differentiation of groups of cells which have multiple roles in eye and hair pigmentation and hearing. *PAX-6* is involved in normal eye formation, especially that of the coloured part of the eye – the iris.

Growth to adult size

Once the fetal limbs are fully differentiated, a period of enlargement follows. Here we have come full circle, as fibroblast growth factors (FGFs) are again required for both shaping and growth.

When considering these stages it becomes apparent that the same sets of molecules can be involved in development at different times and in different places. In this way the numbers of mechanisms required in genetic development are conserved.

2.10 SEX DIFFERENTIATION

One of the first questions the proud parents of a newborn baby are asked is 'is it a boy or a girl?'. Most people know that of our 46 chromosomes, two are called the sex chromosomes – XX for a girl, XY for a boy (see also *Box 2.13*). It seems reasonable to assume therefore that the Y chromosome must contain genes specifically for male development.

A primary male sex determining gene lying proximal to the PAR and inducing testis formation was found in 1991, and is called *SRY* (the sex determining region on the Y). Some books refer to this as the *TDF* (testis determining factor).

Sexual development in the fetus

At six weeks of age embryos contain identical precursors for both the male and female sexual structures, called, respectively, the Wolffian and Mullerian ducts. After six weeks, the pathways of these indifferent gonads begin to diverge, according to the presence or absence of the *SRY* gene on the Y chromosome. It is possible that early in evolution there were two identical chromosomes similar to the present X chromosome. Over time one of the X chromosomes evolved into the present Y chromosome. Vestiges of this heritage are still to be found if we look at certain regions of homology on Xp and Yp, where the DNA sequences are so similar that these regions can pair at meiosis. These are the pseudo-autosomal regions discussed earlier.

The male pathway

By 10 weeks, if the *SRY* gene is present, the Wolffian ducts will begin to develop into the fetal testes. The testes secrete two hormones, one of which inhibits the development of the Mullerian ducts. The other is the male hormone testosterone. Providing the target cells have androgen receptors

Box 2.13 Sex chromosomes in other species

In humans the female is the **homogametic sex** (producing one kind of gamete); the male is the **heterogametic sex** (producing two kinds of gametes). This is not always the case in other species.

In birds, moths and butterflies the sex chromosomes are termed Z and W. The female is the heterogametic sex (ZW) and the males are homogametic (ZZ).

In *Drosophila*, sex determination depends on the X:autosome ratio, i.e. the balance of female determining genes on the X chromosome(s) to the number of haploid sets of autosomes. The Y chromosome plays no part in determining sex. The critical ratio for maleness is 1(X):2(autosome sets). For example:

a normal female = XX + 2*n* autosomes
a normal male = XY + 2*n* autosomes
a 'metafemale' = XXX + 2*n* autosomes
a 'metamale' = X + 3*n* autosomes

which will respond to testosterone, differentiation of internal male structures such as the vas deferens, epididymis and seminal vesicles will follow.

At 15–16 weeks some of the testosterone is converted into 5-hydroxy-testosterone, which is responsible for the development of the external male sexual structures such as the urethra, scrotum and penis.

The female pathway

In the absence of the *SRY* gene, the Wolffian ducts degenerate, and by 10 weeks the fetal ovary is producing oestradiol, leading to the differentiation of the female structures such as the uterus, fallopian tubes and upper vagina.

Other genes involved in sexual differentiation

Besides *SRY*, there are a number of other genes involved in mammalian sexual development. *SF1* is a steroidogenic factor involved in early establishment of male and female gonads. *WT1* is also involved in early development of both gonads and kidneys, and is a possible up-regulator of *SRY* expression in males.

SOX9, acting downstream from *SRY*, is involved in Sertoli cell differentiation and initiation of testis development in normal males. *DAX1* is expressed in females from the short arm of the X chromosome, in a region responsible for dosage-sensitive sex reversal (DSS); see *Chapter 3*.

FGF9 is expressed in males, and is involved with proliferation in the male gonad. *WNT4* is expressed in females and contributes to differentiation in the female gonads. It is now thought that in the early development of the mammalian gonad, *WNT4* and *FGF9* are in equal balance. Disturbing the equilibrium in favour of *WNT4* leads to development of an ovary, while *FGF9* leads to development of a testis. In mice, the presence of *Sry* upregulates *Sox9*, which then activates the *Fgf9* gene. This ensures that the gonad becomes male, and also suppresses *Wnt4*. In the absence of *Sox9* and *Fgf9*, *Wnt4* pushes development of the female pathway, and blocks testis development.

The mechanism of sexual differentiation in non-mammalian vertebrates also appears to rely on the equilibrium between *Wnt4* and *Fgf9*. As only mammals have the *Sry* gene, other factors such as genetics or environment may therefore disrupt the balance in vertebrates who do not possess *Sry*.

2.11 SUMMARY

When genes were first postulated, the definition 'one gene, one polypeptide' or 'one gene, one enzyme' did appear to explain simple linear biochemical pathways. With the concept of gene control came the realization that genes could act in more complex ways, as demonstrated by embryonic development, the immune system, and even by determining which sex we are.

We now know that some genes do not produce proteins – they produce structural RNAs instead, which exert their control by their shape. We cannot even be sure that a gene we receive from one parent will behave in the same way as the allele we receive from the other parent, as demonstrated by imprinting.

Considering the intricate cascades of structural and controlling genes needed for the correct functioning of the human body, it is not surprising that occasionally individuals are born in whom certain genes do not function correctly.

The next chapter looks at some examples of genetic disorders to illustrate how understanding the mechanisms of genetic malfunction helps to understand the normal role of genes.

SUGGESTED FURTHER READING

Carrel, L. and Willard, H.F. (1998) Counting on XIST. *Nature Genetics*, **19:** 211–212.

Colvin, J. *et al.* (2001) Male-to-female sex reversal in mice lacking fibroblast growth factor 9. *Cell*, **104**: 875–889.

Dean, M. (1996) Polarity, proliferation and the hedgehog pathway. *Nature Genetics*, **14**: 245–247.

Evangelista, M. *et al.* (2006) The hedgehog signaling pathway in cancer. *Clinical Cancer Research*, **12**: 5924–5928.

Hammerschmidt, M., Brook, A. and McMahon, A.P. (1997) The world according to hedgehog. *Trends in Genetics*, **13:** 14–20.

Hannigan, B.M., Moore, C.B.T. and Quinn, D.G. (2009) *Immunology*, 2nd Edition. Bloxham: Scion Publishing.

Kim, Y. *et al.* (2006) *Fgf9* and *Wnt4* act as antagonistic signals to regulate mammalian sex determination. *PLoS Biology*, **4**(6): e187.

Mizusaki, H. *et al.* (2003) *Dax-1* (dosage-sensitive sex reversal-adrenal hypoplasia congenital critical region on the X chromosome, gene 1) gene transcription is regulated by Wnt4 in the female developing gonad. *Molecular Endocrinology*, **17**: 507–519.

Ohlsson, R., Hall, K. and Ritzen, M. (1995) *Genomic Imprinting: Causes and Consequences.* Cambridge: Cambridge University Press.

Wolpert, L. (1996) One hundred years of positional information. *Trends in Genetics*, **2:** 359–363.

SELF-ASSESSMENT QUESTIONS

1. Gene A makes a protein that forms the globin of red blood cells. Gene B makes a protein which is a transcription factor that helps to switch on gene C. What is the difference between gene A and gene B; what kinds of genes are they?
2. Why is a primary RNA transcript freshly copied from a DNA template longer than the piece that is eventually translated into protein? Name the process involved.
3. A gene for the enzyme steroid sulphatase is found on the X chromosome very close to the pseudo-autosomal region. Females have approximately twice the

level of this enzyme compared to males. As dosage compensation has not occurred, what explanation is possible?

4. Name a major developmental gene and give an example of its function.
5. Name the gene on the Y chromosome responsible for male development.
6. A genetic condition called Turner syndrome exists; the individual has 45 chromosomes including only one X. What sex will they be and why? Do you think that they will be affected in any way?
7. What is somatic recombination and why is it important in antibody production?

CHAPTER 3

Mechanisms of disease

Learning objectives

After studying this chapter you should confidently be able to:

- **List the major types of DNA mutations and give examples**
  DNA mutations can be divided into categories such as base substitutions, mis-sense, nonsense or frameshift mutations. Such mutations lead to loss or gain of function. Triplet repeats are dynamic mutations which may produce a clinical effect by expansion.
- **Describe how gene control may be altered**
  Gene mutations can also alter the control of gene expression through various mechanisms such as positional change, haploinsufficiency and inappropriate methylation, leading to loss or gain of function. This may occur on a larger scale through chromosomal changes resulting in aneuploidy, polyploidy or structural rearrangements such as translocations or duplications.
- **Explain the consequences of the failure of resetting of imprinting**
  A mutation in an imprinting centre will prevent the correct resetting of a parental imprint in the germ cells. An inherited imprinting centre mutation in a parent may lead to an imprinting disorder in their children, as one parental locus is not functioning normally.
- **Give examples of immunogenetic and mitochondrial disorders**
  Mutations in genes responsible for the formation of B and/or T cells lead to the inability to produce mature or functioning B or T cells, so immunity against infection is compromised. Faults with mitochondrial structure, altered numbers of mitochondrial genes, or mutated mitochondrial genes, may result in mitochondrial DNA disorders.
- **Outline the mechanisms by which DNA repair may fail, and give examples**
  DNA syndromes display increased rates of cancers and immune disorders. Mutations of excision repair genes, mismatch repair genes, or genes involved in DNA replication, lead to uncorrected base changes and failure of DNA repair.
- **Describe the results of gene mutations relating to developmental and sex determining pathways**
  The 46,XY genotype may be expressed as a female phenotype due to mutations or deletions of the *SRY* gene, or a defect in the androgen receptor gene. A 46,XX genotype may appear phenotypically male due to the presence (usually on one of the X chromosomes) of the *SRY* gene, or due to hormonal abnormalities such as those seen in CAH.

3.1 MUTATIONS

Genetic diseases may be caused by a number of different mechanisms (see *Table 3.1* for a summary) but which, at the molecular level, arise from the heritable alteration of a DNA sequence, known as a mutation.

A **mutation** may be defined as any change in the genetic make-up of a cell, a population of cells, or an organism. The mutation originates as an error in one cell, and will only have a determinable effect if it is heritable, i.e. when that cell divides, the mutation becomes distributed throughout the descendants. It is convenient to identify two broad categories, molecular (or gene) mutations, and gross chromosome mutations.

Many mutations are deleterious; some are extremely important in the aetiology and pathology of human genetic diseases, and it is these that comprise the very basis of this book. Some of the possible endpoints of mutation are as follows.

- Accumulation of mutations, occurring as errors of DNA replication, possibly contribute to cellular senescence; some mutations are inevitably lethal to the cell in which they occur.
- Mutations which interfere with normal cellular differentiation and proliferation lead to tumours.
- Mutations occurring in the germ cell line are heritable, and may lead to genetic disease (or handicap) in the offspring.
- The occurrence of advantageous mutations and their contribution to evolution through selection, are beyond the scope of this book.

It is important to understand that DNA is not only a large complex molecule, it is active through replication and transcription, and it is chemically reactive. Inevitably, therefore, changes to DNA occur continually. Mutations or changes in the DNA that become fixed and inherited by daughter cells at cell division, originate either as a straightforward error of a normal cellular process such as replication, recombination or mitosis, or as a result of the influence of an external agent, known as a **mutagen**. Most mutagens are agents reacting with DNA, while others may affect the genetic make-up of a cell indirectly by disrupting cellular processes that control the normal behaviour of the genetic material. Familiar examples include ionizing radiation, ultraviolet light, and certain classes of chemicals such as alkylating agents. Many mutagens have also been proven to be **clastogens** (i.e. they induce breakage of chromosomes, visible at mitosis) and **carcinogens** (i.e. they induce tumours, see *Chapter 8*). The normal cellular responses to these agents, for example, DNA repair, are described in *Chapter 2*.

At the molecular level, an important category of mutation is the **base substitution**, where the replacement of one base by another results in a change in a single coding triplet (see *Box 3.1*). The most common outcome of such a mutation is to alter a single amino acid in that protein for which the affected gene codes. This is known as a **mis-sense mutation**. However, a number of other effects are possible. For example, the substitution may create a chain terminator codon (see *Chapter 1*); this is a **nonsense mutation**, resulting in a truncated protein. Alternatively, a substitution occurring at a

Table 3.1 Mechanisms of change

Normal process	Disorder	Gene	Mechanism
Gene control and expression			
Transcription	β-thalassaemia		Base substitutions or frameshift mutations
	Duchenne muscular dystrophy		Frameshift mutations, duplications
Dynamic mutations	Fragile X	*FMR-1*	Triplet repeat CGG
	Myotonic dystrophy	*DMPK* (myotonin protein kinase)	Triplet repeat CTG
	Huntington disease	*IT15*	Triplet repeat CAG
	Friedreich ataxia	*X25*	Triplet repeat GAA
Chromatin structure	Facioscapulohumeral dystrophy		Positional?
Dosage	Turner, distal 22q11.2 deletion, syndrome		Haploinsufficiency?
X-inactivation	Wiscott–Aldrich syndrome		
Methylation and imprinting	Prader–Willi and Angelman syndrome	*SNRPN?/UBE3A*	Deletion, UPD, imprinting centre mutation
DNA repair	Xeroderma pigmentosum		Mutations in genes involved in excision repair
Immunogenetics	Bruton's agammaglobulin-aemia		B cell defect
	DiGeorge syndrome		T cell defect
	Severe combined immunodeficiency disease		B and T cell defects
Mitochondrial	MELAS, MERRF, etc.		Mutations in the mitochondrial genome
Development			
Growth factors	Achondroplasia Craniosynostosis (Pfeiffer and Crouzon syndromes)	*FGFR3* *FGFR1* and *FGFR2*	Mutations in fibroblast growth factor receptor genes
Hedgehog signalling pathway	Holoprosencephaly	(*SHH*) *Hpe3* 7q36	Haploinsufficiency via deletion/positional silencing via telomere
	Basal cell carcinoma	(Patched) *PTCH*	Haploinsufficiency?
Homeobox genes	Synpolydactyly	*HOXD13*	Gain of function
	Waardenburg syndrome	*PAX3* (2q35)	Haploinsufficiency/loss of function
	Aniridia	*PAX6* (11p13)	Loss of function
Sex determination	Sex reversal	*SRY*	Deletion/mutation
	Complete androgen insensitivity syndrome	Androgen receptor	Deletion
	Congenital adrenal hyperplasia	CAH (*CYP21B*)	Deletion/mutation, gene conversion
	Campomelic dysplasia	*SOX9*	Positional

Box 3.1 Transitions and transversions

A base substitution mutation in which one purine (guanine or adenine) replaces the other, or where one pyrimidine (thymine or cytosine) replaces the other, is called a **transition**. When a purine replaces a pyrimidine, or vice versa, the mutation is known as a **transversion**.

critical location may alter the splicing of an exon to intron junction, or one occurring in the promoter region of a gene may alter the level of expression of that gene.

A classic example of a base substitution mutagen is nitrous acid, which is able to deaminate either adenine to hypoxanthine (the NH_2 group at position 6 (see *Figure 1.4*) is replaced by an oxygen atom), or cytosine to uracil. Hypoxanthine pairs with cytosine; uracil pairs with adenine.

Other molecular DNA mutations involve the addition or deletion of bases, and the result is often a **frameshift mutation**, as commonly found in the disorder Duchenne muscular dystrophy (see *Box 3.2*). Loss or gain of one or two bases means that all the triplet codes downstream of the mutation become out of phase. As the reading frame of the mRNA is based on non-overlapping groups of three nucleotides, the translation will be different from the original mRNA, such that incorrect amino acids will be added to the protein until a stop codon or a splicing signal is encountered. Only if the addition or deletion comprises three bases, or a multiple of three, can the protein continue to be assembled with just the alteration of a small number of amino acids. The most common European cystic fibrosis mutation, p.Phe508del, is an example of a three base pair deletion which removes a single amino acid, phenylalanine, from the protein.

Box 3.2 Frameshift mutagens

Frameshift mutagens include ionizing radiation, one effect of which is to knock out single bases. Certain chemical agents intercalate within the double helix of the DNA, distorting it and making it susceptible to damage. Intercalating agents include some of the fluorescent dyes used in the staining of molecular DNA and chromosomes, for example, ethidium bromide and acridine orange.

Molecular mutations can be conveniently illustrated by considering abnormalities of the blood protein haemoglobin. In the normal adult, most of the haemoglobin is haemoglobin A. This molecule is a tetramer constructed from four protein sub-units, comprising two α-globin chains, and two β-globin chains, coded for by separate genes on chromosomes 16 and 11 respectively. Many abnormal types of haemoglobin have been discovered resulting from various point mutations; some of these are benign, whereas others result in diseases such as sickle cell anaemia, α-thalassaemia, or β-thalassaemia (see *Box 3.3*).

The sickle cell mutation is a point mutation changing the sixth amino acid in the β-globin chain from glutamic acid to valine as a result of a transver-

Box 3.3 Heterozygote advantage

In certain African populations, the sickle cell mutation has reached a high level as a result of the selective advantage of the heterozygous state, or sickle cell trait, which confers resistance to malaria. In the homozygous state, the mutation results in a severe and life-threatening chronic haemolytic anaemia. In some Mediterranean populations, thalassaemia mutations are common, probably also related to malaria resistance, and mutation in the homozygous state is again either lethal or severely life-threatening. In Cyprus, for example, one person in six is heterozygous for a β-thalassaemia mutation.

sion mutation changing the codon from GAG to GTG. The single amino acid difference is sufficient to alter the properties of the haemoglobin molecule.

Some of the common Mediterranean β-thalassaemia mutations involve base substitution at the splice site between the first exon and the first intron, resulting in abnormal processing of the messenger RNA.

The most common β-globin mutation in Sardinia is a transition at codon 39, changing CAG (glutamine) to TAG, which is the amber stop codon. In the homozygote, β-globin is therefore absent.

An example of a frameshift mutation occurring in some Mediterranean β-thalassaemia patients, is the loss of A from the codon 6 triplet GAG.

Mutation at the level of the chromosome generates aneuploidy, polyploidy, and all of the various types of structural chromosome abnormality discussed in *Chapter 5*. The external influences leading to chromosome gain are poorly understood, although various substances are known to interfere with normal spindle fibre formation, and thus affect normal segregation of the chromosomes at cell division.

Structural chromosome changes are induced by ionizing radiation, by ultraviolet light, and by chemical agents, which either damage bases or cross-link them. Chromosome abnormalities (for example, translocations) are generated by incorrect replication, or incorrect repair, in the vicinity of two or more sites of damage, or lesions. Mutagenic and clastogenic agents are commonly used in cancer therapy, where damage to the nuclear DNA of the malignant cells is an effective way of reducing the uncontrolled proliferation of those cells. Radiotherapy with X-rays or radioisotopes is frequently used, while the drugs include alkylating agents such as cyclophosphamide, mitotic spindle inhibitors such as vincristine, and DNA synthesis inhibitors such as hydroxyurea. Patients treated with some types of mutagenic agents have an increased risk of developing a different unrelated, treatment-induced malignancy.

Triplet repeats/dynamic mutations

The human genome has many types of repeat sequences; one type of repeats comprises repetitive runs of three DNA bases called **trinucleotide** or **triplet repeats**. Triplet repeat disorders usually show a dominant pattern of inheritance (see also *Chapter 4*). Normal individuals usually have low numbers of repeats, and they have no clinical effect. As the numbers of repeats rise above a critical level, a pathogenic effect may occur which results in phenotypic expression of that disease. Individuals who inherit larger numbers of triplet repeats are more likely to show a phenotypic effect.

In most of these disorders it has been found that the numbers of repeats increase with each generation, such that the likelihood of disease expression (or severity of disease phenotype) increases down the generations; this is called **anticipation**. Because the triplet repeats change from generation to generation, they are also known as **dynamic mutations**.

The changes in numbers may arise from mispairing of the two DNA strands because there are so many identical short sequences. When replication occurs, the polymerase enzyme copies a longer length of repeats due to the **slippage** and **mispairing** of the strands.

If unequal crossover occurs between the homologues at meiosis, or unequal exchange occurs between chromatids at mitosis (sister chromatid exchange), there will be a duplication (or increase in repeats) on one homologue or chromatid, and a deletion (decrease) on the other (*Figure 3.1*).

The number of diseases known to be caused by triplet repeat **expansion** is currently approaching 20. Huntington disease is caused by an expansion of the triplet repeat **CAG**. Fragile X syndrome is due to an expansion of a **CGG** repeat, and myotonic dystrophy has a **CTG** repeat. The location of triplet repeats (with respect to the gene structure) appears to determine the maximum number of repeats found in an affected person and the disease mechanism (see *Box 3.4*).

An exception to the usual pattern of dominant inheritance is Friedreich ataxia, with the novel repeat **GAA**. The autosomal recessive inheritance pattern suggests a different disease mechanism for the GAA repeat, as two copies of the expansion are required. Fragile X syndrome also has an unusual inheritance pattern which is discussed in *Chapters 4* and *7*.

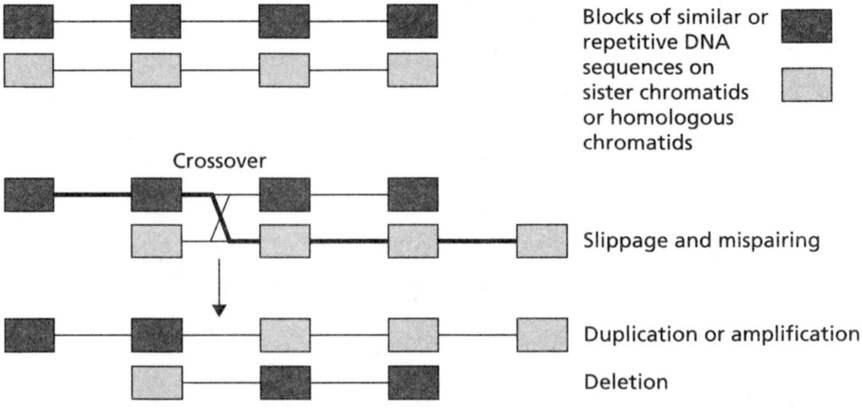

Figure 3.1
Deletions and duplications generated by unequal crossover.

Box 3.4 Location of triplet repeats

If triplet repeats are located in the **coding region** of a gene, they are transcribed along with the unique sequences, and this may limit the expansion size. The maximum number reported so far in Huntington Disease is 121 repeats.

If triplet repeats are in **noncoding regions**, they may show huge expansions up to 2000 repeats, and also very large size changes within a single generation. The expanded repeats may either have no effect on transcription, or they may switch off the gene by, for example, methylation.

In myotonic dystrophy, the triplet repeat CTG is in the 3′ untranslated region of the protein kinase gene on chromosome 19. Affected patients may have very large expansions of up to 2000 repeats. The disease mechanism is as yet undetermined. Its dominant inheritance is difficult to understand, as the mutated RNA is not translated into protein. It is possible that the RNA containing the abnormal expansion exerts a dominant negative effect on normal RNA metabolism. It is also possible that the expansion affects additional as yet uncharacterized genes in this chromosome region.

In the fragile X syndrome, the triplet repeat CGG is found in the 5′ untranslated region. Affected patients have between 200 and 1000 repeats, which affect transcription, as the *FMR-1* gene is silenced by methylation. It appears that the promoter becomes methylated, but the exact mechanism is unknown. It is suspected that both timing of replication and chromatin structure are altered (hence the fragile site on the X chromosome). Under these conditions other genes nearby may also be switched off.

3.2 ALTERATION OF GENE CONTROL

Positional effects

The normal position or location of a gene or part of a gene may be altered directly or indirectly by changes in chromatin structure (see *Box 3.5*), or following a large structural alteration in the chromosomes such as a translocation between two chromosomes, an insertion, or an inversion of one chromosome.

For example, some female carriers were found to be affected with the X-linked condition Duchenne muscular dystrophy (usually only males are affected) because of a translocation between one X and an autosome. One breakpoint was always on the short arm of the X chromosome at Xp21. This was the location of the DMD gene; it was being disrupted in some way by the translocation such that the normal protein dystrophin was no longer being made.

Box 3.5 Functional RNAs

Some genes code for functional RNAs, which may be transcribed but not translated. These are known to be important in such genes as *XIST* on the X chromosome and *H19* on chromosome 11. The transcribed sequences of RNA (transcripts) are necessary for regional transcriptional control, and only work within a particular chromosomal configuration (for example, of chromatin structure).

Therefore, it appears that even though an individual carries an apparently balanced translocation, small positional changes at the molecular level may have unexpected effects when one gene is moved next to a new neighbour.

Other small conformational or positional changes may involve separating a gene from its promoter or enhancer, which might turn it off, or separating a gene from a silencer, which may turn it on. This can lead to:

- **gene enhancing**, in which a gene may continue to function after its usual time span, or in the wrong tissues, or is otherwise turned on inappropriately, can be thought of as a **gain of function**; this is a common mechanism in developmental abnormalities, and is the cause of some cancers (see *Chapter 8*)
- **gene silencing**, in which a gene is turned off at an inappropriate time or in the wrong tissues, can be thought of as a **loss of function** (see *Box 3.6*).

Box 3.6 Gene silencing due to positional effect

Gene silencing may occur if an otherwise active gene is moved to reside in an area that is usually inactive, such as the centromeric heterochromatin or the telomeres of a chromosome. The autosomal dominant disorder facioscapulohumeral dystrophy (FSHD) at 4q35 may be caused by such a position effect.

The FSHD gene maps close to the 4q telomere and is separated from it by a large number of 3.3 kb repetitive sequences. Affected individuals have deletions of some of these repeats, bringing the gene closer to the 4q telomere and thus silencing the gene.

Abnormal dosage and haploinsufficiency

If the ideal state in the nucleus of each cell is the diploid karyotype comprising pairs of chromosomes, then the genes are present in the correct dosage. Problems can arise when there are either more or fewer genes present.

Syndromes due to abnormal gene dosage

Syndromes resulting from excessive dosage may occur when an extra complete chromosome is present, such as in trisomies: Down syndrome (trisomy 21), Edwards syndrome (trisomy 18) or Patau syndrome (trisomy 13). An extreme example is seen in the condition known as triploidy, when a whole extra set of chromosomes is present, resulting in 69 chromosomes.

Haploinsufficiency occurs when, despite the presence of a 50% level of gene product from one normal copy of a gene, this is not enough to prevent phenotypic expression. Examples are DiGeorge syndrome and Turner syndrome.

Distal chromosome 22q11.2 deletion syndrome

Microdeletions of one or more genes at 22q11.2 may cause a variable phenotype, including DiGeorge syndrome and Shprintzen syndrome, or velo-cardio-facial syndrome (VCFS). The DiGeorge phenotype includes several unrelated effects (see *Section 3.6*), so it is thought that haploinsufficiency of one or more genes is involved, with these genes failing to act at a critical early stage such that a cascade of gene expression is disrupted. This disruption results in multiple problems further on in development.

Turner syndrome

In a normal female (46,XX), dosage compensation usually ensures that one of the two female X chromosomes is inactivated, such that the genes on the remaining active X would be expected to produce an equivalent amount of gene product to the single X chromosome of the male.

However, although girls with Turner syndrome (45,X) have a single active X chromosome, which should produce the requisite amount of gene products, they display an abnormal phenotype (see *Appendix: Glossary of disorders*). This implies that Turner females need either extra structural or controlling gene products normally provided by the missing X. It would appear that a 50% level of certain gene products is not enough, and Turner syndrome should also be considered as a type of haploinsufficiency. In a normal female it is known that the tip of the short arm of the inactivated X (Xp22.3) is not inactivated (i.e. the PAR), so that two copies of the genes from this region are required for normal female development.

Skewed X-inactivation patterns

As long as X-inactivation is random and haploinsufficiency is not involved, a carrier female of an X-linked recessive disease should be unaffected (see *Section 4.6*). Most of the time this is the case.

However, in the presence of an abnormal X, it is possible to get skewing (biasing) of the X-inactivation pattern so that the abnormal X is preferentially inactivated. If a female has a deletion of the X chromosome, it is likely that this abnormal chromosome will be inactivated rather than the normal X, thus leaving a comparatively normal genome.

Interestingly, female carriers of the full mutation for the fragile X syndrome may either be normal or mentally retarded. Studies have indicated that the normal females may preferentially inactivate the fragile (abnormal) X, whereas mentally retarded females may preferentially inactivate the normal X.

However, circumstances may arise where, due to an X;autosome translocation, the autosomal genes now attached to part of the X chromosome need to remain active. The genetic balance of the karyotype is also important, so the normal X may be preferentially inactivated. As demonstrated by females affected with Duchenne muscular dystrophy, however, this may lead to positional anomalies and hence disease expression.

Skewed X-inactivation can also be observed in Wiskott–Aldrich syndrome. Symptoms include immune defects, eczema and thrombocytopenia (abnormally long clotting time due to fewer platelets). Some patients develop malignancies and some die of infections or bleeding. As the fault is on the X chromosome (Xp11), males are affected but females are carriers.

Using methylation analysis with DNA probes such as M27β, it was shown that there was a highly skewed X-inactivation pattern in the lymphocytes of female Wiskott–Aldrich syndrome carriers (such that the abnormal X was inactivated). The X-inactivation is so biased in the white blood cells that it can be used as a method of carrier detection using methylation-sensitive restriction enzymes such as *Hpa*II (see *Box 3.7*).

Box 3.7 Skewed X-inactivation in immunogenetic disorders

X-linked agammaglobulinaemia shows non-random X-inactivation in the B lymphocytes of obligate carrier females.

X-linked SCID (severe combined immune deficiency) shows non-random X-inactivation in the T and B lymphocytes and the natural killer cells of obligate carrier females.

Epigenetic pathology

Mutations in genes implicated in the epigenetic processes will lead to an epigenetic pathology. An example is the X-linked dominant Rett syndrome, which almost exclusively affects females. The gene *MECP2* encodes a methyl-CpG-binding protein which selectively binds CpG nucleotides, and is important in assembling transcriptional silencing complexes (i.e. represses transcription).

Mutations of *MECP2* cause a disruption of these silencing complexes such that there is inappropriately activated transcription. The underlying mechanism for the pathogenicity in Rett syndrome is therefore abnormal epigenetic regulation.

Methylation, epigenetic alterations and cancer

Methylation is often employed in the control of gene expression. *De novo* methylation is often associated with tumour suppressor genes (see *Chapter 8*). As a tumour suppressor gene usually acts to restrict growth, if its promoter is methylated, the growth restriction will be de-activated and overgrowth may ensue.

In cases of kidney cancer arising in patients with Von Hippel–Lindau syndrome, it was found that 70% have inactivation of the VHL suppressor gene. The tumour suppressor gene *RASSF1A* is also known to undergo promoter hypermethylation leading to epigenetic inactivation in lung cancer.

Methylation and imprinting disorders

Methylation may also be one of the mechanisms involved in normal imprinting (see *Chapter 2*). Chromosome 15 is known to be imprinted in humans at 15q11-q13. DNA studies have shown that an **imprinting centre** exists in this region, which possibly causes a controlled methylation of the genes in the surrounding area.

A mutation in the imprinting centre will prevent the correct resetting of the parental imprint in the germ cells. For example, if a maternally inherited imprinting centre mutation is present in the father, that maternal 15 will not reset to 'paternal', and if inherited by his children, Prader–Willi syndrome will result, as there is no functioning paternal locus.

3.3 DNA REPAIR DEFECTS IN HUMAN DISEASE

Human disorders arising from defects of DNA repair often display common characteristics such as increased rates of cancers and abnormalities of the immune response. Mutations of excision and mismatch repair genes result in uncorrected base changes which lead to an increased background mutation rate in other genes, predisposing cells to malignant changes (see *Chapter 8* and *Box 3.8*).

Box 3.8 Sister chromatid exchanges

These are exchanges occurring during synthesis between all four strands of the DNA: the two parental strands and the two newly synthesized, or nascent, strands. This is thought to represent a mechanism by which DNA in the S phase is able to replicate around sites of unrepaired damage with a low risk of introducing mutations into the daughter chromosomes.

Bromodeoxyuridine, or 5′BrdU, can be incorporated into replicating DNA. The 5′BrdU molecule is identical to the nucleoside thymidine, except that a bromine atom replaces the CH_3 methyl group (see thymine, *Figure 1.4*). It is said to be a **thymidine analogue**. The presence of 5′BrdU affects the stability of the DNA and thus alters the staining properties of the chromosome at metaphase. In a culture where two cell cycles take place in the presence of 5′BrdU, only one of the two chromatids of each metaphase chromosome contains a grandparental DNA strand without 5′BrdU. Exploiting this phenomenon, it is possible to stain the two chromatids of a metaphase chromosome differently, and see the points at which the pale and darkly staining chromatids exchange.

Xeroderma pigmentosum

Xeroderma pigmentosum (XP) is a very rare (1/250 000) autosomal recessive disorder arising in childhood. It is characterized by extreme photosensitivity to ultraviolet (UV) light affecting exposed areas such as the skin and eyes, resulting in basal cell carcinomas and melanomas, together with abnormal pigmentation and neurological abnormalities.

There are eight XP groups, represented by eight different gene loci which code for the different enzyme subunits in the excision complex. Mutations in the genes *XPA–XPG* result in defects in the initial steps of nucleotide excision repair of the UV-induced damage, while mutations in *XPV* (variant) lead to abnormalities of post-replication repair. Cells from different XP groups can **complement** one another, restoring levels of DNA repair to normal.

Normal cells synthesize DNA in the S phase of the cell cycle, and also exhibit unscheduled DNA synthesis when undergoing DNA repair. In the laboratory XP can be demonstrated by exposing XP cells to UV light; they do not undergo unscheduled DNA synthesis. Although an increase in sister chromatid exchanges and chromatid aberrations can be shown, there is no definitive chromosomal test for XP, although DNA sequencing may reveal a mutation in a particular XP gene.

Other disorders of DNA repair and replication

There are a number of other recessive syndromes in which there is a defect in some aspect of DNA repair or DNA replication. In some of these conditions the exact genetic mechanisms are not fully understood. Although they are clinically distinct syndromes, there are certain features in common; features which might be expected of syndromes in which an abnormally high level of mutation occurs.

Bloom syndrome, ataxia telangectasia and Fanconi's anaemia are often grouped as the 'chromosome breakage syndromes' as they show an elevated level of chromosome damage in culture. They also share a predisposition to tumours and growth retardation (*Table 3.2*). Cockayne syndrome and Werner syndrome are conditions associated with growth retardation and premature ageing.

3.4 IMMUNOGENETIC DISEASES

Disorders of immunodeficiency may affect the B cells (and therefore antibodies), the T cells (and therefore cellular and cytokine response), or both.

Table 3.2 Disorders of DNA repair and replication

Disorder	Frequency	Clinical features	Mechanism	Method of analysis
Ataxia telangectasia (AT)	1/100 000–1/300 000	Sensitivity to ionizing radiation, hence cancer, immune problems, diabetes, ataxia, redness due to dilation of blood vessels	Deficiency in ATM kinase involved in cell cycle and cell signalling	Expose chromosomes to radiation in G_2; look for an increase in chromatid damage
Bloom syndrome	100 living cases	Tumours and leukaemias	Defective helicase in DNA replication	Spontaneous increase in SCEs
Fanconi's anaemia	1/50 000–1/100 000	Sensitive to alkylating (crosslinking) agents. Absent radius and thumbs, pancytopaenia, acute myeloid leukaemia	Molecular basis unknown	Spontaneous increase in chromatid breaks and exchanges; expose chromosomes to alkylating agents

An example of B cell disease

X-linked infantile agammaglobulinaemia (Bruton's agammaglobulinaemia) arises from a faulty gene or genes on the X chromosome such that virtually no immunoglobulins (antibodies) are produced. The fault prevents pre-B cells developing into mature B cells, and is a failure of the variable (V) gene rearrangement mechanism. Absence of an appropriate enzyme results in failure of the V_H (heavy chain) genes to join to the D and J genes.

Male infants of 6 months and over exhibit recurrent bacterial infections, and need injections of IgG.

An example of T cell disease

DiGeorge syndrome is an example of a T cell deficiency disease. The absence of thymus and parathyroids results in hypocalcaemia (low calcium) and patients have virtually no T cells.

Babies are prone to viral, protozoal, fungal and bacterial infections. As there are no helper T cells, no cytokines (actually lymphokines) can be produced to activate the B cells to produce antibodies in the blood.

An example of B cell and T cell deficiency

Severe combined immunodeficiency disease (SCID) results from the failure of stem cells to differentiate into B cells and T cells. There is more than one chromosome location for SCID; one type is on the X chromosome and another is on chromosome 8.

As there are no mature B and T cells, infants are prone to microbial infections. The treatment is usually bone marrow transplantation.

In about half of the patients with autosomal recessive SCID there is a deficiency of the enzyme adenosine deaminase (ADA). Because the gene location is known, the cloned gene has been artificially introduced into a child, and is the first example of successful gene therapy.

3.5 MITOCHONDRIAL MUTATIONS

Mitochondrial disorders (*Table 3.3*) may arise due to faults with mitochondrial structure, altered numbers or impairment of mitochondrial gene function due to DNA mutation. The organs most affected by mitochondrial diseases tend to be muscles (including the heart), kidneys and liver, as they are the most dependent on the energy source derived from the mitochondrial gene processes. The eyes, ears, neurological system and endocrine system may also be affected.

As there are so many mitochondria, if a mutation arises in one mitochondrion it may replicate and exist in the same cell together with normal mitochondria. This is known as **heteroplasmy**. After many cell divisions different ratios of abnormal to normal mitochondria may exist.

Because cloning involves the removal of a nucleus from an egg and the insertion of a somatic cell nucleus into empty cytoplasm, there is the possibility of gene therapy for mitochondrial disorders. If a woman had abnormal

Table 3.3 Mitochondrial DNA disorders

Disorder (abbreviation)	**Full name**	**Clinical symptoms**
MELAS	Mitochondrial encephalomyopathy, lactic acidosis and stroke-like episodes	See left
MERRF	Myoclonic epilepsy and mitochondrial myopathy (with ragged red fibres)	Jerking, hearing loss, ataxia, renal abnormalities, diabetes, cardiomyopathy, dementia – maternal lineage
LHON	Leber's hereditary optic neuropathy	Central vision loss (blindness), sometimes neurological effects on movement. Related through maternal lineage though more males are affected
NARP	Neurogenic muscle weakness, ataxia and retinitis pigmentosa	See left
Leigh		Ataxia, hypotonia, developmental delay, regression, optic atrophy, respiratory abnormalities. Onset 1.5 years

mitochondria, she could theoretically have one of her healthy cell nuclei transplanted into donor cytoplasm, which would contain normal mitochondria.

Besides the finding of mutations in mitochondrial, tRNA and nuclear DNA relating to these conditions, somatic mutations may gradually accumulate over time. Mitochondrial DNA may mutate up to 17 times faster than nuclear DNA, so over a lifetime these somatic mutations may reach a threshold at which normal metabolic function may be impaired.

It is believed that mitochondrial mutations may be implicated in diseases of old age such as Parkinson disease or Alzheimer disease.

3.6 DEVELOPMENTAL CHANGES

Developmental genes tend to act in cascades. When the disruption in one gene causes a multitude of apparently unrelated effects, it is often because that gene is involved in the early development of certain embryonic structures derived from a common origin. A gene which, when expressed, results in multiple phenotypic features is said to exhibit **pleiotropy** (see *Box 3.9*).

Growth factor receptors

Growth factors are required both at the start of embryonic development and towards the end of that process, when finely controlled proliferation and differentiation occur.

Box 3.9 Pleiotropy

Two examples of pleiotropy are the autosomal dominant Marfan syndrome and DiGeorge syndrome:

- **Marfan syndrome** (chromosome 15). Because the faulty fibrillin gene alters the elasticity of the connective tissue, several apparently unrelated symptoms are seen. These include long limbs and fingers, a defective major blood vessel of the heart (the aorta), and dislocation of the lens of the eye. Fibrillin is found in all of these structures.
- **DiGeorge syndrome** (chromosome 22). Migration of neural crest cells cannot be sustained early in embryonic development, and this affects structures called the third and fourth pharyngeal pouches. Structures derived from these are affected in turn, such as the thymus, parathyroids, heart and facial development.

Achondroplasia is an autosomal dominant skeletal dysplasia resulting in dwarfism due to shortening of the limb bones (the 'long' bones) where the gene is expressed. Mutations resulting in achondroplasia have been found in the **f**ibroblast **g**rowth **f**actor **r**eceptor gene *FGFR 3* (found on chromosome 4; *FGFR 1* is found on chromosome 8, *FGFR 2* on chromosome 10, and *FGFR 4* on chromosome 5).

There is another group of autosomal dominant syndromes which result from premature fusion of the skull bones. The name for this abnormal skull growth is **craniosynostosis**. The fusion results in excessive growth elsewhere in the head region, with the result that the head shape is often abnormal. The primitive embryonic cells do not differentiate appropriately into osteoblasts, which make bone. Many of the craniosynostoses also involve limb abnormalities, which implies that face and limb development share at least one common pathway (see *Box 3.10*).

Because of the fusion of the skull bones, the extra digits and syndactyly (fusion of digits), these mutations are considered to result in a **gain of function**. This contrasts with nonsense mutations resulting in haploinsufficiency and a **loss of function**.

Box 3.10 Mutations in FGFR genes

Mutations in *FGFR 1* have been found in Pfeiffer syndrome. Phenotypic features include broad thumbs, great toes, short head, wideset eyes (hypertelorism), and generally a normal IQ.

Mutations in *FGFR 2* have been found in Crouzon syndrome, where the patient has abnormal facial features including premature skull fusion, prominent eyes, beaky nose and poor bite due to jaw abnormalities. They do, however, have normal limbs.

Hedgehog signalling pathway

Sonic hedgehog is one of the segmental polarity genes needed for normal cell growth and development of the notochord (spinal cord/backbone), foregut and limbs. It is also important in ventral midline differentiation.

Mutations have now been found in one sonic hedgehog gene (*HPE3*) at 7q36, which results in **holoprosencephaly**. This is expressed as abnormal facial features such as a single eye or absent nose, the fault arising from aberrant movement of cells around a vertical central section of the head. The mutation mechanism is thought to be haploinsufficiency via deletions or positional silencing (7q36 is near the telomere).

Mutations have also been found in the patched gene (*PTCH*), which controls some aspects of cell growth via the hedgehog signalling pathway. As PTCH normally acts as a growth suppressor, it follows that a mutation in *PTCH* would lead to uncontrolled cell growth, i.e. cancer.

Basal cell naevus carcinoma syndrome (also known as NBCC = naevoid basal cell carcinoma), exhibits mutations in the *PTCH* gene. Besides basal cell carcinomas, there may also be facial and cranial alteration and overgrowth. The mechanism is probably haploinsufficiency.

In *Chapter 8* we will explore the role of *PTCH* as a 'gatekeeper', keeping cancer at bay.

Homeobox genes

These organize the correct anterior and posterior spatial expression of body architecture such as limb position (in the fruit fly larva, loss of function equates with fewer segments, whereas gain of function equates with more segments).

Mutations in the *HOX* or related *PAX* genes would be expected to affect specific areas of the embryo which would be reflected in the clinical symptoms produced. Mutations in *HOXD13* result in synpolydactyly such that there is fusion of the third and fourth digits of the hands or feet, with an extra digit in between. This would be a gain of function.

Mutations in *PAX6* result in the human eye condition aniridia (loss of the iris) and therefore a loss of function. Mutations in *PAX3* result in Waardenburg syndrome type 1. This pleiotropic gene must be important in neural tube development in the brain area, because affected individuals display a variety of symptoms such as deafness, mixed colours in the iris of the eye, and a white forelock. This may result from haploinsufficiency and is therefore a loss of function.

3.7 SEX DETERMINATION

Sex reversal

One particular type of sex reversal is due to unequal crossover between the X and Y chromosome at male meiosis, such that the inheritance of a translocated sex chromosome appears to result in either females with a male karyotype (46,XY), or males with a female karyotype (46,XX).

Although a phenotypically male child may appear to have a 46,XX karyotype at the microscopic level, at a molecular level the presence of the *SRY* gene inherited on the X chromosome results in male development.

In the reciprocal arrangement, phenotypic females appear to have a 46,XY karyotype microscopically, but the *SRY* is not present on the inherited Y chromosome, so female development occurs. *De novo* mutations in the *SRY* gene can also result in 46,XY females.

Evidence of other genes influencing sex determination

There are cases of phenotypic females and abnormal males who have intact *SRY* genes, but duplication of the gene *DAX-1* at Xp21–22.3 (see *Section 2.10*), suggests that gene dosage must also play a part in sex determination.

Intact autosomal genes are also required in the pathway to normal sexual development. Campomelic dysplasia is a skeletal malformation syndrome in which 75% of patients with a male karyotype show sex reversal. In these patients, mutations have been found in a gene called *SOX9* (Sry HMG box) on chromosome 17. *SOX9* shares some homology with the *SRY* gene (a motif called the HMG box) and interacts with *SRY* in the developmental cascade (see *Chapter 2*). There are instances of individuals affected with campomelic dysplasia displaying translocations with breakpoints over 50 kb away from *SOX9*; this suggests that position effect was the most likely mechanism in these patients.

If mice lose *Wnt4* they develop incomplete testes; if they lose *Fgf9* they develop ovaries. It has also been shown that if *Wnt4* is disrupted, there is also a decrease in expression of *Dax-1*.

Other examples of abnormalities of sexual differentiation

Complete androgen insensitivity syndrome

This disorder was formerly known as testicular feminization syndrome (see *Figure 3.2*). Males may display a female phenotype even in the presence of an intact *SRY.* Even when male genes produce the correct male hormones, normal male development depends on the target cells responding correctly.

If the male hormone testosterone targets cells which have a mutation in the androgen receptor gene at Xq11–Xq12, male sexual differentiation does not occur and the individual appears phenotypically female.

Congenital adrenal hyperplasia

Steroid 21-hydroxylase deficiency is just one example of a masculinizing CAH (congenital adrenal hyperplasia). Both boys and girls may inherit the gene, but in the female the deficiency of the hormone 21-hydroxylase results in overproduction of male hormone from the adrenal glands, with differing degrees of male development.

The mechanism by which this is believed to arise derives from the fact that the working copy of the CAH gene (called *CYP21B*) is closely mimicked by a nearby pseudogene (a non-functional gene with a similar DNA sequence) called *CYP21A*. An unequal crossover (or unequal sister chromatid exchange) can result in a product with a deletion of the functional *21B* gene, or sometimes a non-functioning *21A/21B* gene.

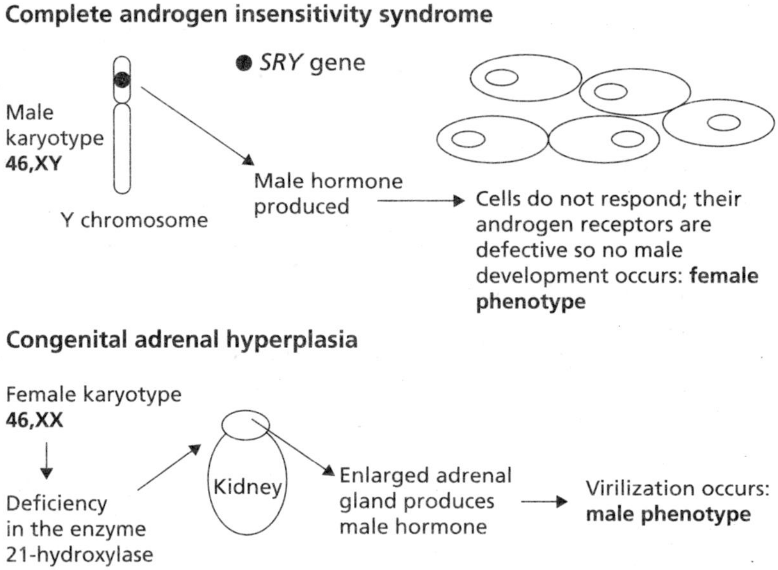

Figure 3.2
Sex reversal due to complete androgen insensitivity syndrome, and congenital adrenal hyperplasia.

CAH also illustrates the mechanism of **gene conversion**. Point mutations in the pseudogene *21A* may be copied into the normal *21B* gene. This non-functional inserted DNA replaces a short stretch of *21B*, and gene conversion occurs.

Another example of gene conversion is adult polycystic kidney disease type 1 (*APKD1*). Mutations in the gene had always been detected with disappointingly low frequency, until it was pointed out that there are three copies of certain sequences near the real gene which resemble it so closely (70% of the sequence is identical) that the potential for unequal crossover or abnormal copying could occur.

3.8 SUMMARY

In order to understand the likely effect of gene mutation, we have to understand both the normal function of that gene, and whether the type of alteration is serious enough to impair that function.

A truncated protein may be less damaging than an altered protein, especially compared to the pleiotropic effect of a mutation in a single developmental gene, resulting in multiple malformations due to its importance in a particular cascade of pathways.

Genes which are expressed in the wrong tissue or at the wrong time may lead to deleterious effects from gain of function, while a deleted gene may lead to loss of function.

There are a large number of disease-causing mechanisms, including well-characterized single base substitutions, the ever-increasing number of disorders due to the expansion of triplet repeats, and the newer concept of epigenetic mutations. Even introns may not be seen as the harmless 'junk DNA' we have supposed them to be.

SUGGESTED FURTHER READING

Bardoni, B., Zanaria, E., Guioli, S., *et al.* (1994) A dosage sensitive locus at Xp21 is involved in male to female sex reversal. *Nature Genetics*, **7:** 497–501.

Foster, J.W., Dominguez-Steglich, M.A., Guioli, S., *et al.* (2002) Campomelic dysplasia and autosomal sex reversal caused by mutations in an *SRY*-related gene. *Nature*, **372:** 525–529.

Goff, D. and Tabin, C.J. (1996) Hox mutations au naturel. *Nature Genetics*, **13:** 256–258.

Koopman, P., Gubbay, J., Vivian, N., *et al.* (1991) Male development of chromosomally female mice transgenic for *Sry*. *Nature*, **351:** 117–121.

Read, A. and Donnai, D. (2007) *New Clinical Genetics*. Bloxham: Scion Publishing.

Roessler, E., Belloni, E., Gaudenz, K., *et al.* (1996) Mutations in the human Sonic Hedgehog gene cause holoprosencephaly. *Nature Genetics*, **14:** 357–359.

Strachan, T. and Read, AP. (2004) *Human Molecular Genetics*, 3rd Edition. Oxford: Garland Science.

SELF-ASSESSMENT QUESTIONS

1. Give examples of two different types of mutations and briefly explain their effects on gene function.
2. What is a triplet repeat? How might a clinical disorder arise from a triplet repeat? Give an example.
3. In the Fragile X syndrome, the expansion of triplet repeats becomes methylated. The affected boys do not produce the FMR-1 protein. Why not?
4. A female with a gene mutation on the X chromosome for the biochemical disorder Hunter syndrome develops that disease despite the presence of a second normal X. What process may be occurring?
5. A baby is born with a particular kind of hydrocephalus (fluid collecting in the brain). When the chromosomes are analysed, it is found that both chromosome 14s come from the mother. Name this phenomenon and explain why the presence of the father's 14 is required for a normal phenotype.
6. An individual's karyotype reveals a balanced chromosome translocation, and yet the patient shows clinical symptoms. It is noted that one of the breakpoints lies very close to a centromere. Why might this be significant?
7. In a famous experiment on the fruit fly *Drosophila* it was observed that instead of antennae growing on the head of the fly there was a leg instead (antennapaedia). Given that the orthologous human genes have a similar function, which group of developmental genes are likely to be involved?

CHAPTER 4

Patterns of inheritance

Learning objectives

After studying this chapter you should confidently be able to:

- **List Mendel's two laws of inheritance**
  Mendel's first law (the principle of segregation) states that during meiosis each of a pair of alleles in a diploid cell segregates into different haploid cells.
  Mendel's second law (the principle of independent assortment) states that genes on different chromosomes segregate independently of one another at meiosis.
- **Outline the principles of linkage and recombination**
  The more closely linked genes are on a particular chromosome, the less chance there is of a crossover and hence genetic recombination. The maximum possible for unlinked genes is 50%.
- **Draw simple pedigrees using the correct nomenclature and symbols**
- **Describe the three major inheritance patterns and give examples of appropriate disorders**
  There are three major modes of inheritance: autosomal dominant, autosomal recessive and X-linked recessive. There are two less common patterns: X-linked dominant and Y-linked.
- **List other problems causing deviations from Mendelian inheritance patterns**
  Within each mode of inheritance, factors such as reduced penetrance or variable expressivity may mask the pattern, or there may be more global problems such as heterogeneity, imprinting, new mutations, or germinal mosaicism. Mitochondrial inheritance is maternal.
- **Define non-Mendelian inheritance**
  This may arise from the combined action of polygenes, or may be due to a varying contribution from the environment, such that a susceptibility gene may produce different phenotypes depending on exposure to a particular environment.
- **Define the Hardy–Weinberg equation in terms of the equilibrium of a population, and list factors disturbing that equilibrium**
  Alleles of genes in balance in a population are distributed according to the Hardy–Weinberg equation. Various factors may disturb this equilibrium, such as random genetic drift, or a change in the mutation level.

■ **Explain how Lod scores can be used in linkage studies, and describe how Bayes' theorem is used in assessing carrier risks**
Lod scores are a means of analysing the extent to which genes are linked. A score of >3 indicates strong linkage. Bayes' theorem is a flexible equation, which weighs up the probability that an individual may be a carrier or affected by a disease.

4.1 CLASSICAL GENETICS

Mendel's laws

The basis for the understanding of inheritance came from the classic work on plants by Gregor Mendel (see *Box 4.1*). By means of meticulous breeding experiments with peas, Mendel explained the basic principles of genetic inheritance, which we now accept to be applicable to all higher organisms with a sexual method of reproduction. Those concepts of heredity, forming the whole basis of the science of genetics, are now commonly referred to by the terms 'Mendelism' and 'Mendelian inheritance'.

Box 4.1 Gregor Mendel

Gregor Mendel was an Austrian monk whose classic experiments used seven sets of well-defined paired characteristics exhibited by pea plants. He first published the data from his experiments with peas in 1866; the full significance to genetics of his two laws was not appreciated until 1900.

As Mendel did not know about chromosomes or genes, he used the term '**factor**'. When considering a factor such as the height of pea plants, for example, he identified the pure **characters** (or traits) 'tall' and 'short', which in modern terminology would now be described as **alleles** (alternatives) of a gene coding for height (see *Box 4.2* for a range of definitions).

Following the hybridization of two pure traits, Mendel termed characters which were transmitted 'entire or almost unchanged' **dominant**, and those which became 'latent' as **recessive**. Mendel's conclusions are most simply expressed as two **principles**; firstly that of **segregation** and secondly that of **independent assortment**.

Mendel's first law: the principle of segregation

- **Factors come in pairs and have different characters or traits**
  Modern interpretation: one gene is inherited from each parent, hence alleles are inherited in pairs.
- **Only one character is passed on from each parent**
  Modern interpretation: although there may be many alleles at a given gene locus, following meiosis only one allele enters the gamete.
- **Equal numbers of gametes inherit each allele**

Box 4.2 Definitions

The following terms are defined with respect to modern human genetics.

Allele: one of several alternative forms of a gene at a specific gene locus; an individual will therefore have 2 alleles (one maternal, one paternal) at each autosomal locus.

Dominant: a genetic trait which is observed (or expressed) in the heterozygous state and refers to the effect of an allele or 'character' at a specific gene locus.

Genotype: the genetic contribution of an individual (or cell). A genotype may also refer to the constitution of alleles at a given locus in an individual.

Hemizygous: having only one copy of a gene (and therefore one allele); for example on the single X chromosome in a male, who is therefore hemizygous for X-linked genes.

Heterozygous: having two different alleles at a specific gene locus.

Homozygous: having two identical alleles at a specific gene locus.

Phenotype: the observable physical effect on an individual (or cell), determined by the genetic constitution.

Recessive: a genetic effect of an allele that is manifest in the homozygous state at a specific gene locus.

Mendel's second law: the principle of independent assortment

Following experiments using unrelated pea plant characteristics such as colour and shape, he proposed that **different characters assort independently**. As long as the genes for colour and shape are on non-homologous chromosomes, the alleles (such as yellow/green or smooth/wrinkled) segregate independently at meiosis.

4.2 LINKAGE

Mendel's second law only works as long as genes are on separate chromosomes, or far enough apart on the same chromosome to assort independently. The closer a pair of genes are (on the same chromosome arm, for example), the less likely there is to be independent assortment – that pair of genes tend to be inherited together, and are said to be **linked**. Whole blocks of genes may be inherited together and the composite genotypes are then known as a **haplotype**.

Genes are sometimes said to be syntenous. **Synteny** means that the genes are on the same chromosome, though not necessarily displaying genetic linkage with each other (i.e. they assort independently). They are known to be on the same chromosome because they are both linked to intermediate genes.

4.3 RECOMBINATION

During meiosis there is at least one crossover (chiasma) between every pair of homologous chromosomes. Sometimes on longer chromosomes there may be more than one chiasma, as there is more room for that to occur (*Figure 4.1*).

Although genes on different chromosomes segregate independently, if genes are well separated on the same chromosome there is still a good probability that a crossover will result in independent segregation. This produces 2/4 parental alleles and 2/4 **recombinants**, i.e. a recombination rate of 50%. This is sometimes called a **recombination fraction** (also known by the Greek letter theta, θ), such that a recombination rate of 50% is expressed as 0.5.

The closer genes are on the same chromosome, the more closely **linked** they will be, and there will be less chance of a crossover occurring between the gene loci, with fewer recombinants and a lower recombination fraction.

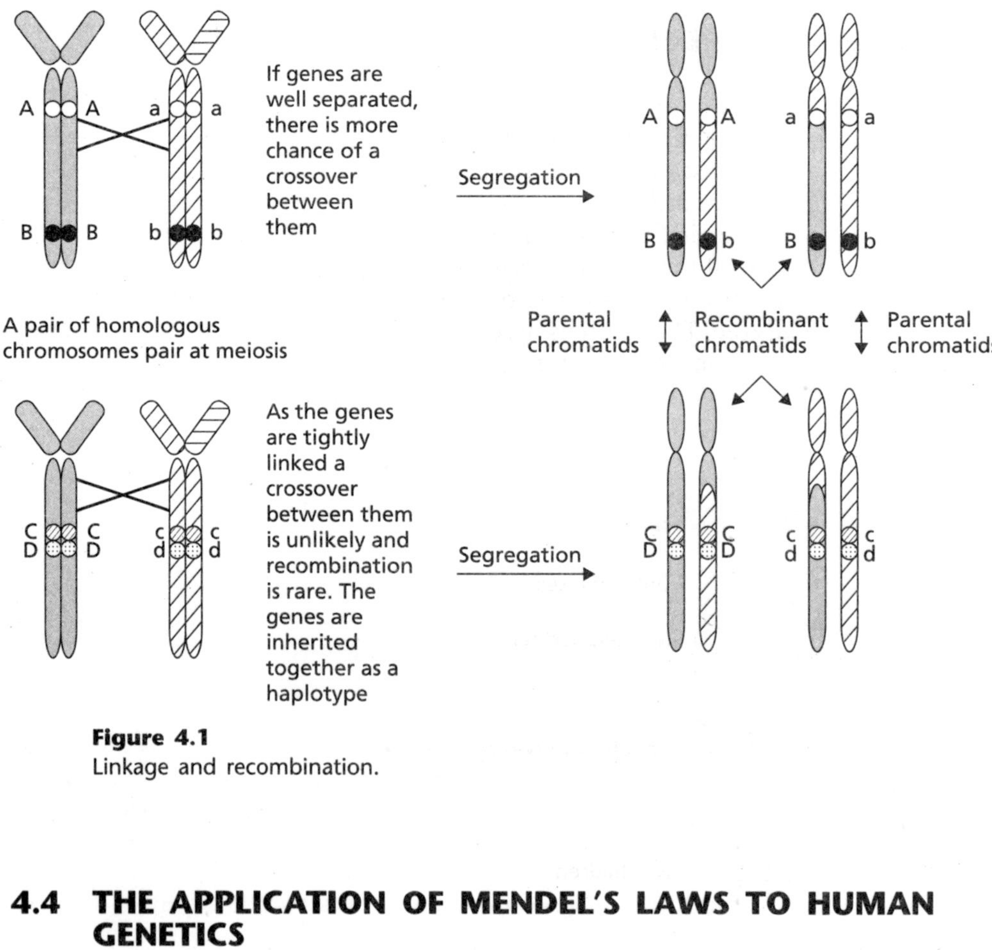

Figure 4.1
Linkage and recombination.

4.4 THE APPLICATION OF MENDEL'S LAWS TO HUMAN GENETICS

Although Mendel formulated his laws using data derived from pea plants, exactly the same principles apply to inheritance patterns in man. Detailed knowledge and understanding of the properties of human genes was harder to obtain simply because it was not possible to undertake the same sorts of breeding experiments that could be carried out on peas, fruit flies and fungi.

To begin with, knowledge and understanding of human genes was drawn mainly from constructing family pedigrees, sometimes to the extent of searching through historical archives and parish records, and subjecting those data to complex statistical analysis. It is only within the past 20 years, since the development of modern techniques of molecular biology, that substantial progress into the intricacies of human genetics has been made, but deduction of basic patterns of inheritance often still requires pedigree analysis.

Around 5000 human diseases are now traceable to mutations in single genes. These are the **single gene disorders** and the pattern of their inheritance is often well characterized.

4.5 PEDIGREES

A family with a genetic disorder will usually be referred to a clinical geneticist, a medical consultant with specialist training in the identification of genetic diseases. It is important to elucidate the **mode of inheritance** of the disease, in order to give the family an idea of the probability that the disease will recur in future generations – the **recurrence risk**.

One of the easiest ways to start this investigation process is to take a family history and pictorially express this in the form of a family tree, known as a **pedigree**. The symbols used in such pedigrees can be seen in *Figure 4.2*. Each generation (grandparents, parents and children, for example) is given a

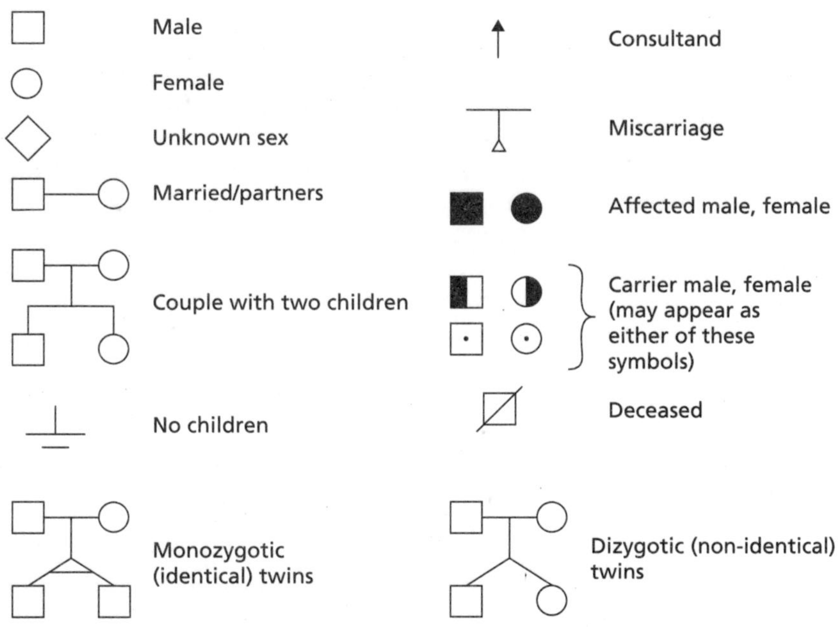

Figure 4.2
Pedigree symbols.

Roman numeral (I, II, etc.). Each individual within that generation is sequentially numbered from left to right using Arabic numerals. Miscarriages and pregnancies are also included. The affected individuals are indicated on the pedigree and by studying how frequently the disease was passed on and in which sex, the geneticist can often deduce the **pattern of inheritance**.

4.6 MENDELIAN INHERITANCE PATTERNS

There are three major and two less common types of inheritance pattern. These are classified according to whether the gene responsible for a particular characteristic or disorder resides on a sex chromosome or an autosome, and also whether that gene is expressed in its homozygous or heterozygous state.

Autosomal dominant inheritance

Autosomal dominant inheritance (see *Table 4.1* for examples of some autosomal dominant genetic diseases) follows the general rules given below.

- A dominant gene is found on one of a pair of homologous **autosomes.** This means that either the father or the mother can pass on a dominant disorder.
- Only one parental gene is required to display the associated phenotype – it is **dominant** to the allele on the homologous chromosome. Although two alleles are present at that gene locus (the individual is heterozygous), the dominant allele is able to be **expressed** in this heterozygous state.
- The children of a parent with an autosomal dominant disorder have a **50% chance** of inheriting that disorder, **irrespective of sex**. If an individual is phenotypically (and hence genotypically) unaffected, his or her children will also be unaffected.
- An autosomal dominant **pedigree** has a high chance of having affected family members in each generation, and is sometimes said to have a '**vertical**' appearance (*Figure 4.3*).

Table 4.1 Examples of autosomal dominant genetic diseases

Disease	Chromosome locus
Huntington disease	4p16.3
Myotonic dystrophy	19q13.2–q13.3
Adult polycystic kidney disease type 1	16p13.3–p13.12
Waardenburg syndrome type 1	2q35
Neurofibromatosis type 1	17q11.2
Familial adenomatous polyposis coli	5q21–q22

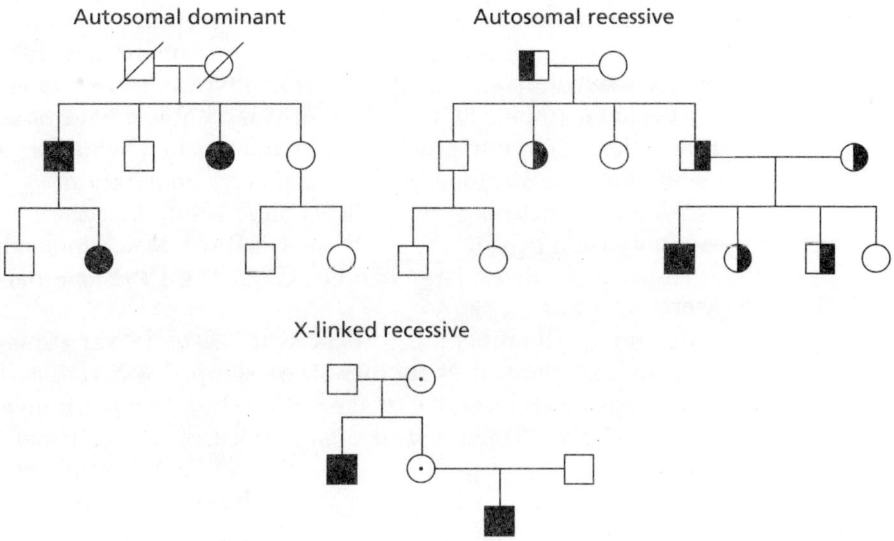

Figure 4.3
Inheritance patterns.

Other features of autosomal dominant disorders

Co-dominance. This occurs when the characteristics of each of two alleles are expressed in the heterozygous state. The most well-known example is the ABO blood groups, such that a person with blood group AB shows antigens for both group A and group B.

Reduced penetrance. In most autosomal dominant diseases, **a certain proportion of individuals** will not show the expected abnormal phenotype, sometimes to the extent of appearing completely phenotypically normal (non-penetrant). Depending on other factors such as the environment, the proportion of individuals with specific diseases but not showing the abnormal phenotype may be predictable.

For example, in retinoblastoma (a childhood tumour of the eye) the penetrance is 90%, such that in any 10 individuals carrying the *RB1* gene, one will not develop the tumour. The mechanism of this particular disorder is now well understood and is described in *Chapter 8*.

Biochemical disorders such as porphyria and hereditary pancreatitis are examples where certain carriers may only show biochemical but no overt external symptoms.

Variable expression. This refers not to the individual in isolation, but the observation that there can be a great deal of phenotypic variation between different members of the same family, who all carry the same mutation in a dominant gene. **This variation is the degree to which the gene is expressed.**

A uniform degree of expression is the exception in autosomal dominant disorders. Achondroplasia (dwarfism) is one of the few disorders in which the heterozygous state usually results in a typical recognizable phenotype.

Affected members of a family with Waardenburg syndrome, however, may show different combinations of the characteristic phenotypic features such as a white forelock, deafness and unusual eye pigmentation.

Neurofibromatosis type 1 (NF1) may result in either a nearly normal phenotype comprising a few café-au-lait spots, which look a little like freckles, through to many large benign tumours on the external surface of the skin.

Autosomal dominant disorders with extreme variation in expression result in the apparent phenomenon of 'skipped generations'. The abnormal gene is passed through every generation, but one key individual may have been so mildly affected that he or she appeared to be normal.

Problems arise in predicting the phenotype of an unborn child who has inherited the abnormal gene. A parent who is very mildly affected with NF1, for example, has to be made aware of the full range of abnormalities that may be expressed in their child, and also that the abnormalities may be considerably worse than those of the parent.

Late onset. The most notable example of a late onset disorder is Huntington disease. This neurological disorder affects the brain leading to severe motor disturbances, both physical and mental. It is often not expressed until the third or fourth decade of life, when an affected individual may have already had a family. Each of those children will then have a 50% risk of having inherited Huntington disease.

Anticipation. Although anticipation is found in other types of inheritance patterns (such as X-linked inheritance for fragile X), some of the original examples were noted in autosomal dominant diseases. As the faulty gene is passed on, the phenotype appears to become more severe (or exhibit more clinical symptoms) in later generations.

Myotonic dystrophy, for example, may show minimal symptoms in a grandfather, who might have a slight drooping of the eyelid muscles, baldness, cataracts, but a normal life span. His daughter may have a more pronounced weakening of the facial muscles including the typical downturned mouth and hand myotonia. This disease is usually more severe if it is passed through the mother, and her newborn child may then have congenital myotonic dystrophy, which results in a severe whole body muscle weakness, mental retardation and poor survival. In myotonic dystrophy there is an inability to relax muscles and patients have a typical handshake: once they have gripped the clinician's hand they find it very difficult to let go.

The underlying molecular genetic basis for myotonic dystrophy is now known, and is due to a number of triplet repeats (CTG) on chromosome 19, which expand from generation to generation until a critical number is reached, resulting in a severe phenotype (see *Chapters 3* and *7*).

Autosomal recessive inheritance

Autosomal recessive inheritance follows the general rules given below (see also *Table 4.2*).

- A recessive gene is found on **both** of a pair of homologous **autosomes.** The father and mother each pass on a recessive gene to their child.
- Both parental genes are required to display the associated phenotype, which is only revealed when the mutant alleles are in the **homozygous state.**
- The children of parents who are each heterozygous for an autosomal recessive disorder have a **25%** chance of being **affected**, a **50%** chance of being a phenotypically unaffected **carrier**, and a **25%** chance of being **normal, irrespective of sex**. A healthy sibling of a known affected child has a 2/3 chance of being a carrier (see *Box 4.3*).
- An autosomal recessive **pedigree** has less chance of having affected members in every generation, as any carriers would have to meet another carrier with the same defective gene from the general population. These pedigrees have a '**horizontal**' look, or even just one apparently 'sporadic' case (see *Figure 4.3*). This is especially true in small families.
- These pedigrees also give rise to the notion of 'missing a generation', when a carrier may not have any affected children, but their phenotypically normal children may be carriers who will meet another carrier in the next generation.

Table 4.2 Examples of autosomal recessive disorders

Disease	Chromosome locus
Cystic fibrosis	7q31.2
Phenylketonuria	12q24.1
Spinal muscular atrophy type 1	5q12.2–q13.3
Tay–Sachs disease	15q23–q24
Congenital adrenal hyperplasia	6p21.3

Box 4.3 Carrier risk calculation

The reason why a healthy sibling of a child affected with an autosomal recessive disorder has a 2/3 chance of being a carrier is as follows:

- the children of heterozygous parents have genotypes in the proportion **1 normal:2 carrier**:1 affected
- the affected genotype is not relevant to the calculation
- the proportion of normal phenotypes are therefore 1 normal:**2 carrier**
- therefore the carrier risk to the healthy sibling is **2/3**

Other features of autosomal recessive disorders

Consanguinity. In order for an affected child to be born, the parents must be two asymptomatic carriers of the same disorder. The chance of them meeting is dependent on the frequency with which that gene mutation is found in the general population. Usually for most autosomal recessive conditions the gene frequency is low, but if a couple are genetically related, the presence of a recessive mutation in a common ancestor make it more likely that they may both have inherited the mutation (see *Box 4.4*).

The exact risk that a related couple both carry the same inherited mutation depends on their genetic relationship and hence the number of genes they would share. First cousins share 1/8 of their genes; if their common grandparent passed on a mutation, the chance that their child will inherit both mutations is 1/32, despite the fact that the frequency of the disease may be very low (say 1/40 000 for a typical biochemical disorder) in the general population. This works as follows:

- each cousin has a carrier risk of 1/4 from the common grandparent
- the chance of them both passing this on to a child is 1/2 each; that is $(1/4 \times 1/2) \times (1/4 \times 1/2) = 1/64$
- because they share two common grandparents the risk is twice as great: hence $1/64 \times 2 = 1/32$

Box 4.4 Carrier risks

For a rare disease, a gene frequency of 1/200 in the population would result in a carrier frequency of about 1/100 and hence a disease incidence of 1/40 000 (see the Hardy–Weinberg equilibrium in *Box 4.8* and *Section 4.10*).

However, some disorders have a much higher gene frequency and hence a higher population carrier risk. Examples are cystic fibrosis with a British carrier risk of about 1/24, and haemochromatosis (a treatable disorder of iron metabolism) which has a carrier risk of 1/10.

X-linked recessive inheritance

X-linked recessive inheritance follows the general rules given below (see *Table 4.3* for examples).

- An X-linked gene is found on the **X chromosome**.
- **Males**, who receive an abnormal X from their mother, will be **affected**.
- **Females** who carry one normal X and one abnormal X will usually be **unaffected** carriers.
- Affected male children are usually born to unaffected parents; the father (who contributes his Y chromosome) will usually be normal and the mother will be an asymptomatic carrier. Such a couple will have a **25%** risk of an **affected son**, a **25%** risk of a **normal son**, **25%** risk of a **carrier daughter**, and a **25%** risk of a **normal daughter**.

Table 4.3 Examples of X-linked recessive disorders

Disease	Chromosome locus
Duchenne muscular dystrophy	Xp21.2
Becker muscular dystrophy	Xp21.2
Haemophilia A	Xq28
X-linked ichthyosis (steroid sulphatase)	Xp22.32
Adrenoleucodystrophy	Xq28
Hunter syndrome (MPSII)	Xq28

- An X-linked **pedigree** is characteristic in that the **affected individuals will all be male** (see *Figure 4.3*). Female carrier status may be inferred in a woman with both an affected brother and son.

Other features of X-linked recessive disorders

Reproductive ability of the affected male. Normally an X-linked disorder is passed on by a phenotypically healthy female carrier. Some X-linked diseases result in a lethal outcome for an affected male before the age of reproduction, so there is no chance of his abnormal X being passed on. This is the case with Duchenne muscular dystrophy and the severe form of Hunter syndrome.

However, some disorders are now treatable, or are mild enough to allow the affected males to marry and reproduce. Haemophilia A can now be treated with infusions of Factor VIII. Becker muscular dystrophy (which is an allelic form of Duchenne muscular dystrophy) can have a very mild phenotype.

If an affected male can reproduce (for example in haemophilia), **all his daughters will be carriers**, as they receive his single abnormal X (together with a normal maternal X). **All his sons will be normal**, as they will receive his normal Y together with their mother's normal X.

X-linked dominant inheritance

X-linked dominant inheritance is rare, but follows the general rules given below (see *Table 4.4* for examples).

- An X-linked dominant gene is carried on the **X chromosome.**
- **Both males and females** will be **affected**.
- The **females** may have a more **variable expression** as they also have a normal X (the phenotype may depend on the randomness of the X-inactivation).
- **If a female is affected, each child** will be at **50% risk of being affected**, irrespective of sex. If a male is affected, all his daughters (but none of his sons) will be affected.

Table 4.4 Examples of X-linked dominant disorders

Disease	Chromosome locus
Incontinentia pigmenti	Xq28
Rett syndrome	Xq28

Fragile X – a special case?

Fragile X was originally thought to be an X-linked recessive disorder, until it was found that around 33% of obligate carrier females had clinical symptoms. However, this is still too low a percentage to be a fully penetrant X-linked dominant disorder (all female carriers would then be affected).

As explained in *Chapter 3*, the molecular basis of fragile X is now known to be caused by a DNA triplet repeat (CGG). The numbers of CGGs increase down the generations until a critical number of repeats are reached (usually around 200). At this point the gene becomes methylated, such that it cannot produce the protein FMR-1, with the consequent phenotypic results. The disease therefore appears to display **anticipation**.

Y-linked inheritance

There are no well-characterized clinical disorders found on the Y chromosome and, apart from azoospermia, there are very few Y-linked characteristics, but they follow the general rules of inheritance given below.

- Y-linked characteristics are found on the **Y chromosome**.
- All male children would inherit that characteristic.
- If a male did have a Y-linked clinical disorder, **all his sons would inherit it**.

Two examples of Y-linked disorders/characteristics are given in *Table 4.5*.

4.7 OTHER PROBLEMS WITH INHERITANCE

Heterogeneity

Locus heterogeneity

There are two different kinds of heterogeneity. If a mutation in a gene produces a specific phenotype, and a mutation of a completely **different gene** (perhaps on a different chromosome) produces a **similar phenotype**, the disease is said to show **locus heterogeneity** (see *Box 4.5*).

- Example 1: there are several kinds of polycystic kidney disease. Adult polycystic kidney disease type 1 (APKD1) is dominant; the gene is found

Table 4.5 Examples of Y-linked disorders/characteristics

Disorder/characteristic	Chromosome locus
Sex determining region on Y (*SRY*)	Yp11.3
Azoospermia/Y microdeletion syndrome	Yq11.2

Box 4.5 Complementation and compound heterozygotes

Locus heterogeneity can be detected when a couple who each appear to have the same disorder, marry and have normal children. Deafness is one well-known example where several different genes may interact in the complex processes leading to the normal function and anatomy of the ear.

A deaf couple may each have a mutation on a different chromosome; the normal child may in fact be heterozygous for each autosomal recessive form of deafness at two unrelated gene loci.

When genes react in this way they are said to be **complementary**; the interaction of two non-allelic gene products such that a normal phenotype is produced is called **complementation.** In certain circumstances the interaction of gene products may be demonstrated by mixing cell lines from two individuals and revealing the presence of a functional protein.

The corollary of complementation is the **compound heterozygote.** In this case, each parent carries a different intragenic mutation which is passed on to the child, but there is no complementation and the child is affected. An example would be cystic fibrosis, where any mutation in the *CFTR* gene usually leads to disruption of the protein function. The child inherits two different cystic fibrosis mutations, but the two proteins are dysfunctional in different ways and do not complement. Thus cystic fibrosis behaves in a conventional recessive manner.

on chromosome 16. There is another dominant form (APKD2) with virtually the same phenotype found on chromosome 4. A third form is infantile autosomal recessive polycystic kidney disease (ARPKD), found on chromosome 6.

- Example 2: there are many kinds of muscular dystrophies and muscular atrophies. There are nearly 30 types of spino-cerebellar ataxia, for example, found on different chromosomes.

It is important that the clinician correctly identifies the locus of that family's disease correctly, otherwise DNA tests, which depend heavily on the correct gene location, will be inaccurate.

Allelic heterogeneity

Sometimes different mutations within the **same gene** produce clinically **different phenotypes.** These are then **allelic forms**, one example being the severe phenotype found in Duchenne muscular dystrophy, compared to the much milder Becker muscular dystrophy. These are found at the same gene locus (Xp21) but tend to have different allelic mutations. Although both diseases may be caused by deletions, those in Duchenne muscular dystrophy destroy the reading frame of the gene such that no protein (or a completely non-functional protein) is produced. Becker muscular dystrophy has in-frame deletions which result in a smaller protein with the functional ends intact.

Parental origin

In *Chapter 2*, we saw that parental **imprinting** played a natural part in some forms of gene expression. Inheritance of some normal gene functions may therefore depend on the **parent of origin**, as demonstrated by diseases such as Prader–Willi syndrome, Angelman syndrome and Beckwith–Wiedemann syndrome.

New mutations

Not all mutations are inherited. Some are new mutations which arise in the parental germ cells and appear for the first time in the affected individual.

In those diseases which are lethal before reproductive age, the generation of new mutations serves to replace the old mutations leaving the population. In lethal dominant, or X-linked, diseases therefore the new mutation rate tends to be fairly high. In Duchenne muscular dystrophy, for example, approximately 1/3 of all affected boys result from a new mutation on the mother's X chromosome.

In diseases where affected individuals live to reproduce (for example, late onset Huntington disease), the natural mutation rate is low, as the inherited mutation is maintained at a higher level in the population.

Germline mosaicism

Occasionally a small group of mutant cells may arise in the male or female germ cells as a clone. Although normal germ cells are also present, one of the mutant germ cells may be inherited, resulting in an affected child. For example, if a woman has two sons with Duchenne muscular dystrophy and she has no other family history, there are two possibilities:

- she may be a carrier as she inherited an abnormal X from her mother
- she could be a germline mosaic, the mutation having arisen in her germ cells

The carrier risks to various members of the family are very different with these two scenarios. The sisters of a germline mosaic would have a low carrier risk, but if their mother is an obligate carrier, they would each be at 50% risk of being carriers. Extended studies of the family history and molecular genetics studies in the laboratory can be undertaken to determine carrier risks.

4.8 MITOCHONDRIAL INHERITANCE

As the cytoplasm of our cells is derived from the cytoplasm of the maternal egg cell, **mtDNA is inherited from the mother by cytoplasmic inheritance**.

As there is cooperation between the nuclear and mitochondrial genomes (see *Chapter 1*), some mitochondrial diseases are not maternally inherited, but show a Mendelian pattern. Examples are Leigh syndrome, which is autosomal recessive in 7–20% of cases, and autosomal dominant progressive external ophthalmoplegia.

Sperm have very little cytoplasm and very few mitochondria, so although a fertilized cell has both a maternal and paternal set of chromosomes, it has only maternal mitochondria. It therefore follows that if mutations occur in the mitochondrial genome, they will be passed to children of both sexes by the mother (see *Box 4.6*).

Box 4.6 Tracing families using mitochondrial DNA

As mitochondria are maternally inherited, it was thought in the 1980s that a female ancestor could be traced back over 200 000 years; she would have been the 'Eve' of the present day population. Although the conclusion of the example above may be exaggerated, mitochondrial DNA has been used, for example, in proving the identity of the remains of the Russian royal family (the Romanovs) who were shot after the Russian Revolution in July 1918. Their mtDNA was compared to that obtained from a blood sample given by Prince Philip, the Duke of Edinburgh, who shared a common female ancestor.

Sometimes, however, a particular mutation is not present in every mitochondrion. There will then be a mosaic population comprising normal and abnormal mitochondria, known as **heteroplasmy**. Where heteroplasmy exists, the proportion of normal to mutated mitochondria may determine the severity of the disease, and this is known to be variable in mitochondrial disorders. Examples of mitochondrial disorders are given in *Section 3.5*.

4.9 GRADATION OF INHERITANCE

The most straightforward patterns of inheritance are the Mendelian disorders, which are usually based on the premise of 'one faulty gene, one disorder'. Chromosomal abnormalities may be incorporated into the parental germline, and will then follow Mendelian patterns of segregation.

Some characteristics, however, such as skin colour or intelligence depend on the additional small effects of each of a number of genes. These are called **polygenes**. Height was thought to be typically polygenic until the discovery of the *SHOX* gene on the X chromosome, which probably accounts for around 70% of our final height. There are, however, other genes for height, one of which is thought to be located on Yq.

Prediction for such characteristics is therefore much more uncertain. When the effects of environment are also taken into consideration, it becomes very difficult to isolate the genetic from the environmental components of a disease. There are certain disorders which appear to be more likely to express themselves if the individual has certain genetic markers known to be associated with the disease. These are **susceptibility** genes or markers (*Table 4.6*), and sometimes may need an extra environmental factor in order to develop the disease phenotype. Not everyone with the marker DQA1 will develop multiple sclerosis, for example; the other factors necessary to instigate the disease may range from viruses to a faulty immune system or even diet.

There is no true distinction between 'polygenes plus an environmental effect' and the term multifactorial disorders, which was the name formerly applied to diseases with no clear inheritance pattern, but with an element of genetics.

Heart disease covers a multitude of different heterogeneic anatomical and phenotypic problems, yet we now know of a particular susceptibility if an individual carries the gene for familial hypercholesterolaemia. The males are prone to heart attacks in their 30s and 40s, yet if they can follow a suitable diet from childhood and have cholesterol-lowering drugs, this risk is considerably lowered.

Identical twins, who have identical genotypes (they are essentially clones), are a naturally occurring phenomenon which allows us to study the different effects of nature (genetics) and nurture (environment). Although genetically identical, their environmental experiences may be quite different. There are cases of twins being separated and adopted at birth into different family environments. By comparing whether one twin develops a particular disorder compared with the other one also developing the same disease (**concordance**), we are able to estimate the genetic contribution towards that disease compared with the environmental effect.

Table 4.6 Susceptibility markers

Disorder	Gene/marker	Allele
Ankylosing spondylitis	HLA (tissue) type	B27
Multiple sclerosis	HLA (tissue) type	DQA1
Alzheimer disease	Apolipoprotein	E4
Rheumatoid arthritis	HLA (tissue) type	DR4

4.10 POPULATION GENETICS

Each gene may have two or more alleles, which may express themselves as harmless polymorphisms (differences) such as eye colour or unimportant DNA sequences in introns; alternatively they may represent a mutant gene on one chromosome and a normal gene on the homologue.

In a large heterogeneous population many alleles will be represented. Normal and abnormal alleles will be present in the heterozygous and homozygous form. There appears to be a natural balance of alleles in large populations; they are said to be in **equilibrium**.

The Hardy–Weinberg equilibrium

The Hardy–Weinberg equilibrium (see *Box 4.7*) is a mathematical way of representing the balance of alleles in a population. Each **allele** (note that these are alleles of genes, not the genes themselves) is present in a population at a particular **frequency** (*Figure 4.4*). If we take a gene that has two alleles **A** or **a**, then there are three combinations possible at the gene locus on a pair of autosomes: **AA**, **Aa** or **aa**.

If we say that **A** has an allele frequency of $\boldsymbol{p}$ (i.e. A=p)
and that **a** has an allele frequency of $\boldsymbol{q}$ (i.e. a=q)
and that in a population all the alleles must add up to **100%** (i.e. 100%=**1**)
then $\boldsymbol{p+q=1}$
This also means that $\boldsymbol{p=1-q}$ and that $\boldsymbol{q=1-p}$

However, the Hardy–Weinberg equation describes the **distribution of alleles in a population:**

i.e. how many **AA**s (i.e. $\boldsymbol{p \times p = p^2}$)
i.e. how many **Aa**s and **aA**s (i.e. $\boldsymbol{2 \times p \times q = 2pq}$)
i.e. how many **aa**s (i.e. $\boldsymbol{q \times q = q^2}$)

When numbers are squared in this way, the allele distribution is represented by a **binomial equation**, which also shows the balance (**equilibrium**) of the three combinations of alleles:

$$p^2+2pq+q^2$$

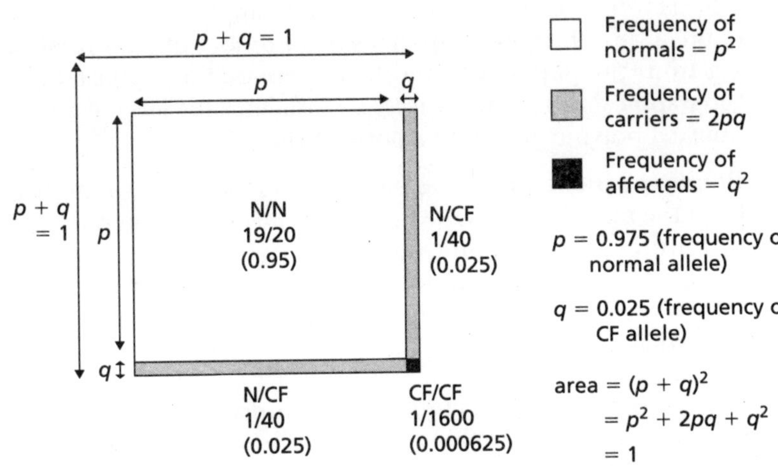

Figure 4.4
The Hardy–Weinberg equilibrium using the allele frequencies of cystic fibrosis, where N = normal allele and CF = cystic fibrosis allele.

Box 4.7 Using the Hardy–Weinberg equation to calculate carrier frequency in cystic fibrosis

If $p+q=1$, then $(p+q)^2=1$, so expanding the equation: $p^2+2pq+q^2=1$.
This can be illustrated by the recessive disease cystic fibrosis.
If **A** = the normal allele, then the **normal allele frequency** = p
if **a** = the CF (abnormal) allele, then the **abnormal allele frequency** = q

p^2 = the proportion of **normal** people in the population
$2pq$ = the proportion of **carriers** in the population
q^2 = the proportion of cystic fibrosis **affected** people in the population

We can use the Hardy–Weinberg equation to calculate various useful pieces of information such as carrier rates and incidence of the disease. If the **incidence** of newborns with cystic fibrosis in the Caucasian population is approximately **1/1600**:

then $q^2 = 1/1600$
so $q = \mathbf{1/40}$ (0.025) which is the **cystic fibrosis gene frequency** (or abnormal allele frequency)
As $p = 1-q$
then $p = \mathbf{0.975}$ (1–0.025) which is the **normal gene frequency** (approximately equals 1)
so $2pq = 2\times0.975\times0.025 = 0.04875 = \mathbf{1/20}$ which is the **carrier** frequency

Another way of considering carrier risk is to say that each chromosome 7 (where the cystic fibrosis gene locus is found) has a 1/40 risk of carrying cystic fibrosis. As there are two chromosome 7s, the risk is 1/40 + 1/40 = 1/20.

If we just know the carrier risk, the incidence is calculated as follows. The chance of a female (carrier risk 1/20) meeting a male (carrier risk 1/20) and having an affected child (affected risk 1/4) is $\mathbf{1/20\times 1/20\times 1/4 = 1/1600}$

Factors disturbing the Hardy–Weinberg equilibrium

The Hardy–Weinberg equation only applies to large randomly mixed populations. It depends upon a small number of abnormal genes disappearing from the population and being replaced at a constant rate. Physical or geographical factors may therefore influence the level of an abnormal gene or unusual polymorphism in a population.

Random genetic drift. A small number of people may become separated from the main population either voluntarily (by emigration) or involuntarily by features of the terrain (mountains, rivers, etc.).

There is a chance that in a small population (who may be more closely related anyway) an imbalance of genes may exist, such that there are more abnormal genes for a particular disease than expected by the Hardy–Weinberg equation.

If an imbalance in a clinically insignificant polymorphism occurs, it may increase in frequency (by chance due to genetic drift) to become the only allele carried in that population (Native American Indians do not have blood group B, for example).

Sometimes the origin of a particular gene abnormality can be traced back through the ancestors of the affected family members to a single individual.

- There is only one mutation for the second breast cancer gene (*BRCA2*) in Iceland. Researchers at the Icelandic Cancer Society discovered that a 16th century cleric called Einar was responsible for nearly every case of this type of breast cancer in Iceland today. It was possible to trace him, as historically there was a relatively small gene pool on this remote island.

  The individual bearing the first mutant gene found in a population is responsible for the phenomenon of **the founder effect.**

Change in mutation level. The stability of allele frequencies will alter in a population if there is a change in mutation level resulting in an increase or decrease of q^2. An increase or decrease in population size may influence the frequency of an allele. If a population becomes smaller, for example, due to illness, an abnormal allele may remain in the survivors at a higher frequency than previously. This allele will be perpetuated as that population grows again. With any inbred population there is an increased risk of autosomal recessive disorders.

An abnormal allele can sometimes be maintained at an abnormally high level due to the phenomenon of **heterozygote advantage**. Here the **carrier state for a particular autosomal recessive disorder** confers a physical advantage over the homozygous normal state and is selected for (see *Box 4.8*).

Box 4.8 Heterozygote advantage of cystic fibrosis

It has long been a puzzle why the fatal disease cystic fibrosis is so common in the Caucasian population. In the homozygous affected state, the thick mucus produced on the surface of the body's epithelial cells clogs the lungs and intestines. This is due to a fault in the chloride channels of the cell membrane. Because chloride ions cannot pass out of the cell, water is also prevented from exiting into the lungs or intestines, and the mucus normally present there is insufficiently hydrated.

This means that the body fluids are not as dilute as they should be. It has now been suggested that there may be a heterozygote advantage such that in the past a carrier of cystic fibrosis may have had more resistance to cholera, which led to death from dehydration due to rapid loss of water through diarrhoea. Perhaps the carrier had more resistance to water loss than the normal person.

As described in *Chapter 3*, sickle cell anaemia is common in countries where malaria is found. Although HbS red cells carry oxygen normally, in the deoxygenated state the red blood cells are shaped like sickles. Sickled cells are deformed and therefore cannot traverse capillaries, leading to downstream tissue hypoxia, necrosis and pain. Malarial parasites invade the red blood cells via mosquito bites and consume intracellular oxygen, triggering sickling and the selective removal of parasitized cells in the heterozygous carriers. Those individuals with the sickle cell trait therefore appear to be more resistant to malaria.

4.11 LINKAGE AND LOD SCORES

Two genes close together on the same chromosome are said to be linked. Before DNA testing it was often noticed that two separate characteristics could run in the same family such that the affected individuals also had the same allele for an unassociated physical characteristic.

Sufferers of the autosomal dominant nail patella syndrome (oddly shaped fingers, toenails and kneecaps) in a particular family were sometimes found to have the same allele from the gene that determines ABO blood groups, whereas the normal individuals had a different allele. The reason was that both nail patella syndrome and the genes for the ABO blood group are found close together on chromosome 9; they are linked. Although coding for completely separate characteristics, the alleles are passed on together as a haplotype. Another example of linkage is between the dominant disease myotonic dystrophy and the secretor locus (i.e. positive or negative for a substance in the saliva) on chromosome 19. In a small nuclear family it might be that all the affected people had the negative secretor allele, whereas those unaffected carried the positive secretor allele. The locus for secretor was so close to the locus for myotonic dystrophy that there was very little chance of a crossover of alleles at meiosis, so the two alleles remained linked.

With the advent of DNA testing which can distinguish different polymorphisms (often called **markers**), it becomes useful if it can be established whether the polymorphisms are linked to any genetic disorders. If a marker allele can be shown to be linked to a disorder in a family, the disease can be **tracked** through that family even if we do not know the exact gene locus of the disease (see *Chapter 7*).

By using the amount of recombination between a disorder and a marker, the distance between the two loci can be estimated. This can be done empirically, i.e. by looking at existing cases in families and collecting data. There is an association between the amount of recombination and map units which determine the distance apart of two loci: one **centimorgan** is equivalent to a 1% recombination rate.

It is obviously important to know how far away a polymorphic marker is with respect to a disorder where the gene locus is not known, because the closer the marker, the more tightly linked they will be and the less will be the recombination. This would mean that a close, tightly linked marker would more accurately track the disorder in a family, and the predictive value would increase.

A statistical method can be used which estimates not only the chance that a disease might be linked to a marker, but even the most likely recombination frequency between the disease and the marker. This involves calculating **Lod scores** (Log of the odds on linkage, see *Box 4.9*). Taking the example of the family in *Figure 4.5*, it can be seen that some members are affected with an autosomal dominant disease. A marker suspected to be linked to this disease has six alleles: A, B, C, D, E or F. Chromosomes come in pairs, so there is always a choice of two alleles out of these six.

Notice that most of the affected members in this family carry allele A, suggesting that this allele is linked to the disease, and is inherited along with the abnormal disease gene. One affected member of the family has not inherited allele A; there has been a recombination between the marker locus and the disease locus. The number of recombinants in a family gives us an estimate of the recombination risk: in the family depicted in *Figure 4.5* there is only one

Box 4.9 Working with logarithms

The example below uses **logarithms (logs) to the base 10** (log_{10}).

- Large numbers are expressed as powers of 10, so for example $1000 = 10^3$, so $\log_{10} 1000 = 3$.
- Using the same principle, $\log_{10} 100 = 2$, $\log_{10} 10 = 1$ and $\log_{10} 1 = 0$.
- Where two numbers are ordinarily multiplied, the logs of each of those numbers are added (e.g. $M \times N = MN$; $\log(MN) = \log M + \log N$).
- Powers of numbers can be multiplied with logs (e.g. $\log_{10} 4^7 = 7\log_{10}4$).

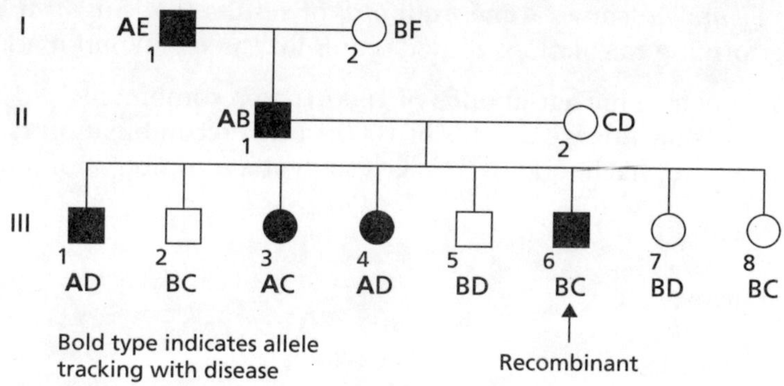

Figure 4.5
Calculating Lod scores.

recombinant in eight people; if these loci were unlinked we would expect 50% recombination ($\theta = 0.5$), i.e. four recombinants out of eight.

However, we have to look at the family structure as a whole to calculate the Lod scores at different values of θ. This is because we have to take into account the fact that two loci could be linked by chance.

It can be deduced that allele A tracks along with the disease because affected members of three generations show this pattern (with the exception of the recombinant). This is called knowing the **phase**, that is, in II_1 allele A tracks with the disease and allele B tracks with the normal state. Lod scores become more difficult to calculate if some crucial family members have died and phase is not definitely known. The Lod score for the family structure is called Z and the recombination fraction (rate) is represented by θ.

In this family 7 people have not recombined $(1-\theta)^7$
1 person has recombined (θ)

The chance of no linkage in eight people is 50% $(0.5)^8$

The ratio (or likelihood) of linkage to non-linkage is therefore:

$$Z = \log_{10}[(1-\theta)^7 \times \theta \text{ divided by } (0.5)^8]$$
$$Z = 7\log_{10}(1-\theta) + \log_{10}\theta + 8\log_{10}0.5$$

Z is calculated for each of several recombination fractions (usually a range from θ = 0 to θ = 0.5):

θ	0	0.1	0.2	0.3	0.4	0.5
Z	– ∞	1.0874	1.0308	0.8009	0.4575	0

Usually there is said to be linkage if **Z > 3**, and no linkage if **Z < –2.**

Notice that in our example the family is not big enough to reach a score of 3; often the results of several families have to be combined to be statistically significant.

Figure 4.6 shows some examples of results that might be obtained after performing calculations of Lod scores for three different markers:

A. probable linkage at odds of 1000:1, no recombinants
B. possible linkage at odds of 1000:1 but a recombination rate of 0.3
C. a lower likelihood of linkage (10:1) at a recombination rate of 0.3

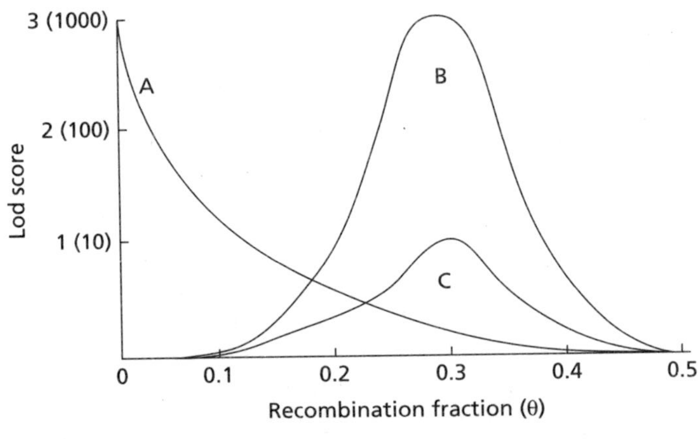

A: Lod = 3 (1000:1 odds on linkage) with θ = 0
B: Lod = 3 (1000:1 odds on linkage) with θ = 0.3
C: Lod = 1 (10:1 odds on linkage) with θ = 0.3

Figure 4.6
Using Lod scores to estimate recombination risk.

4.12 BAYES' THEOREM

Another very useful mathematical equation was named after its inventor, Reverend Bayes, in 1763.

We may wish to know the chances of someone being a carrier of a disease, but we must also take into account the chance of them not being a carrier. In a similar manner we can compare the chance that someone is affected with a disease versus the chance that they are not affected. The advantage of Bayes' theorem is that it can incorporate new information contributing to the final risk.

We usually start with the initial estimated risk taken from pedigree information. This is called the **prior risk.** This risk can then be modified, making it more or less likely that someone is a carrier or is affected. Many different factors may be taken into consideration; these are called the **conditional risks.** If there is more than one condition they can be multiplied together. Examples of such conditions include:

- numbers of normal children
- age (with a late onset disease)
- enzyme levels
- mutations already tested for by DNA

The consultand (see *Box 4.10*) from the pedigree in *Figure 4.7* is asking if she is a carrier of Duchenne muscular dystrophy. This is an X-linked disorder, and her sister has had an affected boy, suggesting that she is a carrier. *Table 4.7* shows the risks as assessed using Bayes' theorem.

Box 4.10 Terms used in clinical genetics

An **index case** is the first affected individual in a family to be investigated for a particular disorder. A **consultand** is any individual attending a clinic for a **consultation** with a medical specialist, in this case a **consultant** clinical geneticist.

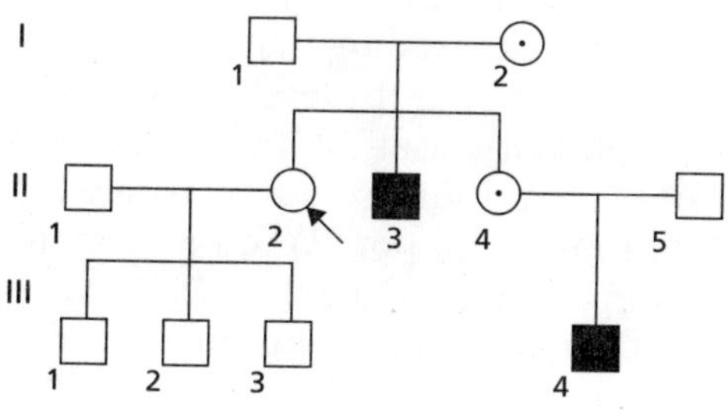

Figure 4.7
Use of Bayes' theorem to calculate a DMD carrier risk.

Table 4.7 Bayes' theorem used in Duchenne muscular dystrophy

	Carrier	Not a carrier
Prior risk	1/2	1/2
Conditional risk	1/2×1/2×1/2=1/8	1
Joint risk	1/16	1/2

The grandmother has one abnormal X chromosome and one normal chromosome. There is therefore a one in two chance that she has passed the abnormal X to her daughter (prior risk of being a carrier = 1/2). The chance she has passed on the normal chromosome (i.e. not a carrier) is also 1/2.

The consultand has already had three normal sons. Common sense tells you that this makes her more likely not to be a carrier. If she is a carrier, the chance of her having a normal son is 1/2 (i.e. the chance of her passing on her normal X). As she has had three normal sons this would have happened three times; these chances are multiplied together giving a conditional risk of $(1/2)^3$ (i.e. $1/2 \times 1/2 \times 1/2 = 1/8$).

If the consultand is not a carrier she can only have normal sons – a certainty of 1.

The prior risk is multiplied by the conditional risk to give a **joint risk**. This is the initial risk modified by the extra information we have incorporated. These figures are now put into the Bayes' theorem to give a final carrier risk. This will be expressed in words as:

$$\frac{\textbf{the joint risk of being a carrier}}{\textbf{the joint risk of being a carrier + the joint risk of not being a carrier}}$$

Putting in the above figures:

$$\frac{1/16}{1/16 + 1/2} = \frac{1/16}{9/16} = 1/9$$

Her risk has therefore been reduced from 1/2 to 1/9.

It would also be possible to incorporate the results of testing for the enzyme creatine kinase, which is present at a higher level in female carriers of Duchenne muscular dystrophy. The reading may modify the risk towards or away from being a carrier, depending if the level is high or low, respectively.

In the next example, a phenotypically normal individual is wishing to know their carrier risk for the autosomal recessive disease cystic fibrosis. The carrier risk in the Northern European population is around 1/20. The prior risk of that person not being a carrier is therefore 19/20 (1 – 1/20).

A DNA laboratory tests their DNA for a range of cystic fibrosis mutations, known to account for around 90% of all cystic fibrosis mutations in the local population, and finds them negative (i.e. normal) for those mutations. The only way that person can still be a carrier is if they carry a mutation present in the 10% not tested for; this is the conditional risk. Someone who is not a carrier will by definition not carry any cystic fibrosis mutations – with a certainty of 1. *Table 4.8* shows the risks as assessed using Bayes' theorem.

Applying the formulation as for the first example, the final carrier risk will be:

$$\frac{0.005}{0.005 + 0.95} = \frac{0.005}{0.955} = 1/191$$

The carrier risk for this individual after a negative DNA test has been reduced from 1/20 to 1/191.

Table 4.8 Bayes' theorem used in cystic fibrosis

	Carrier	Not a carrier
Prior risk	1/20 (0.05)	19/20 (0.95)
Conditional risk	10% (0.1)	1
Joint risk	0.005 (0.05 × 0.1)	0.95 (0.95 × 1)

4.13 SUMMARY

Classical genetics includes Mendel's laws and the major Mendelian inheritance patterns, and is the basis upon which prediction of familial single gene disorders is based.

Risks formerly calculated solely from pedigrees can now be augmented by batteries of medical tests, and statistical calculations have been developed in order to incorporate these results into a more accurate risk estimate.

Although thousands of single gene disorders are already characterized, a vast range of diseases such as autoimmune disorders and heart defects still remain, for which the genetic basis is strongly modified by environmental factors.

SUGGESTED FURTHER READING

Harper, P.S. (2004) *Practical Genetic Counselling,* 6th Edition. London: Hodder Arnold.

Young, I.D. (2007) *Introduction to Risk Calculation in Genetic Counselling,* 3rd Edition. Oxford: Oxford University Press.

SELF-ASSESSMENT QUESTIONS

1. The recombination fraction for two genes is 0.1. What does this tell you about these genes? Explain your reasoning.
2. Name the three major types of inheritance patterns and give an example of a disorder for each one.
3. Albinism (a lack of pigment in the skin and eyes) is an autosomal recessive disease. A man has a sister who has this disorder. What is the chance that he is a carrier of albinism?
4. A man and a woman both affected with achondroplasia decide to marry and have children. Given that the homozygous dominant form is lethal prenatally, what proportion of their children will be of normal height?
5. A man who has azoospermia (produces no mature sperm) wishes to undergo infertility treatment by intracytoplasmic sperm injection (ICSI) surgically, using one of his mature sperm to fertilize an egg. It is found by molecular

analysis that he has a deletion at the azoospermia locus on his Y chromosome. What are the prospects for his male children?

6. An adult male had a paternal grandmother who died of Huntington disease. His father is 65 years old and has not yet developed the disease. Using Bayes' theorem, and given that by 65 years of age, 85% of people carrying the Huntington disease gene would have developed the disease, what is the risk that his father is carrying the gene? What is the risk of the son having Huntington disease?

CHAPTER 5

Cytogenetics

Learning objectives

After studying this chapter you should confidently be able to:

- **Describe how chromosomes are prepared**
  Chromosomes are only visible at mitosis, therefore most analysis is undertaken on cells that have been grown in tissue culture, such as blood lymphocytes, amniotic fluid cells and skin cells. The cells are arrested at metaphase, swollen with hypotonic salt solution, fixed in a mixture of methanol and acetic acid, and then spread on a slide. The chromosomes are then stained and examined using high power oil immersion optics.

- **Describe the normal human karyotype**
  The human karyotype consists of 46 chromosomes, 22 autosomes numbered 1–22 in decreasing size, and the sex chromosomes: XX in females and XY in males.

- **Describe the different types of chromosome abnormalities**
  Chromosome abnormalities may affect number (aneuploidy and polyploidy) or structure of chromosomes. Structural abnormalities include translocations, inversions, rings, deletions, duplications, isochromosomes and extra structurally abnormal chromosomes. Chromosome abnormalities may be balanced, with material rearranged but with no net loss of material, or unbalanced, with extra or missing material.

- **Outline the system for describing chromosomes and interpret a simple karyotype**
  Chromosomes are described using the International System for Human Cytogenetic Nomenclature, ISCN. The system describes each chromosome as a series of bands with each chromosome in the human somatic cell complement considered to consist of a continuous series of bands, with no unbanded areas. There are standard abbreviations for describing abnormalities.

- **Describe the effects of unbalanced and balanced karyotypes**
  An unbalanced karyotype is generally responsible for serious clinical effects. Common unbalanced karyotypes include trisomy, deletion, and products of segregation of a translocation. The carrier of a balanced structural chromosome abnormality frequently has a risk of miscarriage or birth of a physically and mentally handicapped child. Common balanced structural chromosome abnormalities conferring a reproductive risk include Robertsonian and reciprocal translocations, and pericentric inversions.

- **Name some well recognized chromosomal syndromes and the chromosome defect responsible**
  The common chromosomal syndromes include Down syndrome (trisomy 21) Edwards syndrome (trisomy 18), Turner syndrome (monosomy X), Klinefelter syndrome (47,XXY), and cri-du-chat syndrome (deletion 5p).
- **Understand the impact on sexual differentiation and development**
  The Y chromosome contains the male determining gene essential for male differentiation. Sex chromosome abnormalities are associated with their own particular syndromes, and imbalance of the sex chromosomes is tolerated to a greater extent than autosomal imbalance.

Cytogenetics is a branch of genetics that is concerned with the microscopic study of chromosomes and cells. Clinical cytogenetics involves the processing and analysing of different sample types with the purpose of detecting and interpreting chromosome abnormalities in the karyotype (see *Box 5.1*).

The work of the cytogeneticist falls into three main categories:

1. analysis of blood from individuals with a variety of problems including congenital abnormalities, learning difficulties, reproductive difficulties (recurrent miscarriages and infertility), and sexual development problems
2. prenatal diagnosis of chromosomal abnormalities from amniotic fluid or chorionic villus samples (see *Chapter 9*)
3. analysis of samples, usually bone marrow or blood, from patients with known or suspected cancers such as leukaemia, to aid in the diagnosis and management of the disease (see *Chapter 8*)

The results of cytogenetic investigations have a huge impact on the patient and often their families, depending on the abnormality, and careful genetic counselling is paramount (see *Chapter 10*).

Box 5.1 What is a karyotype?

Karyotype means chromosome complement. It can be used in the sense of the chromosome complement of a species, a single organism, or even an individual cell. The word is also applied to a picture in which the chromosomes are cut and pasted into matching pairs, see *Figure 1.10*.

5.1 THE HUMAN KARYOTYPE

Humans have 46 chromosomes as 23 pairs. All chromosome pairs except the X and Y sex chromosomes are identical with respect to length, centromere position and genetic loci. There are 22 pairs of homologous **autosomes** (those chromosomes not directly involved in sexual differentiation) and one pair of **sex chromosomes** that are either homologous XX in females or heterologous XY in males. The autosomes are numbered from 1 to 22 in descending size and according to their overall length (i.e. chromosome 1 is the largest and 22 the smallest).

Each normal somatic cell of the body contains the same **diploid** ($2n$) complement of chromosomes with two copies, **disomy**, of each chromosome: one of paternal and one of maternal origin. **Aneuploidy** refers to any chromosome complement with one or more additional or missing chromosomes compared to the full set, usually the diploid chromosome number.

Deviation from this normal balanced complement results in clinical abnormality. It is important to understand that, in most cases where an abnormal phenotype is associated with an abnormal karyotype, the phenotype is usually attributable to extra or missing chromosome material that is an imbalance, or to a dosage effect of **normal** genes rather than to particular **defective** genes.

This is an important difference between cytogenetic syndromes and single gene disorders (*Chapters 4* and *7*). The commonest types of unbalanced karyotype involve **monosomy** (one copy) or **trisomy** (three copies) for all or part of a chromosome, resulting from a numerical or structural abnormality.

Trisomy 21 (Down syndrome) is the commonest trisomy seen in humans at birth. Currently in the United Kingdom, about 60% of trisomy 21 pregnancies are diagnosed prenatally (see *Chapter 9*), although detection rates are increasing as the methods for screening improve. Most of the remainder are detected at birth where the baby will display characteristic features familiar to the paediatrician (see *Box 5.2*).

Nullisomy (no copies) and **tetrasomy** (four copies) are rarely encountered. There are also other structural abnormalities of the chromosomes, which, while balanced and therefore compatible with a normal phenotype, generate unbalanced forms at the meiotic division of gametogenesis, resulting in an abnormal phenotype in the following generation.

Box 5.2 Trisomy 21 – Down syndrome

Down syndrome, the result of trisomy for chromosome 21, is the most familiar chromosome syndrome and is a typical example of a recognizable specific facial dysmorphism, with the characteristic small nose, flat profile both to the face and the back of the head, and distinctive shape to the eyes resulting from epicanthic folds. The ears are small and low-set. The stature is short. Mental handicap is universal, and heart defects common. Although individuals can now live into middle age, three-quarters of conceptions with Down syndrome are lost during pregnancy but, of those that are born, a number of them will do well and manage to lead independent lives.

Ploidy

Ploidy refers to the number of chromosome sets in a karyotype. The single set of chromosomes in the sperm or the egg is known as **haploid**. The two haploid gametes, each with 23 chromosomes, fuse to give a diploid zygote with 46 chromosomes. Other levels of ploidy are occasionally seen: **triploidy,** which is three complete chromosome sets, occurs most commonly when an ovum is fertilized by two sperm, the result being an abnormal pregnancy that usually miscarries. **Tetraploidy**, which is four chromosome sets, occurs when the nucleus and cytoplasm of a diploid cell fail to divide following normal replication and division of the chromosomes. It is seen sometimes in tissue culture, in the placenta, or in tumours.

With a few exceptions, gain or loss (trisomy or monosomy) of a specific region of the genome is responsible for the expression of a particular spectrum of clinical abnormalities. There are therefore many identifiable chromosomal syndromes attributable to specific gain or loss. Most chromosome syndromes include mental handicap, and many have general features in common such as growth retardation, skeletal malformations, heart and other organ defects, and midline defects such as cleft palate. Often a syndrome is associated with distinctive dysmorphic facial features peculiar to that syndrome. However, patients with a syndrome, although having similar phenotypes, can display varying degrees of effects and no two patients will be exactly the same.

There is a limit to the extent of imbalance compatible with life, and many conceptions with an unbalanced karyotype abort spontaneously. There are some chromosome regions, probably containing vital developmental genes, for which imbalance cannot be tolerated.

5.2 PREPARING CHROMOSOMES

Individual chromosomes are not distinguishable in the interphase nucleus, but become visible only as they condense immediately preceding mitotic or meiotic division. Cell culture is usually required for conventional cytogenetic analysis, as chromosomes can only be seen during those stages of mitosis and

Table 5.1 Tissues cultured for cytogenetic studies

Tissue	Type of culture	Culture duration	Reason for referral
Blood (lymphocytes)	Suspension	2–3 days	Wide range of post-natal reasons
Amniotic fluid	Monolayer	7–14 days	Prenatal diagnosis
Chorion	Direct	0–24 hours	
	Monolayer	7–14 days	
Skin/organ	Monolayer	7–21 days	Abortion/post-mortem, investigation of mosaicism
Bone marrow	Suspension	0–72 hours	Leukaemia
Tumour	Suspension or attached	Varies	Solid tumours

meiosis when the genetic material is supercoiled and condensed into discreet chromosomes of distinctive length and shape; these stages are prophase and metaphase of mitosis and stages 1 and 2 of meiosis (see *Section 1.4*).

In the majority of somatic tissues there is insufficient mitotic activity to retrieve chromosome preparations directly from the tissue, and it is necessary to resort to tissue culture to produce the necessary dividing cells (see *Table 5.1*). Two exceptions are the chorion of the placenta, where the cytotrophoblast cells display spontaneous activity (see *Chapter 9*), and the bone marrow (see *Chapter 8*). A variety of tissue culture methods are appropriate for different tissues.

In general terms, cultured cells require a sterile environment and a supply of nutrients for growth. In addition, the culture environment should be stable in terms of pH and temperature. Various defined basal media are commercially available which will support the growth of a wide range of cell types. Types commonly used in cytogenetics laboratories includes RPMI 1640, Hams F10 and TC199 (see *Box 5.3*).

The basic constituents of media include:

- inorganic salts
- carbohydrates
- amino acids
- vitamins
- fatty acids and lipids
- proteins and peptides
- serum

pH

Cells are very sensitive to pH and must be kept close to neutrality (pH 7.0) – they will only grow at about pH 6.8–7.4. Close control of pH is essential for optimum culture conditions: natural buffers such as bicarbonate/CO_2 systems maintained in an atmosphere of 5–10% CO_2 in air, or chemical buffers such as HEPES, are added to the media to maintain the required conditions. The pH of the medium changes as the cells grow, use the nutrients and release waste products. Phenol red is usually included in the media so that the pH status of the media can been seen by the colour – yellow indicates acid and purple alkali.

Temperature

Cells are usually grown at 37°C. They are very sensitive to changes in temperature and are easily killed at a few degrees above normal body temperature. Lower temperatures can be tolerated but growth will be slower. Tissues to be used for cell culture will survive for a few days at room temperature which enables the samples to be sent to the laboratory in the normal post.

Box 5.3 Constituents of culture media RPM11640

Basal media contains a complex and carefully controlled mixture of nutrients, some in minute quantities. Cells often require additional supplements such as serum or growth factors. One litre of the basal media RPMI 1640 always contains the following nutrients.

Compound	Amount (grams/litre)
Inorganic salts	
Calcium nitrate	0.1
Magnesium sulphate	0.04884
Potassium chloride	0.4
Sodium bicarbonate (for buffering)	2
Sodium chloride	6
Sodium phosphate	0.8
Amino acids	
L-Argine	0.2
L-Asparagine	0.05
L-Aspartic acid	0.02
L-Cysteine	0.0652
L-Glutamic acid	0.02
L-Glutamine (0.3 g/l must be added)	none
Glycine	0.01
L-Histidine	0.015
Hydroxy-L-proline	0.02
L-Isoleucine	0.05
L-Leucine	0.05
L-Lysine	0.04
L-Methionine	0.015
L-Phenylalanine	0.015
L-Proline	0.02
L-Serine	0.03
L-Threonine	0.02
L-Tryptophan	0.005
L-Tyrosine	0.02883
L-Valine	0.02
Vitamins	
D-Biotin	0.0002
Choline chloride	0.003
Folic acid	0.001
Myo-inositol	0.035
Niacinamide	0.001
D-Aminobenzoic acid	0.001
D-Pantothenic acid	0.00025
Pyridoxine	0.001
Riboflavin	0.0002
Thiamine	0.001
Vitamin B12	0.000005
Other	
D-Glucose	2
Glutathione	0.001
Phenol red (pH indicator)	0.0053

Cells are grown in sterile liquid medium (*Figure 5.1*). Depending on the types of cells being cultured (see *Table 5.1*) they will grow in either suspension, in a culture tube (e.g. for blood cells), or by attaching to a substrate in a tissue culture flask (e.g. amniotic fluid and fibroblast cells).

Depending on the length of the cell cycle, the cells are typically grown for several days until sufficient mitotic activity is occurring. The cells are then arrested at metaphase of mitosis by the addition to the culture of colchicine or its synthetic derivative colcemid. The effect is to destroy the mitotic spindle fibres so that the chromosomes, randomly distributed within the cell rather than on the metaphase plate, are unable to proceed to anaphase. The cells are then exposed to a hypotonic salt solution, for example potassium chloride or sodium citrate, in order to swell the cells and facilitate separation of the chromosomes. Finally, they are fixed in a mixture of methanol and acetic acid. The chromosome preparation is made by placing a drop of concentrated fixed cell suspension onto a microscope slide and simply allowing the fixative to evaporate. As the alcohol evaporates and the slide dries, the chromosomes spread out. The ideal is to produce chromosomes that are long and well-defined with little overlapping and which can then be analysed under the microscope.

A wide range of tissues can be used as a source of chromosomes for analysis. The quality of the preparation varies depending on the cell type, with lymphocyte culture generally giving the best results. Lymphocytes are the cells of choice for most post-natal chromosome studies (see *Box 5.4*). They are easily obtained from a small blood sample of just 1–2 ml, they are easily grown, and they provide a high quality result.

Amniotic fluid cells or chorionic villus samples are most commonly used for prenatal diagnosis (see *Chapter 9*), and bone marrow or tumour samples for cancer studies (see *Chapter 8*). Skin or placental samples can be used to investigate pregnancy loss, miscarriages and stillbirths.

Box 5.4 Lymphocyte culture

Lymphocytes are a type of white blood cell concerned with immune response. They can be cultured in various media such as RPMI1640 basal media with 20% fetal calf serum, antibiotics and glutamine (an essential amino acid that cells cannot synthesize).

A particular sub-type, the T cell, does not normally divide actively, but can be stimulated to do so by adding a substance called phytohaemagglutinin to the culture medium. In the standard blood culture, the lymphocytes are not separated from the blood sample, but a few drops of whole blood are simply added to the culture medium.

Lymphocyte cultures are often synchronized by temporarily blocking the passage of cells through the DNA synthesis phase of the cell cycle by the addition of a high concentration of thymidine or an antimetabolite drug such as methotrexate. Following release of the S-phase block, it is possible to harvest a high yield of cells in late prophase to early metaphase where the chromosomes are relatively uncontracted.

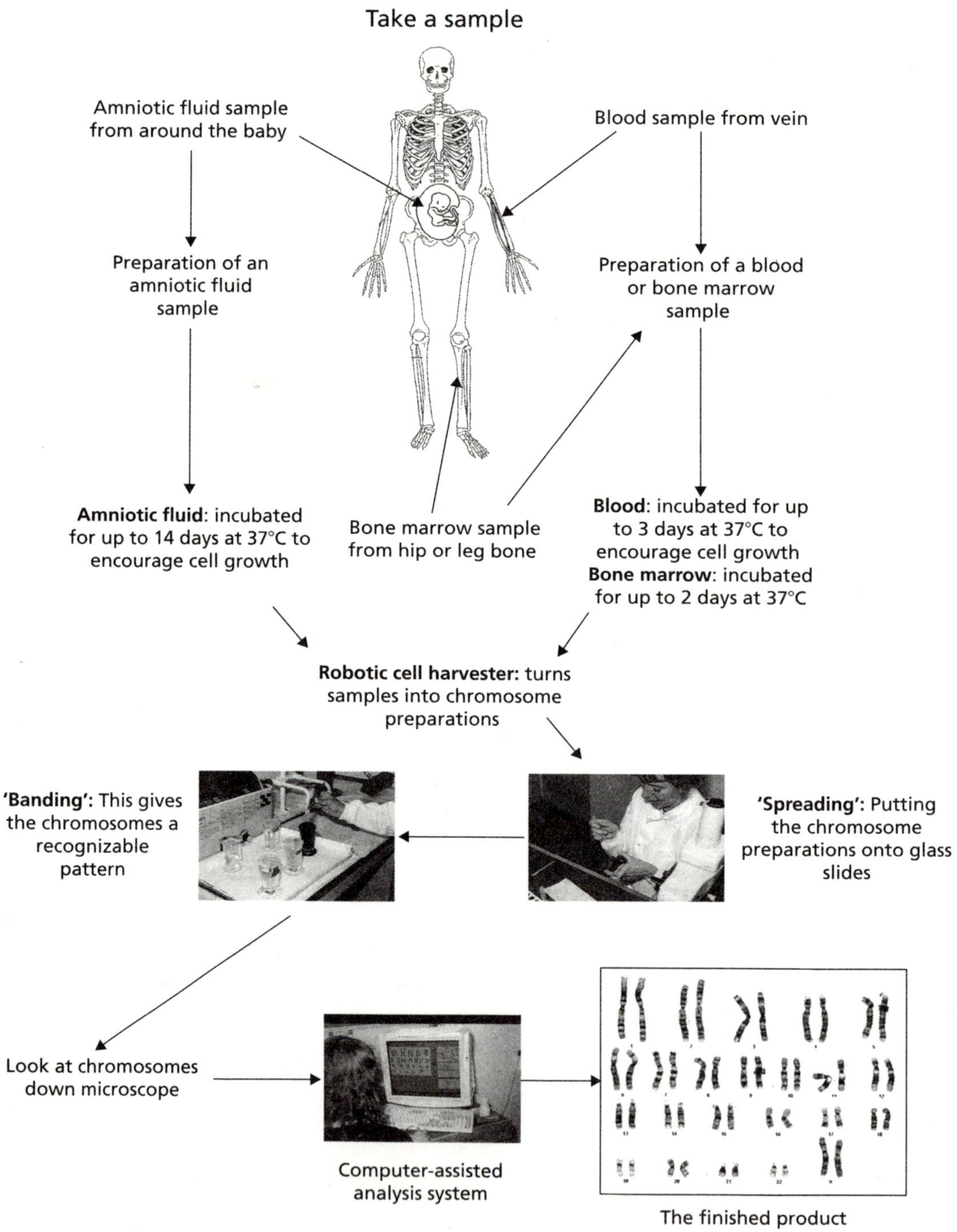

Figure 5.1
How to prepare chromosomes.

5.3 CHROMOSOME ANALYSIS

When undertaking chromosome analysis, the total chromosome number, the morphology of each chromosome with respect to its length and centromere position, and the staining pattern, are all taken into account.

Chromosomes with the centromere close to the mid-point are called **metacentric**, those with a noticeably shorter short arm are **submetacentric**, and those with a very short short arm are **acrocentric**. Some other species (mice for example) have **telocentric** chromosomes where the centromere is located right at the end; this shape is not seen in normal human chromosomes.

Block or solid staining stains the entire chromosome evenly. It is not possible to identify individual chromosomes, but they can be assigned to seven groups (A–G) based on size, centromere position and shape (see *Box 5.5*):

- A comprises chromosomes 1–3, which are large and metacentric
- B comprises 4 and 5, which are large submetacentric
- C comprises 6–12 and X, which are medium-sized submetacentric
- D comprises 13–15, which are medium-sized acrocentric
- E comprises 16–18, which are smaller submetacentric
- F comprises 19 and 20, which are small metacentric
- G comprises 21 and 22 and Y, which are small acrocentric

Chromosome analysis also shows features such as 'fragile sites'. These appear as gaps within a chromosome arm. Fragile X syndrome is associated with a fragile site at Xq27 (see *Figure 5. 2a*). Fragile X mental retardation occurs in about 1 in 2000 males and is due to multiple repeats of the sequence CGG (see *Chapter 3*).

For most routine investigations, analysis makes use of the **G- (Giemsa) banding** technique, a method in which a cross-banding pattern unique to each chromosome is achieved by a variety of treatments. The most widely used and simplest method entails dipping the slide in a solution of the enzyme trypsin before staining the slide in Giemsa or Leishman's stain. Using this method, the trained cytogeneticist is able to identify each pair of chromosomes and spot abnormalities of the banding pattern.

Box 5.5 The groups of human chromosomes

Group	Chromosomes	Size and shape
A	1–3	Large and metacentric
B	4–5	Large and submetacentric
C	6–12 (and X)	Medium and submetacentric
D	13–15	Medium and acrocentric
E	16–18	Small and submetacentric
F	19–20	Smaller and submetacentric
G	21–22 (and Y)	Small and acrocentric

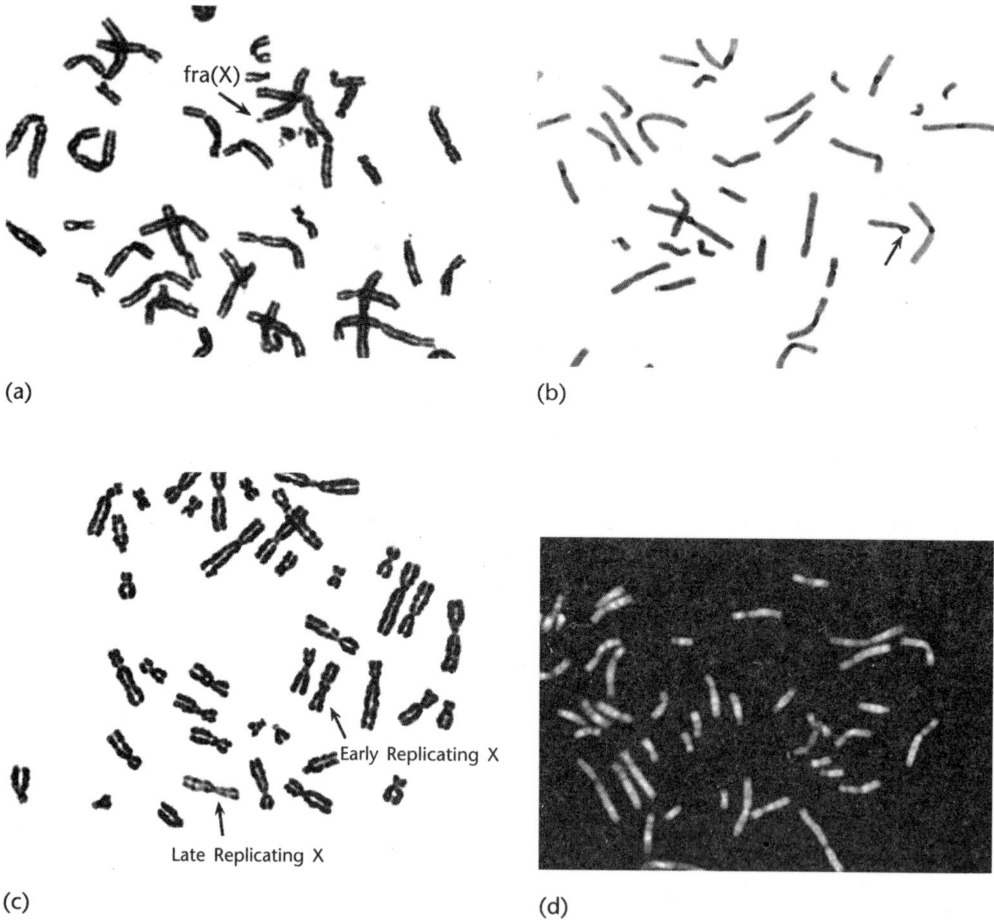

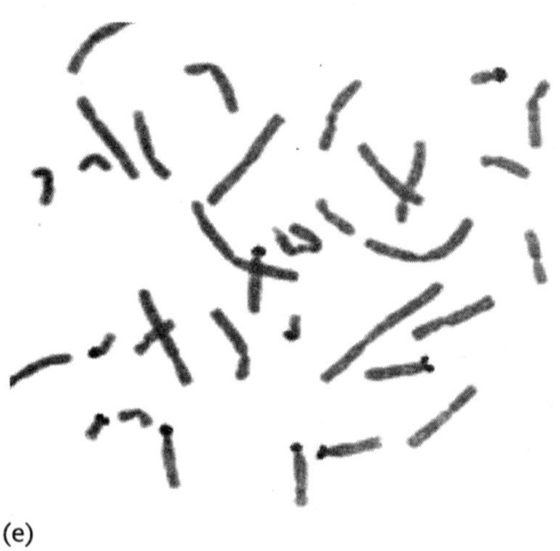

Figure 5.2
Staining of chromosomes. (a) Solid or block staining is used to observe chromosome shape and size. Note the fragile site on the end of the X chromosome. (b) C-banding is used to stain heterochromatin at the centromere. Note the additional heterochromatin on the short arm of one acrocentric chromosome – this is a normal variant. (c) Replication banding is used to study later replacing regions. Note the pale late replicating X chromosome. (d) Q-banding is used to stain Y chromosomes and other regions of heterochromatin. Note the Y chromosome with a large bright heterochromatin region on the long arm. (e) Silver staining is used to stain the nucleolar organizer regions on the acrocentric chromosomes. Note that not all NORs stain – in this cell nine are stained.

Various other staining methods, outlined in *Table 5.2* and illustrated in *Figure 5.2*, may be used to highlight specific regions of the karyotype, occurring at particular chromosomal locations, to help in interpretation of changes in the chromosomes seen down the microscope. These include **heterochromatin** at the centromere region and **nucleolar organizer regions** in the short arms of the acrocentrics.

Analysis using a microscope entails first scanning the slide using a low power objective lens in order to locate suitable chromosome spreads. A high power objective lens (giving a magnification of 1000×) is then used to allow the cytogeneticist to count the number of chromosomes. Finally, the chromosomes are visually assigned to their pairs. Microscopic analysis can be supplemented by the production of a permanent karyotype using automated image processing systems, with which the cytogeneticist can manipulate digitized images of the chromosomes to produce a karyotype of photographic quality (see *Figure 5.1*).

A minimum of four or five cells need to be analysed in order to be confident of excluding an abnormality. In some circumstances in which there is

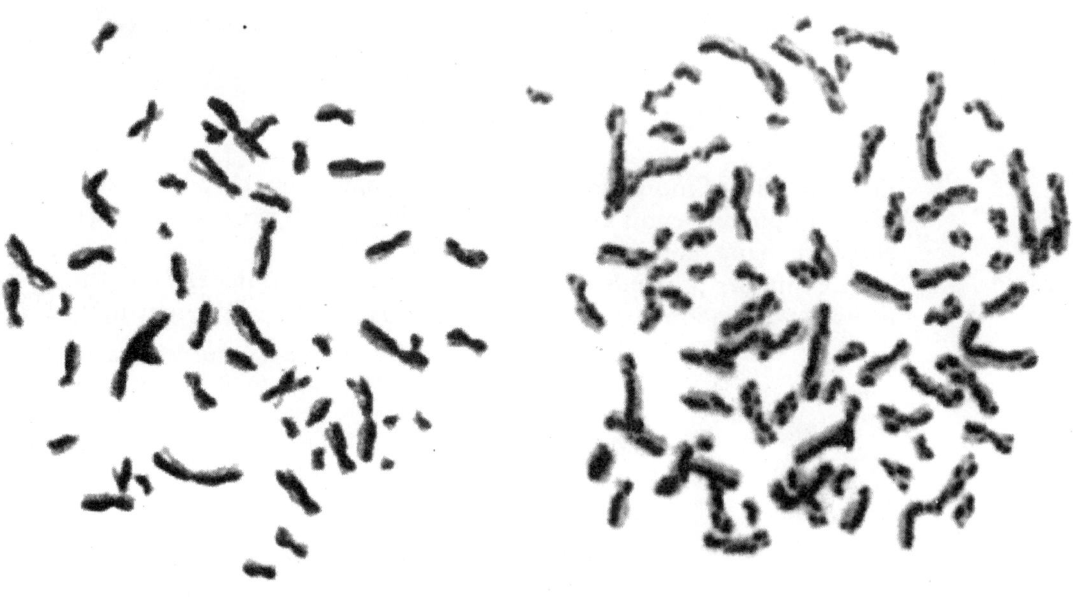

Figure 5.2 (*contd*)
(f) Harlequin staining is used to show differential staining of sister chromatids and to investigate chromosome damage. Note the sister chromatid exchanges (SCEs) are shown by the switch from light to dark. The normal cell has 6–10 SCEs and the abnormal cell 50+.

Table 5.2 Staining methods for chromosome analysis

Method	Displays	Use
G-banding	Unique pattern for each chromosome G dark bands are AT-rich (gene poor)	Routine analysis
Block (or solid) staining (see *Figure 5.2a*)	Length and centromere position only	Some aspects of morphology; aberrations, e.g fragile sites
C-banding (see *Figure 5.2b*)	Constitutive heterochromatin (area of no active genes)	Defining centromeres, centromeric and Yq variants, markers
R-banding (see *Figure 5.2c*)	Late replicating areas stain dark, e.g. late replicating X	Sex chromosome abnormalities, X;autosome translocations
Q-banding (see *Figure 5.2d*)	Yq chromosome, some other regions of heterochromatin	Y chromosome abnormalities or mosaicism
Silver staining (see *Figure 5.2e*)	Nucleolus organizer regions on acrocentric short arms	Defining acrocentrics and markers
Harlequin staining (see *Figure 5.2f*)	Differential staining of chromatids, sister chromosome exchange	DNA repair in chromosome breakage syndromes, e.g. Bloom syndrome
In situ hybridization	Specific region of probe homology	Microdeletions (see *Chapter 6*)

the possibility of more than one cell line with different chromosome complements (a phenomenon known as mosaicism which is explained later in this chapter), many more cells have to be examined.

The number of bands visible along a chromosome depends on the extent of its condensation (or contraction) (see *Figure 5.3*). Different numbers of

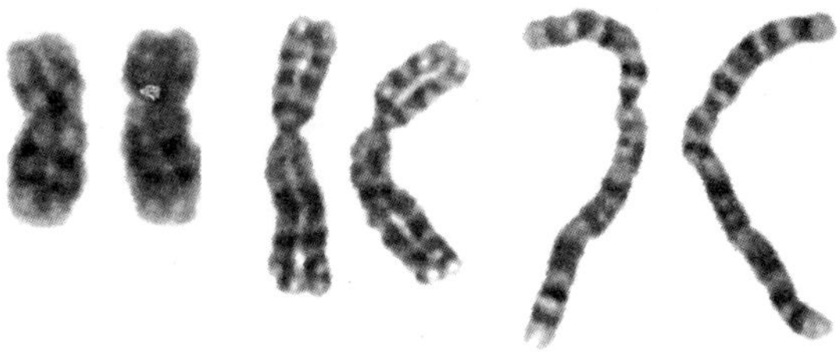

Figure 5.3
One chromosome number 2 with different numbers of bands visible due to difference in DNA chromosome contraction. The greater the number of bands the smaller the cytogenetic abnormality that can be observed.

bands tend to be seen in typical metaphase cells of different tissues, and this is probably the result of cell cycling characteristics and basic cell morphology. Because of these differences, certain cell types tend to yield more highly condensed chromosomes.

The limit of resolution of chromosome analysis is the detection of the gain or loss of a single band. A good quality lymphocyte preparation contains 1100–1600 dark and pale bands per karyotype; the smallest bands contain approximately 3–5 Mbp, and so potentially many hundreds of genes may be lost or duplicated in even the smallest unbalanced structural chromosome abnormality. Chromosomes from amniotic fluid cell cultures, the usual choice for prenatal diagnosis, usually yield 700–1100 bands, while chromosomes of malignant tumour tissues, frequently poorly defined and difficult to identify with the same level of accuracy as chromosomes from other tissues, may have as few as 100–400.

The techniques of molecular cytogenetics, such as fluorescence *in situ* hybridization (FISH), can be used to increase this resolution further. These techniques are described in more detail in *Chapter 6*.

Reliable detection of subtle abnormalities of the banding pattern also depends on the clinical information accompanying the request for analysis. It is much easier to detect, for example, a small deletion if the clinical features of the patient suggest a specific deletion syndrome.

Describing chromosomes – band nomenclature

Chromosomes are described using the International System for Human Cytogenetic Nomenclature (ISCN). The system describes each chromosome as a series of bands, with each chromosome in the human somatic cell complement considered to consist of a continuous series of bands, with no unbanded areas. A band is defined as a part of a chromosome clearly distinguishable from adjacent parts by virtue of its lighter or darker staining intensity.

The bands are allocated to various regions along the chromosome arms, and the regions are delimited by specific landmarks. which are distinct morphological features important in identifying chromosomes, such as the ends of chromosome arms, the centromere, and certain bands.

The centromere divides each chromosome into two arms, with the symbols p and q being used to designate the short arm and long arm, respectively. The centromere is designated 10, with the part facing the short arm being p10 and the part facing the long arm being q10. Regions and bands on each chromosome are numbered consecutively from the centromere outward along each chromosome arm. The regions adjacent to the centromere are labelled as 1 in each arm; the next more distal region as 2 and so on. The band numbers are designated similarly, with the first band in a region (moving away from the centromere) being 1, then 2, etc. Four items are required to direct a person to a specific band:

- the chromosome number
- the arm symbol (p or q)

- the region number
- the band number

These items are given in order without spacing or punctuation, for example, 9p21 indicates chromosome 9, short arm p, region 2, band 1 (see *Figure 5.4*). Whenever an existing band is subdivided, a decimal point is placed after the original band designation and is followed by the number assigned to each sub-band. The sub-bands are numbered sequentially from the centromere outwards, for example, 1p31.1, 1p31.2 and 1p31.3.

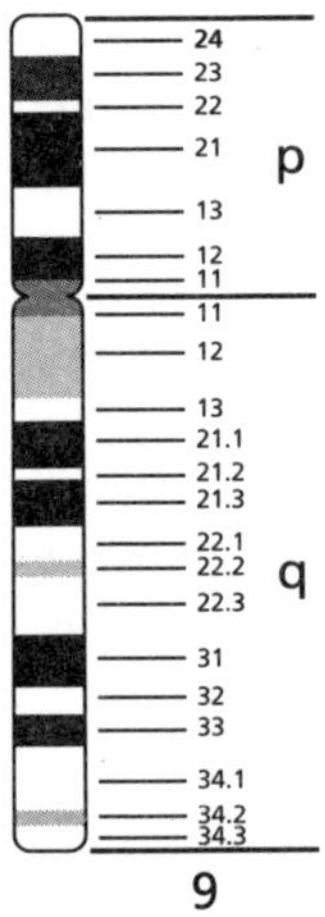

Figure 5.4
Diagram of one chromosome 9 showing the system for the numbering of bands.

A karyotype can be described using a series of conventions described in the ISCN. The total number of chromosomes are recorded first, followed by a comma and then the sex chromosomes. So 46,XX is a normal female karyotype and 46,XY a normal male karyotype. The autosomes are only specified when they are abnormal. A series of standard abbreviations are used to describe different types of chromosome abnormalities (see *Box 5.6*).

5.4 TYPES OF CHROMOSOME ABNORMALITIES

Chromosome abnormalities can affect the number of chromosomes (aneuploidy), the number of sets of chromosomes (polyploidy), or the actual structure of the chromosomes. Aneuploidy is the most commonly identified chromosome abnormality in humans and most aneuploidy results from meiotic nondisjunction.

DNA polymorphisms have been used extensively to study the origin of aneuploid conditions. These studies have shown that there is variation with respect to the origin of the additional/missing chromosome; however, maternal meiosis I errors predominate amongst almost all trisomic conditions.

Box 5.6 Examples of symbols used to describe karyotype abnormalities

Symbol	Description
add	Additional material of unknown origin
Single colon (:)	Chromosomal break
Double colon (::)	Chromosomal break and reunion
Comma (,)	Separates chromosome numbers, sex chromosomes and chromosome abnormalities
del	Deletion
der	Derivative chromosome
dic	Dicentric chromosome
dup	Duplication
i	Isochromosome
ins	Insertion
inv	Inversion
mat	Maternal origin
minus sign (–)	Loss
p	Short arm of chromosome
parentheses ( )	Surround structurally altered chromosome
pat	Paternal origin
q	Long arm of chromosome
r	Ring chromosome
rec	Recombinant chromosome
t	Translocation

Altered genetic recombination has been shown to be a significant contributing factor. Most aneuploidy in humans results from meiotic non-disjunction (see *Chapter 1*).

Autosomal aneuploidies

The common viable aneuploidies are trisomy 21 (Down syndrome), trisomy 18 (Edwards syndrome), and trisomy 13 (Patau syndrome). Trisomy for almost all the other chromosomes has been demonstrated in tissue from spontaneous abortions. An example of standard nomenclature for trisomy would be 47,XX,+21 – this would represent a female with Down syndrome.

Edwards syndrome occurs in 1:3000 live births. Birth weight is low, the chin is small, the ears malformed, and there is a distinctive shape to the back of the head. The hands are clenched with overlapping index and fifth fingers. The feet are distinctive, with a shape usually described as 'rocker-bottom'. Only a small proportion survive infancy, and these children are profoundly delayed. The karyotype would be either 47,XY,+18 or 47,XX+18, for a male or female, respectively.

Patau syndrome occurs in 1:5000 live births and can be identified by cleft lip and palate, small eyes and polydactyly. Congenital heart defects are common, and survival beyond infancy is rare.

If it is suspected at birth that a baby has trisomy 21, 18 or 13, the chromosome result to confirm this suspicion can be given to the paediatrician within 24–48 hours after receipt of the blood sample in the laboratory.

Except for rare cases of mosaicism (see later), no other autosomal aneuploidies are found in live births, although they contribute significantly to spontaneous abortions.

Chromosome abnormalities in spontaneous abortions

Chromosome non-disjunction is a frequent occurrence; however, most of them are lost through spontaneous abortion or miscarriage. About 20% of recognizable pregnancies abort spontaneously, usually in the first three months of pregnancy. Of these, it is estimated that up to half may have a chromosome abnormality. Amongst those with an abnormal karyotype, about 50% have an autosomal trisomy, 20% have 45,X (Turner syndrome), and 16% have triploidy (69 chromosomes from three whole sets) (see *Figure 5.5*). The most common trisomy seen in miscarriage tissue is trisomy 16 which occurs in more than one-third of the trisomic abortions. This abnormality is incompatible with survival to birth.

Many other factors can cause miscarriage. The incidence of chromosome abnormalities decreases as the pregnancy progesses, with only 20% in losses between 12 and 20 weeks, and 5% in stillbirths in the later part of pregnancy.

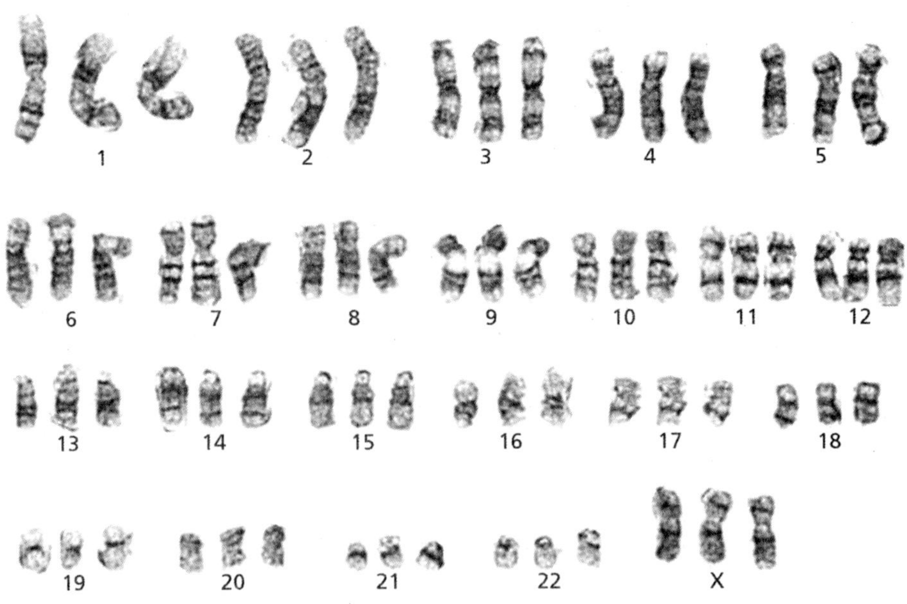

Figure 5.5
A female triploid karyotype with 69 chromosomes, 69,XXX. Triploidy occurs in about 1% of all recognizable pregnancies; it mainly arises when two sperm fertilize one egg. The pregnancy is not viable and will be lost, usually during the first trimester.

The risk of autosomal trisomy increases with maternal age, a factor which is related to the increasing incidence of Down syndrome in older mothers, as explained in *Chapter 9*. There is no paternal age effect for autosomal trisomy. Population studies indicate that the risk of recurrence after one autosomal trisomy conception is low at approximately 1% above that for maternal age.

Unbalanced structural abnormalities may also be found and these may have arisen from a parent carrying a balanced rearrangement. It is important to analyse the chromosomes from couples who have had three or more miscarriages, because in as many as 5% of these couples one partner will carry a balanced rearrangement which is the cause of the miscarriage. Carriers may then be at risk of the birth of chromosomally unbalanced offspring (see later in this chapter).

Structural abnormalities

Structural chromosome abnormalities result from changes in number and/or position of segments within one or more chromosomes, and arise due to breakage and incorrect rejoining of chromosome segments. There are many factors which may cause chromosome breaks including environmental factors, or simply errors during replication. The body usually repairs these effectively using the DNA repair enzymes.

Rearrangements are described as **balanced** when the chromosomal segments have changed their position but no significant genetic material has been lost. They are described as **unbalanced** when genetic material has been lost (partial monosomy) or gained (partial trisomy).

Each rearrangement is almost always unique to an individual or family and very rare. Having a child with a rare chromosome disorder can be a huge shock for parents and family and can stir up a whole range of emotions. Careful support and counselling is essential (see *Chapter 10*) and consideration should be given to prenatal diagnosis in any further pregnancies (see *Chapter 9*).

Individually, chromosome disorders are rare but collectively they are common. One in every 200 babies born has a rare chromosome disorder, which may be balanced or unbalanced, with one in every 1000 babies having abnormalities noted at birth or in early childhood (see *Box 5.7*). The effects of rare chromosomal disorders can be very varied. The vast majority of carriers of balanced rearrangements will have no symptoms but might have problems in reproduction.

Where there has been loss or gain of chromosomal material, the symptoms arising might include a combination of physical problems, general health problems, learning difficulties and/or challenging behaviour. The combination and the severity of effects very much depend upon which parts of the chromosomes (and hence which genes) are involved. The outcome for affected children can be different even if they carry abnormalities that appear cytogenetically similar.

In general, loss of a segment of a chromosome is more serious than the presence of an extra copy of the same segment. Defects of the autosomes,

Box 5.7 Incidence of chromosome abnormalities in the unselected newborn

Abnormality	Incidence per 1000 births
All abnormalities	9.1
Autosomal trisomies	1.4
Trisomy 13	0.07
Trisomy 18	0.12
Trisomy 21	1.5
Balanced autosomal rearrangements	5.2
Unbalanced autosomal abnormalities	0.6
Sex chromosome abnormalities	
In phenotypic males	1.2
XXY (Klinefelter syndrome)	1.5 of male births
In phenotypic females	0.75
45,X (Turner syndrome)	0.4 of female births
47,XXX	0.65

Based on Jacobs, P.A. *et al.* (1992) *J Med Genet,* **29**: 103–108.

chromosomes 1–22, tend to be more serious than those of the sex chromosomes X and Y.

Translocations

There are basically two types of translocation, Robertsonian and reciprocal.

Reciprocal translocation

A **reciprocal translocation** arises when a two-way exchange of chromosome segments takes place between two chromosomes, resulting in two **derivative** chromosomes. The exchanged portions are the translocated segments, while the remainder of each derivative chromosome is often called the centric segment. The reciprocal translocation is balanced, and therefore generally occurs in normal healthy people. This type of chromosome rearrangement is most often discovered when the carrier seeks medical advice following miscarriages, or when a handicapped child is born, having inherited an unbalanced form of the translocation.

At meiosis, pairing and recombination takes place as normal, but there are homologous segments on four chromosomes, two normal and two derivatives, which therefore come together as a **quadrivalent or pachytene cross.** (*Figure 5.6*). When the four chromosomes part company at anaphase of meiosis I, they can enter the two daughter cells in almost any combination. The segregation patterns are called **alternate, adjacent 1** and **adjacent 2.**

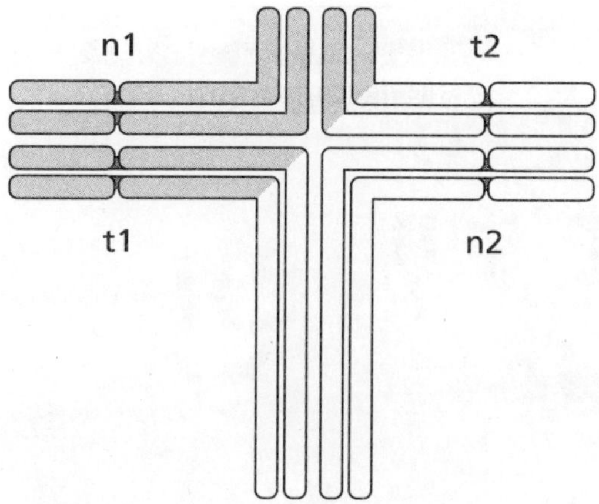

Figure 5.6
Meiotic pairing in a reciprocal translocation heterozygote results in a quadrivalent cross formation, the pachytene cross. Contribution of alternate chromosomes n1 and n2 to the gamete is normal, and of t1 and t2 is balanced. Contribution of adjacent chromosomes, for example t1 and n2, is one example of an unbalanced gamete.

In alternate segregation, the alternate centromeres segregate together to produce balanced gametes with either both parts of the translocation t1 and t2, or both normal chromosomes, n1 and n2, (see *Figure 5.6*).

Adjacent 1 segregation occurs when non-homologous centromeres segregate together and adjacent 2 occurs when the two homologous centromeres segregate to the same poles at anaphase. Only unbalanced gametes result. With extreme imbalance, the offspring is unlikely to be viable, and for most translocations there are only one or two unbalanced types that can form recognizable pregnancies. These are usually where the two **adjacent non-homologous** chromosomes segregate together, that is n1 with t2, or n2 with t1.

In the pachytene cross, if you draw the longest line the most likely method of segregation is that which gives rise to the least imbalance. Thus when the translocated segments are short this will be adjacent 1 segregation. When the centric segment is short the most likely method is adjacent 2 segregation. When one of the derivative chromosomes is small this will be 3:1 segregation. In general adjacent 1 has the least imbalance if the translocated segments are shorter than the centric ones.

Adjacent 2 has the least imbalance when the centric segments are shorter than the translocated ones. The 3:1 segregation occurs when one derivative is very small. Although it is possible to predict the most likely method of segregation, it is important to consider all types where viability is possible.

Figure 5.7a shows a translocation between the short arm (p) of one chromosome 8 and the long arm (q) of one chromosome 11. This pairs to form a quadrivalent (see *Figure 5.7b*) and adjacent 1 segregation will give rise to the unbalanced karyotype shown in *Figure 5.7c*. The rearranged chromosome is called a derivative (der) chromosome and is identified according to which centromere it possesses.

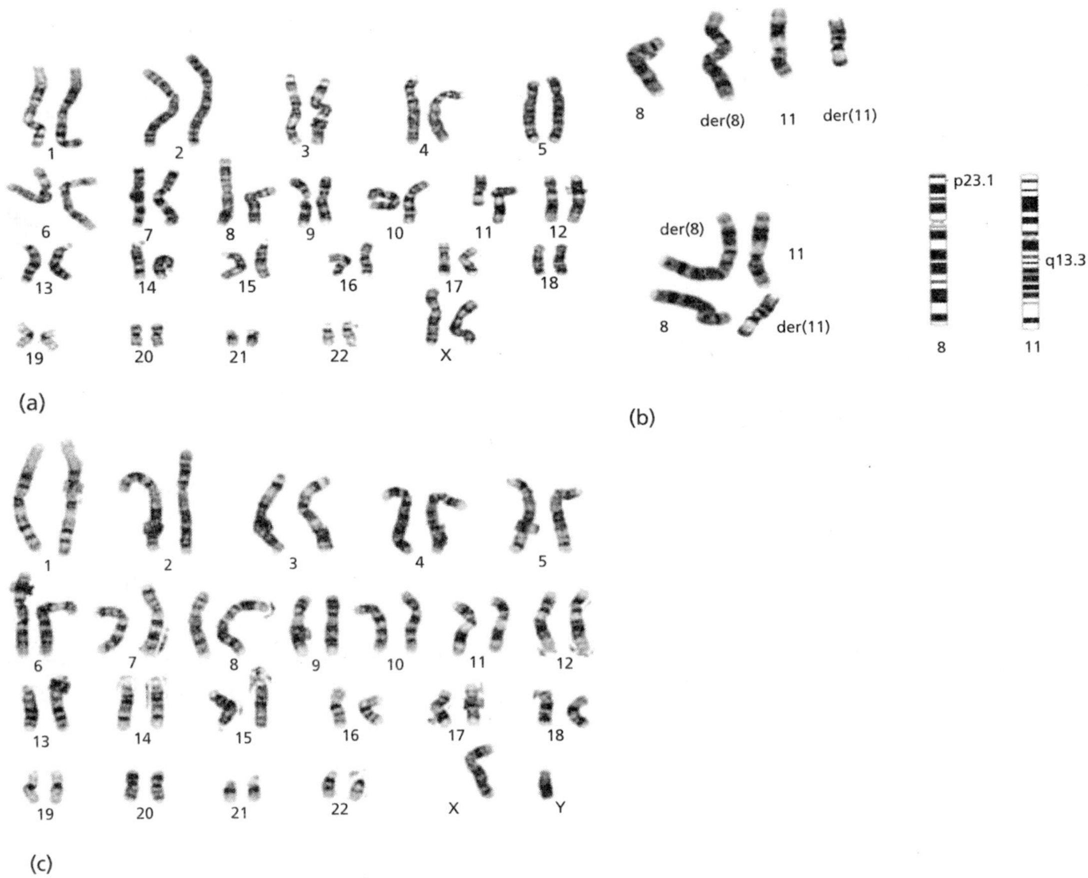

Figure 5.7
(a) G-banded female karyotype of a balanced reciprocal translocation between the short arm of one chromosome 8 and the long arm of one chromosome 11. The ISCN for the karyotype is: 46,XX,t(8;11)(p23.1;11q13.3). (b) Pairing of the translocation as a 'quadrivalent'. (c) Male karyotype with the derivative chromosome 8 but two normal chromosomes 11s. This karyotype is unbalanced. The ISCN for this karyotype is written as: 46,XY,der(8)t(8;11)(p23.1;q13.3)mat. The mat denotes it has been inherited from the balanced translocation carried by the mother.

Reproductive risks of reciprocal translocation carriers

The question that carriers of a balanced translocation often want answered is "what is the risk of having a chromosomally abnormal offspring?". A broad estimate of risk may be derived from a consideration of four factors.

1. If the mode of ascertainment is through a live-born unbalanced offspring, this is evidence that this can occur again. If it is only through miscarriages then the risk is less clear.
2. The predicted type of segregation leading to a particular type of gametes. This can be deduced from the shape of the quadrivalent and the chromosomes involved. Some general rules can be applied. Alternate segregation

is the most frequent mode of segregation. If the translocated segments are small in genetic content, adjacent 1 is the most likely type of malsegregation capable of giving rise to viable abnormal offspring. If the centric segments are small in content, adjacent 2 is the most likely segregation.

3. The sex of the transmitting parent, as the risks are higher in female carriers than male carriers for reasons that are not fully understood but may relate to problems in spermatogenesis.
4. The assessed imbalance of potentially viable gametes. It is possible to use the size of imbalance as an indicator of risk because different amounts are tolerated differently depending on the chromosomes involved.

Translocations are usually unique to a family. One exception is the translocation between chromosomes 11 and 22, t(11;22)(q23;q11.2) (see *Figure 1.10*) which has been described in a number of families and for whom the reproductive risks are well described. This translocation can undergo 3:1 segregation, with three centromeres moving to one pole and one to the other, and again unbalanced offspring occur. This results in 47 chromosomes with an extra derivative chromosome 22 which appears as a supernumerary marker chromosome (see below) and results in either miscarriage, or a child with an abnormal phenotype and including growth retardation, heart defects, microcephaly, muscular hypotonia and profound mental deficiency.

Apparently balanced karyotypes associated with phenotypic abnormality

Occasionally what appears cytogenetically to be a balanced translocation is found in someone with mental handicap or some kind of congenital or heritable problem. This can at times be coincidental. However, in some cases there is evidence of disruption of a gene at one of the breakpoints. For example, the location of the neurofibromatosis gene (*NF1*), was eventually shown to be on chromosome 17 after an affected person was shown to have a translocation involving that chromosome. In one family, adult polycystic kidney disease (APKD) was shown to segregate with a translocation involving chromosome 16. Duchenne muscular dystrophy is an X-linked disorder that affects males. It rarely affects girls and some of those who are affected have been shown to have a translocation involving the critical locus on the X chromosome.

With increasing use of the more sensitive molecular cytogenetic technologies, it is being found that in some cases where a rearrangement appears to be balanced cytogenetically, it is in fact unbalanced. The use of array CGH (see *Chapter 6*) has shown that this may be the case in as many as 40% of these individuals.

Robertsonian translocation

In the **Robertsonian translocation**, two V-shaped acrocentric chromosomes join at the centromere into a single X-shaped structure. As two centromeres fuse into one, the balanced translocation karyotype has only 45 chromosomes (see *Figure 5.8*).

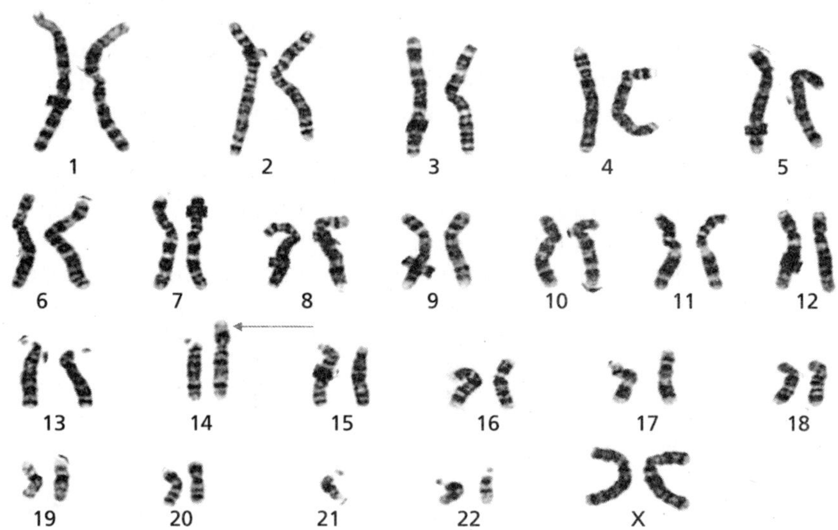

Figure 5.8
A female karyotype with a Robertsonian translocation between chromosomes 13 and 14. Note that there are only 45 chromosomes. The ISCN for this karyotype is written as: 45,XX,rob(13;14)(q10;q10) or 45,XX,der(13;14)(q10;q10). The band q10 designates the centromere. The short arm material has been lost.

It is the human acrocentric chromosomes 13, 14, 15, 21 and 22 which become involved in Robertsonian translocations. They each have a very small short arm with no vital unique genes although, as explained already, they do have nucleolar organizers. The short arms are lost in the fusion which creates the translocation, but the imbalance has no clinical effect. The resulting product is in fact only part of the rearrangement so it is called a derivative. The karyotype for a male carrier of a Robertsonian translocation involving chromosomes 13 and 14 would be written as 45,XY,der(13;14)(q10;q10). Band 10 is the centromere; note that there are only 45 chromosomes.

Reproductive risks of Robertsonian translocation carriers

About one person in 1000 carries a Robertsonian translocation, more than half of which involve chromosomes 13 and 14. The majority of the remainder involve 14 and 21. As these translocations are among the commonest structural chromosome rearrangements found, the reproductive risks for them are well established.

Carriers of a 14;21 translocation are phenotypically normal and will usually have offspring that are chromosomally normal or who have the same balanced translocation themselves. However, the presence of the balanced translocation does confer a risk of having a child with Down syndrome. During pairing at meiosis, instead of the normal bivalent formation, there are three partially homologous components forming a **trivalent** (*Figure 5.9*).

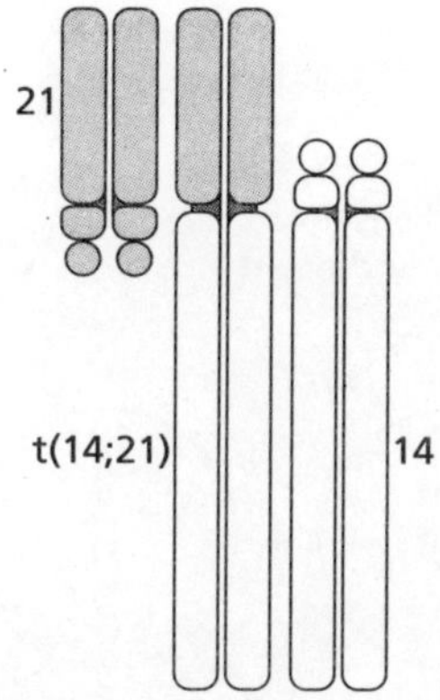

Figure 5.9
Meiotic pairing in a Robertsonian 14;21 translocation heterozygote results in a trivalent configuration. If the translocation chromosome and the chromosome 21 both go into the gamete, the end result will be a Down syndrome pregnancy.

Segregation from this trivalent can result, in theory, in six types of gametes (see *Figure 5.10*). Only normal, balanced or trisomy 21 are viable, with trisomy 14 and monosomy 14 and 21 being unviable. In a family with more than one affected child a 14;21 translocation is often the cause.

Some carriers have a history of miscarriage; possibly some of their pregnancies had the non-viable trisomy 14. Carriers of a Robertsonian translocation would be offered the opportunity of prenatal diagnosis by amniocentesis or chorion biopsy.

Empirically the risk of an unbalanced outcome is relatively small and sex-dependent. A female carrier has a 15% risk of an unbalanced conception and a male carrier of the same translocation a 1.4% risk. The frequency at birth will be lower as 75% of Down syndrome pregnancies are lost before birth. Translocation Down syndrome can occur to a carrier at any age and is not associated with increasing maternal age, unlike those with a free trisomy 21.

Similarly, a carrier of a 13;14 Robertsonian translocation is at risk of having a child with trisomy 13. The risk at birth is very low (less than 1%) due to the severe nature of the condition which rarely survives to birth.

Deletions

A deletion involves loss of material from a chromosome. Two types of deletion are identifiable, the **terminal deletion**, a single break resulting in loss of the end of a chromosome, and the **interstitial deletion**, where two breaks result in a segment being lost from within a chromosome.

All deletions result in an unbalanced karyotype, with partial monosomy, and therefore the result is almost inevitably one of serious clinical effect. A number of clinical syndromes have been ascribed to specific deletions (see *Box 5.8*). Deletions range in size from large ones readily detected in standard G-banded chromosome preparations, to those very small ones requiring

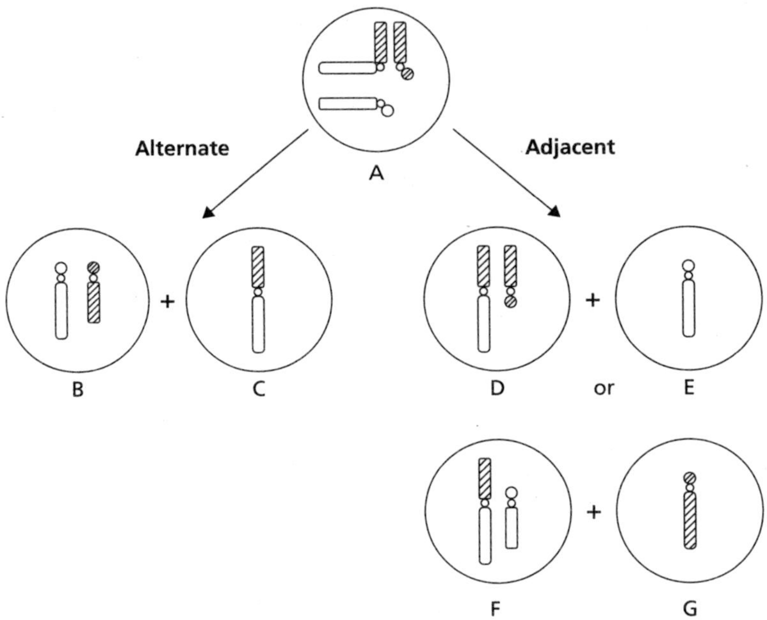

Figure 5.10
Meiotic behaviour of a 14;21 Robertsonian translocation. The three chromosomes pair as a trivalent (A) and six possible gametes can be produced: two are normal/balanced and four are unbalanced. Alternate segregation results in normal (B) or balanced carrier (C) gametes. Adjacent segregation results in trisomy 21 (D), monosomy 21 (E), trisomy 14 (F) or monosomy 14 (G). Gametes (E), (F) and (G) are not viable.

Box 5.8 Nomenclature for deletions

The ISCN nomenclature easily distinguishes between terminal and interstitial deletions. For example, a terminal deletion of the long arm of chromosome 10 in a male might be 46,XY,del(10)(q26), while an interstitial deletion might be 46,XY,del(10)(q24;q26).

molecular techniques such as fluorescence *in situ* hybridization (FISH) to detect (see *Chapter 6*).

Even smaller deletions within a gene may only be detectable using the molecular methods described in *Chapter 7.*

Chromosome microdeletion syndromes

The first 'microdeletion' syndrome described (cri-du-chat syndrome) involved a deletion of part of the short arm of chromosome 5. Cri-du-chat syndrome is so called because of the cat-like cry that the infants make (see *Appendix: Glossary of disorders*).

The identification of these often tiny deletions (see *Table 5.3* and *Figure 5.11*) needs a combination of excellent technical and analytical skills on the

Table 5.3 Some autosomal deletion syndromes

Syndrome	Deletion
Wolf–Hirschhorn syndrome	4p16.3
Cri-du-chat syndrome	5p15
Williams syndrome	7q11.2
Langer–Giedion syndrome	8q24.11
9p syndrome	9p13
WAGR syndrome	11p13
Jacobsen syndrome	11q23
Retinoblastoma	13q14
Prader–Willi syndrome	15q11.2
Angelman syndrome	15q11.2
ATR-16 syndrome (α-thalassaemia with mental retardation)	16p13.3
Miller–Dieker syndrome	17p13.3
Smith–Magenis syndrome	17p11.2
18p syndrome	18p11
18q syndrome	18q21
DiGeorge/VCFS Shprintzen syndrome	22q11.2

part of the cytogeneticist, and the critical expertise of the specialist clinician to recognize the phenotypic features and relate them to a known syndrome. There are dysmorphology databases that collect together clinical features from published reports and other sources to continually improve the knowledge in this area. Some deletions are sub-microscopic and only detectable by FISH as described in *Chapter 6*.

New rare deletion syndromes continue to emerge. Recent examples include terminal deletion of 1p36 (the tip of the short arm of chromosome 1) and of 2q37 (the tip of the long arm of chromosome 2). Even more are being discovered using the sensitive array CGH techniques (*Chapter 6*) and more families are able to be given an explanation of the problems they or their offspring have.

Deletions and telomeres

If a terminal deletion were as simple as the loss of the end of a chromosome, there would be no telomere to stabilize the tip of the deleted chromosome. It is widely accepted that any stable chromosome arm *must* have a telomere, including an arm which under the microscope appears to have a straightforward terminal deletion. Where a deleted chromosome arm gets its telomere is not known for certain but the possibilities are that:

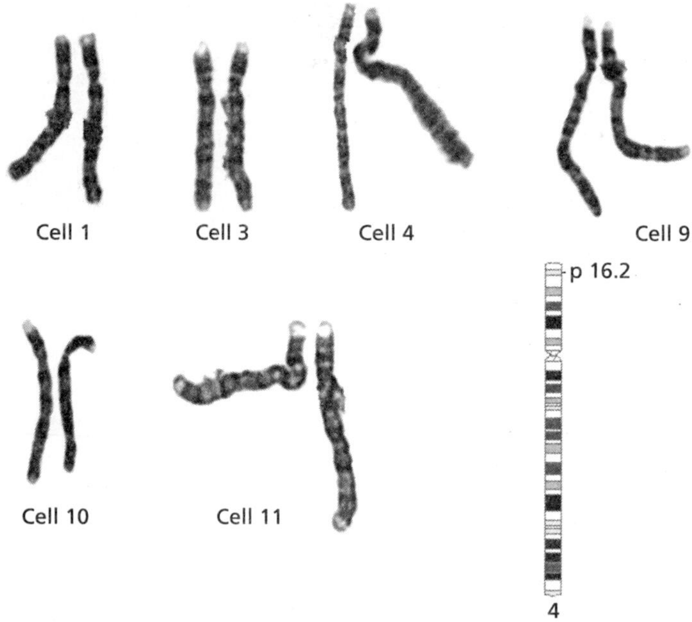

Figure 5.11
Homologues of chromosome 4 with deletion of the tip of the short arm in a child with the Wolf–Hirschhorn microdeletion syndrome; the abnormal chromosome is on the right. Note that it can take the analysis of several cells in good quality preparations from blood samples to clearly see this subtle abnormality.

- there is activation of an ancestral interstitial telomere located at the breakpoint
- a new telomere is manufactured by normal cellular processes
- the rearrangement actually involves a second breakpoint close to a telomere; in other words, it is a cryptic interstitial deletion or a translocation product

Isochromosomes

An isochromosome is a chromosome composed of two copies of the same arm. In other words it is a mirror image about the centromere. Visually it is not possible to distinguish between a true isochromosome and a centromere to centromere translocation involving two homologous chromosomes. A Robertsonian translocation between two homologous chromosomes would appear as an isochromosome.

Isochromosomes can also be present as an extra or supernumerary chromosome. Examples of this include the 'iso-12p' found in Pallister–Killian syndrome, and the 'iso-18p' which results in tetrasomy for the short arm of chromosome 18, and is associated with a defined syndrome of dysmorphism and mental handicap.

An isochromosome may also be confined to certain tissues. The isochromosome for 12p is only found in fibroblast cells and is not seen in blood cells, a fact that is important when looking for this abnormality.

The isochromosome for the long arm of the X chromosome, the iso-Xq (see *Figure 5.12*), is found in about 10% of females with Turner syndrome (see *Section 5.5*). It is often found in mosaic form with a mixture of some normal and some abnormal cells, and has been reported to be inherited in a small number of families.

An isochromosome in the karyotype can also **replace a normal chromosome.** In most chromosomes such a situation would result in an extreme and unviable imbalance, with monosomy for one complete chromosome arm and trisomy for the other arm of the same chromosome.

An isochromosome of the long arm of chromosome 21 is found in about 1–2% of cases of Down syndrome. Here the chromosome number is 46, rather than the expected 47, with the trisomy attributable to the presence of a normal 21, plus two copies of the long arm of 21 in the form of an isochromosome. Carriers of an isochromosome 21 have a 100% risk of having a Down syndrome baby.

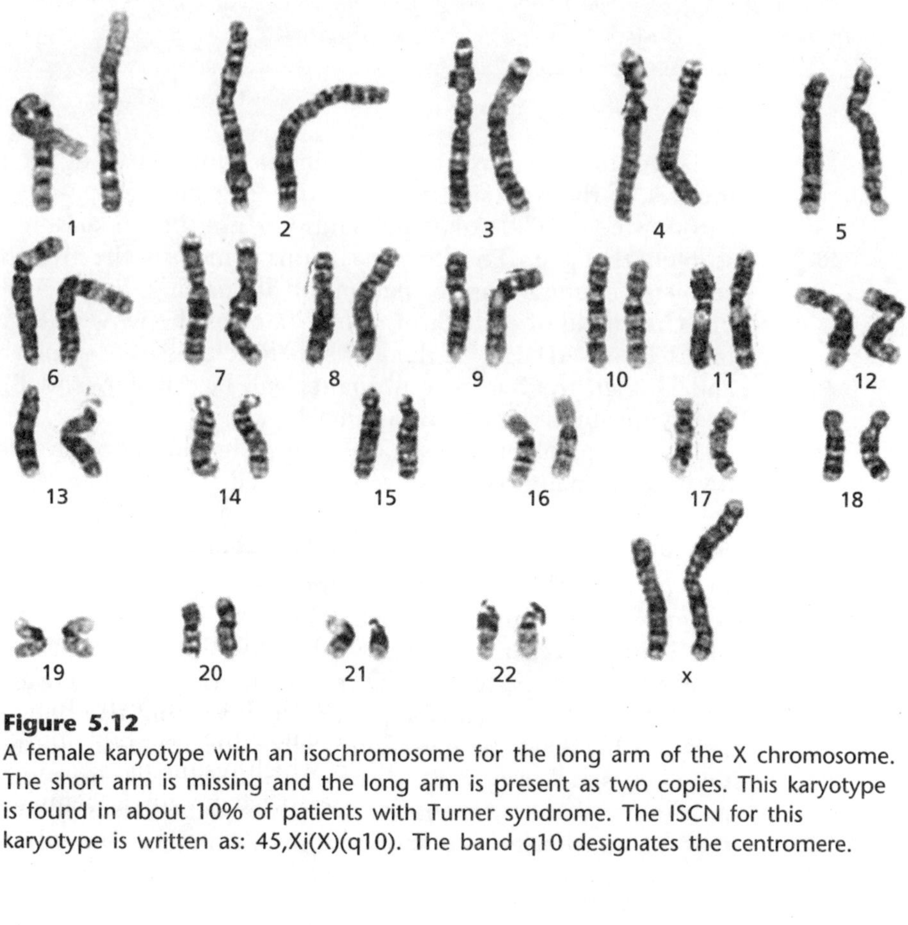

Figure 5.12
A female karyotype with an isochromosome for the long arm of the X chromosome. The short arm is missing and the long arm is present as two copies. This karyotype is found in about 10% of patients with Turner syndrome. The ISCN for this karyotype is written as: 45,Xi(X)(q10). The band q10 designates the centromere.

Inversions

An **inversion** is the reversal of a segment of a chromosome. It requires two breakpoints and, as it is balanced, it is generally compatible with normal development. The carrier of an inversion may be at risk of producing gametes with an unbalanced karyotype and, depending on the structure of the inversion (i.e. the location of the breakpoints), the result may be reduced fertility, miscarriage, or the live birth of a handicapped child.

There are two types of inversion, **paracentric** and **pericentric**. In a paracentric inversion the reversed segment is within one chromosome arm while in a pericentric inversion the reversed segment includes the centromere. The consequences of meiotic recombination and the reproductive risks are very different for these two rearrangements. The ISCN nomenclature can distinguish between paracentric and pericentric inversions (see *Box 5.9*).

Box 5.9 Nomenclature for inversions

The ISCN nomenclature easily distinguishes between paracentric and pericentric inversions.

A paracentric inversion of the long arm of chromosome 10 in a male might be 46,XY,inv(10)(q22;q26), while a pericentric, which has a short arm (p) and a long arm (q) breakpoint might be 46,XY,inv(10)(p12;q26).

There are two ways to maintain maximum pairing of the bivalent at meiosis. If the inversion is very large (i.e. the breakpoints are near to the chromosome ends), only the ends do not match and it is the inverted segment that pairs. For other inversions at meiosis, the inverted chromosome segments adopt a loop formation for full pairing. *Figure 5.13* illustrates just one chromatid of each chromosome, a normal one ABCDE, pairing with an inverted one ADCBE. If there is a meiotic crossover somewhere between B and D, then the two **recombinants** will be ABCD**A** and **EDCB**E, both of which are abnormal and unbalanced.

It is the imbalance caused by an uneven number of crossovers within the inverted segment that needs to be assessed.

Reproductive risks of inversion carriers

In a paracentric inversion, the centromere is at A (see *Figure 5.13*), so the recombinant chromatids are either dicentric or acentric, and therefore unstable. Thus paracentric inversions generally have a low reproductive risk.

However, in a pericentric inversion with the centromere at C, the products are stable and a carrier of a pericentric inversion may have a high risk. The risks are higher if the inversion is large and the non-inverted segments are therefore small, especially if the duplication/deletions combination of the non-inverted end segments is viable. Risk can be assessed in a similar way to that for translocations described above, including method of ascertainment,

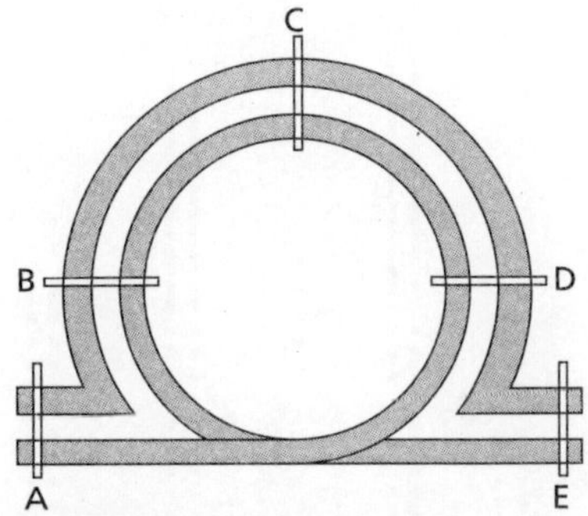

Figure 5.13
Pairing of an inverted chromosome segment ADCBE with its normal homologue ABCDE. Crossing over within the loop results in recombinants ABCDA and EBCDE. Note that the two ends of the recombinant are the same AA or EE. This can help to identify recombinants chromosomes.

size of imbalance, sex of carrier, and prediction of viability of unbalanced combinations.

The maximum theoretical risk of imbalance is 50%. If there has been a history of a previous imbalance or if a literature review suggests viability, then a risk of 5–10% is probable. If not, then the risk is more likely to be 1–2%.

Figure 5.14 shows a pericentric inversion of one chromosome 3, written as inv(3)(p25;q12). It is a large inversion so it is almost certain that there will be recombination in the inverted segment. The risk of abnormality is estimated to be about 33% and there are two possible recombinants. One recombinant will have duplication (trisomy) of the short (p) arm distal to the breakpoint and deletion (monosomy) of the long (q) arm distal to the breakpoint. This amount of chromosome missing is not viable and will result in miscarriage very early in pregnancy. The other will have duplication of part of the long arm and deletion of part of the short arm. This is a much smaller imbalance and has been described in the literature, so we know that it is possible that a pregnancy with this chromosome abnormality could surive to birth, but the child will have major phenotypic abnormalities.

Of course any pregnancy could have inherited only the normal chromosome 3 from the carrier parent, or could have inherited the balanced inversion. The outcome in both of these cases would be normal, although the child carrying the inversion will also be at risk of chromosomally unbalanced offspring in the same way as its parent.

Inversions as variants

Not all inversions are clinically significant. Pericentric inversion of a small region of inert heterochromatin located around the centromere of chromosome 9 occurs in about 1% of the human population. It is completely benign. Other rarer small pericentric inversions are also considered to be benign variants, including those involving chromosomes 1, 2, 10, and 16.

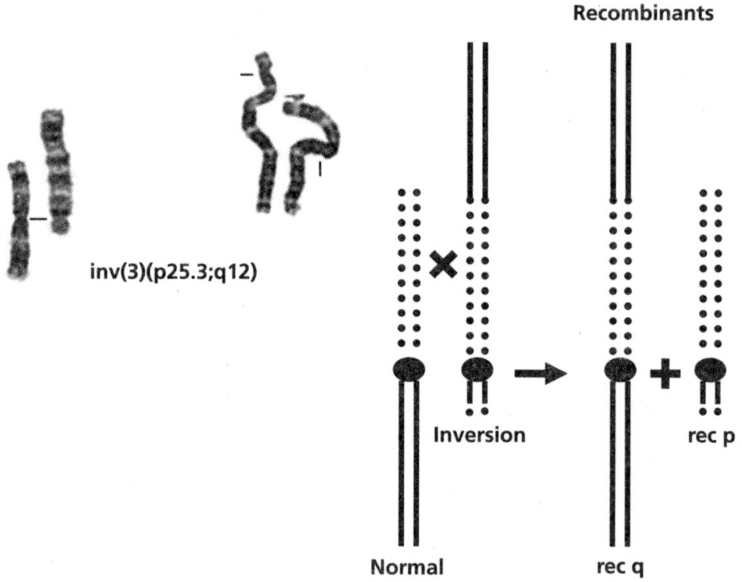

Figure 5.14
Inverted chromosome 3 with breakpoints in each arm, i.e. pericentric. The diagram shows the recombinants that would be produced from a crossover within the inverted region and would have either duplication of the end of the short arm or the end of the long arm. The karyotype for a male with a recombinant chromosome involving duplication of the q arm distal to the breakpoint would be written as: 46,XY,rec(3)dup(3)inv(3q)(p25.3;q12).

Ring chromosomes

A ring chromosome is another manifestation of the incorrect rejoining of two breakpoints, one on each arm of a chromosome, such that a **circular chromosome** containing the centromere is formed (see *Figure 5.15*). Ring chromosomes are inevitably unbalanced because the tips of each arm distal to the breakpoints are lost. They are usually associated with partial monosomy and, like other unbalanced chromosome abnormalities, have a serious clinical effect. A further problem with a ring chromosome is that it may be somewhat unstable, with double-sized dicentric structures forming at replication; at mitosis the chromosome complement of the daughter cells may become even more unbalanced and eventually non-viable. Because normal cell division is compromised, growth retardation and failure to thrive may be noted in patients who have a ring chromosome.

Reproductive effects of a ring chromosome

There have been occasional cases of a clinically normal carrier of a ring chromosome. In such cases the breakpoints are so close to the ends of the chromosome that there is no significant loss of important genetic information. The carrier is generally identified because of reproductive problems,

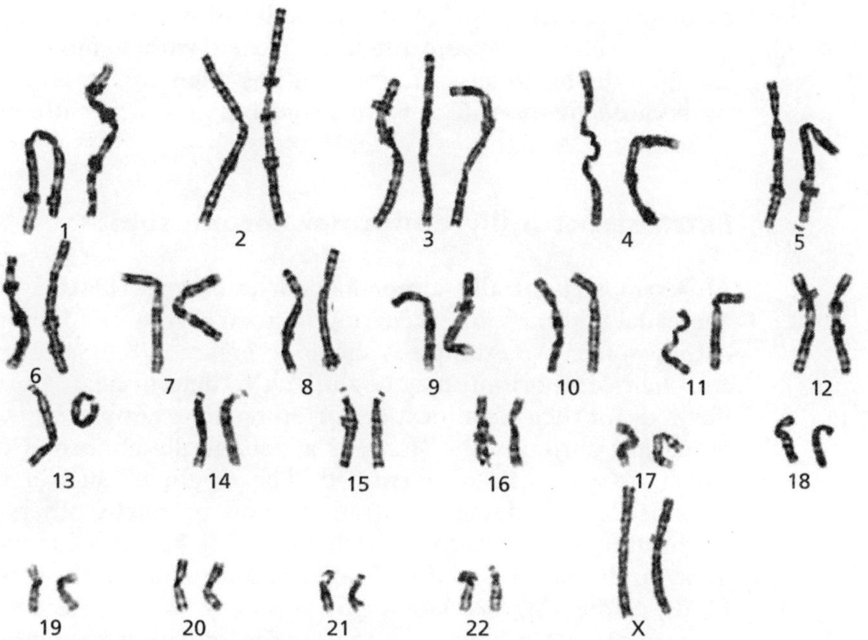

Figure 5.15
A female karyotype with a ring chromosome 13. Rings are formed by breaks at each end which then join and a small amount of chromosome is lost. The ISCN for this karyotype is written as: 46,XXr(13)(p11;q34).

since there is a high probability of abnormal chromosomes being generated by meiotic recombination between the ring chromosome and its normal partner.

Duplications

In another rare unbalanced type of chromosome abnormality, which is almost never familial, the duplication has the structure ABCDE**CDE**FGH, with a segment of chromosome repeated in tandem. Occasionally duplications may be inverted, that is ABCDE**EDC**FGH. As with other types of unbalanced abnormality, the patient is usually identified because of mental handicap, or physical dysmorphic features.

Complex chromosome abnormalities

Some rare structural chromosome abnormalities are more complex than those discussed in this chapter. They include balanced and unbalanced rearrangements which involve three or more breakpoints. Those which are balanced include three- or four-way translocations, insertions of interstitial chromosome segments in incorrect locations, and composite translocation/inversions. They are often associated with a high reproductive risk owing to the much larger number of possible unbalanced forms, or infertility because of the difficulty of achieving complex patterns of pairing at meiosis.

Extra structurally abnormal chromosomes

An **extra structurally abnormal chromosome** (ESAC) is a structurally abnormal chromosome occurring in a karyotype in addition to 46 normal chromosomes. An example is shown in *Figure 5.16*. ESACs are mostly small, and their composition may be difficult to determine, requiring a battery of methods for their identification including, for example, *in situ* hybridization. A **marker** chromosome (mar) is a structurally abnormal chromosome in which no part can be identified. The origin of such a chromosome is variable; some originate as translocation products, others are tiny rings, isochromosomes or dicentric structures. The karyotype in which it occurs is always unbalanced, and the clinical effect depends on the composition of the ESAC. Some tiny accessory chromosomes which effectively have only a centromere and contain no important genetic information are benign, while

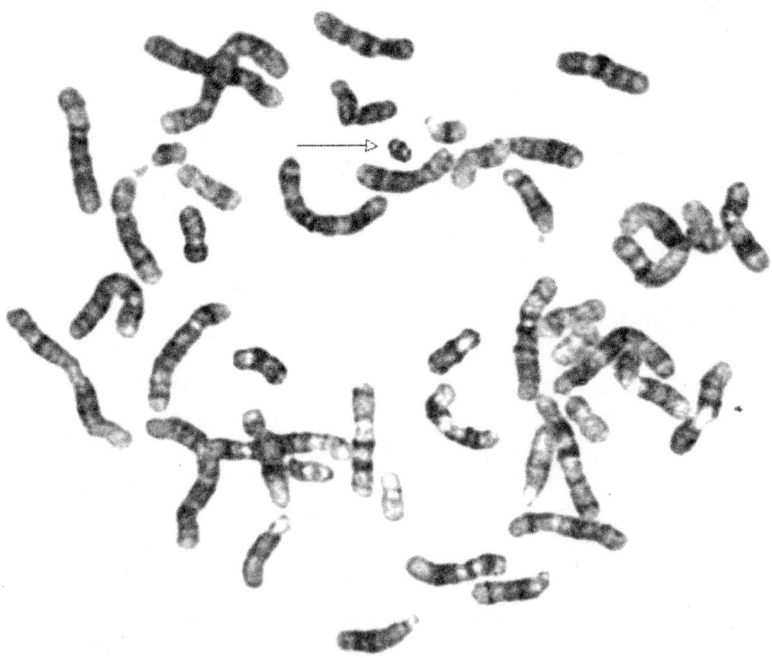

Figure 5.16
A G-banded metaphase with an extra structurally abnormal chromosome (ESAC) or marker. Additional staining tests showed this one to be derived from heterochromatin and was present in a normal individual.

others result in serious mental handicap or physical malformation. The identification of an ESAC at prenatal diagnosis poses a difficult dilemma, since the clinical effect is often impossible to predict.

Markers derived from chromosome 15

Many ESACs identified to date originate from the region around the centromere of chromosome 15, and basically fall into two types, those which are small and benign, and those which are larger and occur in patients with a recognizable syndrome of mental handicap and physical dysmorphism. This latter type of chromosome is usually dicentric and contains the centromeres of both maternal chromosome 15s. It is believed that some kind of anomalous meiotic pairing and recombination gives rise to these abnormal chromosomes.

5.5 THE SEX CHROMOSOMES

In addition to the 22 autosomes, the normal human karyotype has two sex chromosomes, X and Y.

Normal females have two X chromosomes and normal males have one X and one Y chromosome.

Unlike partial or total loss or gain of autosomes, imbalance involving the sex chromosomes is relatively well tolerated.

XY pairing

Although the X and Y chromosomes are very different in morphology and genetic composition, they do have some genes in common – the pseudo-autosomal region. This region is 2.6 Mb long and is located at the tip of Yp and Xp. At meiosis, these regions pair end to end and there is recombination between these homologous DNA sequences. The obligatory crossover between the X and Y is an important factor in controlling the segregation of the two sex chromosomes into separate daughter cells at the reduction division, meiosis I.

Sex determination

Sex determination is regulated by the sex chromosomes. The Y chromosome contains the sex-determining region (*SRY*) gene located at the tip of the short arm. It has a single exon and codes for a 204 amino acid DNA-binding protein. It initiates a cascade of activation of other genes elsewhere in the genome (present on the X chromosomes and on autosomes) which cause the testes to develop. The testes secrete hormones which cause the development of the male reproductive system. Hence male sexual differentiation in the embryo is controlled by the Y chromosome.

It is the absence of the Y chromosome (or absence of the *SRY* gene) which results in female fetal development. For normal female development, post-natal growth and secondary sexual differentiation, the presence of at least two X chromosomes is required.

Turner syndrome

Turner syndrome occurs in approximately one in 2500 live-born females, although 95% of Turner conceptuses fail to reach term. Those that do survive have short stature and a lack of secondary sexual development, including rudimentary 'streak' ovaries, and they are often diagnosed in puberty because of primary amenorrhoea (failure to start menstruation); they are infertile.

People suffering from Turner syndrome may be diagnosed during pregnancy because of the presence of **cystic hygroma** on ultrasound scan; this fluid-filled cyst is a consequence of fetal oedema leading to a build-up of fluid in the lymphatic system, particularly in the neck region. They may also be diagnosed at birth because of excess skin at the nape of the neck (webbed neck) and lymphoedema (puffiness) of hands and feet. About 20% have a heart defect. Intelligence and lifespan are normal.

The most common karyotype in Turner syndrome, accounting for 55% of cases, is 45 chromosomes including a single X, (45,X). In 25% of cases there is one normal X and one abnormal X. These can be deletions, rings or an isochromosome for the long arm of the X chromosome, i.e. 46,X,i(X)(q10) (see above). Approximately 15% have a mosaic karyotype.

Klinefelter syndrome

Approximately one in 1000 males has the karyotype 47,XXY, which gives rise to Klinefelter syndrome. XXY males do not produce adult levels of testosterone, and often have poorly developed secondary sexual characteristics and small testes. Some have gynaecomastia (breast development), and as a group they are tall with long limbs. The patient is often identified through the infertility clinic, although approximately 20% display mild mental handicap which may be noted during childhood. Again, they are often mosaic with a 46,XY cell line in addition, and some of these mosaic individuals may be fertile.

Other aneuploidies of sex chromosomes

A **47,XYY** karyotype occurs in about one male in 1000; it is often asymptomatic and fertility is normal. These males tend to be tall. In surveys of high-security criminal hospitals in the 1960s, more of these males were found than would be expected by chance. At the time, those observations led to a popular misconception that the XYY karyotype was found in big men prone to aggressive and violent behaviour. This has turned out to be untrue, although when these boys are compared to their brothers, there does appear to be a tendency towards relatively low IQ, poor social adaptation, disruptive behaviour, and so on.

It is possible for an individual to have multiple copies of the sex chromosomes and karyotypes such as 47,XXX, and 49,XXXY have been described. The more X chromosomes that are present the greater the incidence of mental retardation.

Genes on the X chromosome

The X chromosome contains a number of genes which are not related to sexual development, including the genes at:

- Xp11.2–p22.1; ovarian failure
- Xq13; X inactivation centre (XIST)
- Xq13–Xq26; the 'critical region' which must be present for ovarian function

Most of the Y long arm contains a large block of inactive heterochromatin which stains brightly with Q-banding and which can be very variable in length between individuals. There are other genes present in the long arm some of which when deleted can result in infertility.

Females are disomic for many X-linked genes for which males are monosomic and, in order to compensate for this difference, one X chromosome in the normal female becomes genetically 'switched off' or inactivated early in embryonic development. This occurs approximately 2 weeks post-fertilization. It occurs randomly in all of the 5000 or so cells present at this stage and all descendant cells maintain the same X inactivation pattern from then onwards. A well-known example of this phenomenon is the tortoise-shell cat. It carries one black allele on one X and one ginger allele on the other X chromosome. Random X inactivation results in patchy expression of different coat colour alleles.

This process of inactivation is sometimes referred to as **Lyonization**, after the cytogeneticist Mary F. Lyon who first proposed the mechanism. The inactive X chromosome replicates at the end of the S phase of the cell cycle and, providing both X chromosomes are normal, X-inactivation is random, with half of the cells expressing the maternal X-linked genes and half the paternal. The gene controlling the Lyonization process, the X-inactivation centre (X-inactivation-specific transcript, XIST), is located on the long arm of the X. Inactivation spreads in both directions along the chromosome from the XIST. It is *cis*-acting, i.e. can only influence the chromosome it is on. The inactive X then takes on the properties of heterochromatin and replicates late in the cell cycle (see *Table 5.4*).

X-inactivation is not complete; it affects most but not all genes on the X chromosomes. This includes the genes in the pseudo-autosomal regions on Xp.

X-linked diseases such as Duchenne muscular dystrophy usually only manifest themselves in males. This is because they have only one X chromosome which carries the mutant gene and no normal X. Carrier females are usually unaffected as they have two X chromosomes, one with the mutant gene and one with the normal gene.

Table 5.4 X-inactivation in sex chromosome abnormalities

Phenotypic sex	Chromosome complement	No. of inactive X chromosomes	Pattern
Male	XY	None	
	XXY	One	Random
	XXXY	Two	Random
Female	XX	One	Random
	XXX	Two	Random
	X,del(X)	One	Abnormal X-inactivated
	X,i(X)	One	Abnormal X-inactivated
	Balanced t(X;autosome)	One	Normal X-inactivated
	Unbalanced t(X;autosome)	One	Usually abnormal X-inactivated

Structurally abnormal chromosomes which incorporate X chromosome material can only be inactivated if the inactivation centre is present (*Chapter 2*). When the X chromosome is involved in a structural rearrangement, X-inactivation is frequently non-random, since those cells which are effectively more genetically balanced have a selective advantage.

Replication banding

Replication banding (see *Figure 5.2c*), a valuable method for visualization of the late replicating X chromosome, exploits the properties of the substance 5′-bromodeoxyuridine (or 5′BrdU), which has been mentioned before in relation to sister chromatid exchange (see *Chapter 3*). The G_2 phase is known to last about 4–5 hours, so addition of 5′BrdU to a culture for the final 6–7 hours before harvest ensures that it is incorporated only into the late-replicating regions of the genome, which can then be differentiated using particular stains and buffers.

X;autosome translocations

This type of translocation is very rare, especially in males. In females, the balanced carrier may be infertile if the X chromosome breakpoint occurs in a segment of the long arm known as the **critical region**. Sometimes a female who has a partial autosomal trisomy as a result of an unbalanced X;autosome translocation may display relatively mild clinical features, since the

Lyonization of the abnormal X spreads into the segment of the autosome and partially inactivates it. Depending on which X chromosome is inactivated, balanced carriers of X;autosome translocations may have an abnormal phenotype (unlike balanced autosome–autosome translocations) and unbalanced carriers of X autosome translocations may have a normal phenotype if the material is inactivated. An example is shown in *Figure 5.17*.

Sex reversal

Sex reversal is the term used to describe individuals where the karyotypic sex does not correlate with the phenotype (see *Chapter 3*). It can occur due to abnormalities of the sex-determining genes or due to abnormalities of the biochemical pathways involved in sexual development.

Females with a 46,XY karyotype include those with testicular feminization syndrome (now known as complete androgen insensitivity syndrome, or CAIS), where the gonads are testes but a deficiency of cell binding of testosterone inhibits male differentiation. Males with an XX karyotype often have the *SRY* gene present as a cryptic translocation.

Congenital adrenal hyperplasia (CAH) is due to a deficiency of 21 hydroxylase resulting in 'mascularization' of external genitalia in the female fetus. A true hermaphrodite has both ovarian and testicular elements and is very rare.

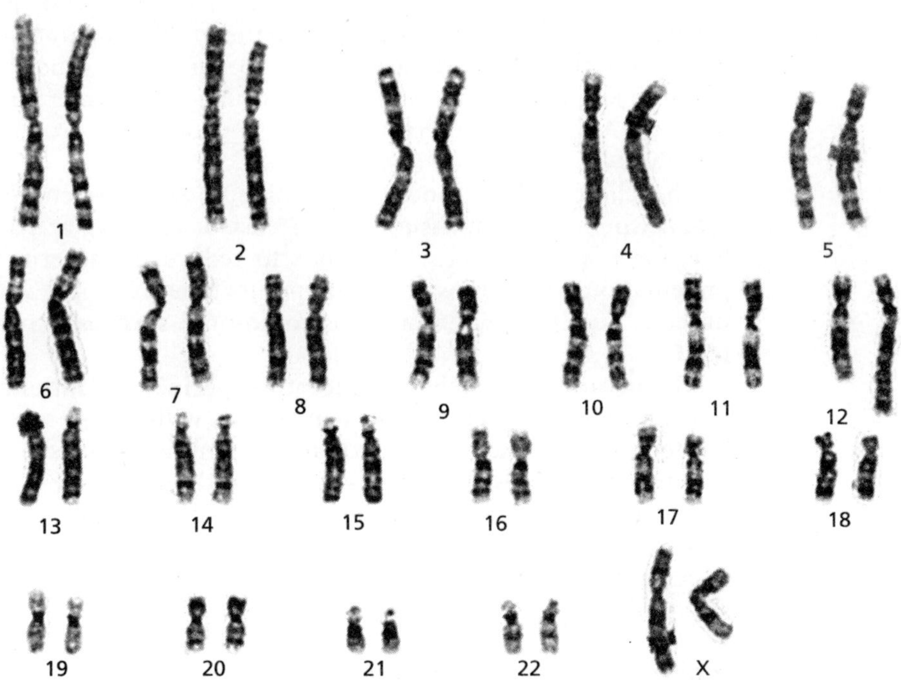

Figure 5.17
A female karyotype with an X autosome translocation involving chromosome 12; autosome translocation – X;12.

Individuals may also have ambiguous genitalia externally. This may be due to mosaicism, or chimerism with a mixed 46,XX/46,XY karyotype, or to mutations in the genes in the pathway downstream of *SRY*.

Mosaicism

Mosaicism is the term applied to the situation where two or more chromosomally different cell lines occur in a single individual. This is in contrast to chimaerism, which is very rare, in which more than one cell line is present in an individual due to the union of more than one separate conception. The major cause of mosaicism is mitotic (i.e. somatic) non-disjunction in an initially non-mosaic zygote.

If non-disjunction occurs on the first mitotic division, a non-mosaic karyotype results which appears to be due to meiotic error. If non-disjunction occurs in any division after the first mitotic division, then three cell lines result. In most cases growth of the monosomic cell line is severely disadvantaged, this cell line is lost, and the resulting zygote therefore has normal and trisomic cell lines. The exception to this is when there is non-disjunction of an X chromosome. In this case both abnormal cell lines may persist, e.g. one missing an X and one with an extra X giving a karyotype 45,X, 46,XX and 47,XXX.

Alternatively, mosaicism can arise in an initially trisomic conceptus. This can be either due to non-disjunction or anaphase lag. Anaphase lag occurs when a chromosome fails to connect to the spindle apparatus or 'lags behind' when chromosomes are segregating to the poles. The chromosome in question fails to be included in the reforming nuclear membrane and instead it is lost. Although numerical mosaics are common, structural mosaics are very rare.

The clinical significance of mosaicism is dependent on the level of mosaicism in different tissues. Coexistence of a normal and an abnormal cell line, for example trisomy 21, tends to reduce the severity of the clinical presentation. Other mosaic karyotypes include a cell line with a mitotically unstable structurally abnormal chromosome such as a ring, dicentric or ESAC.

Mosaicism is relatively rare except in specific circumstances. 45,X/46,XX mosaicism is found in up to 10% of females with Turner syndrome. Trisomy 8 is found in mosaic form in patients presenting with a syndrome involving specific dysmorphism. Pallister–Killian syndrome involves mosaicism for a supernumerary isochromosome composed of two copies of the short arm of chromosome 12, an abnormality never seen in lymphocyte preparations; patients suspected of having this syndrome are investigated by analysis of a skin biopsy.

Interphase FISH and mosaicism

Increasing use of *in situ* hybridization may mean that mosaic abnormalities, especially those confined to particular tissues, can be confirmed more easily by using interphase nuclei recovered from buccal scrapes. For example, use of a chromosome 8 centromere probe may show two populations of nuclei, one with two signals, one with three (see *Chapter 6*).

A search for chromosome mosaicism may be made when a patient presents with asymmetrical growth, or hemi-hypertrophy: different proportions of normal and abnormal cells on one or other side of the body which may lead to different growth rates.

Mosaicism in the gonadal tissue is suspected after the birth of two siblings with the same *de novo* abnormality. Mosaicism may be confined to extra embryonic tissue during embryogenesis and this is thought to occur in 1–2% of placentas. This confined placental mosaicism may cause problems in prenatal diagnosis of chromosome abnormalities, as explained in *Chapter 9*.

SUGGESTED FURTHER READING

Gardner, R.J.M. and Sutherland, G.R. (2004) *Chromosome Abnormalities and Genetic Counselling*, 3rd Edition. Oxford: Oxford University Press.

Rooney, D.E. and Czepulkowski, B.H. (2001) *Human Cytogenetics: A Practical Approach; Volume I: Constitutional Analysis*. Oxford: Oxford University Press.

Schinzel, A. (2001) *Catalogue of Unbalanced Chromosome Aberrations in Man*, 2nd Edition. Berlin: Walter de Gruyter.

Shaffer, L.G. and Tommerup, N. (eds) (2009) *ISCN 2009, An International System for Human Cytogenetic Nomenclature*. Basel: S Karger AG.

SELF-ASSESSMENT QUESTIONS

1. What are the key requirements for tissue culture?
2. Which chromosome abnormality would you expect to find in: (a) a Down syndrome child born to a 40-year-old mother, and (b) a Down syndrome child with a similarly affected brother and maternal aunt?
3. How many chromosomes are most commonly found in the diploid karyotype in Edwards, cri-du-chat, and Turner syndromes, respectively?
4. If a normal chromosome has the structure ABCDEFG, and an inverted one is ABEDCFG, what are the consequences of recombination in the heterozygote?
5. What is: (a) a microdeletion, and (b) the difference between terminal and interstitial deletions?
6. What sort of chromosome abnormalities contribute to the high level of spontaneous abortion?
7. Explain the difference between a reciprocal and a Robertsonian translocation.
8. A woman was referred to a genetics centre because of a family history of early miscarriages. Cytogenetic analysis revealed an inversion of chromosome 3 involving bands 3p21 and 3q26. Write the ISCN and state what type of inversion this is.
9. What factors determine the risk of abnormal offspring?

CHAPTER 6

Molecular cytogenetics

Learning objectives

After studying this chapter, you should confidently be able to:

- **Explain the principles of fluorescence *in situ* hybridization (FISH)**
  A sequence of DNA, called a probe, is labelled with a coloured dye and mixed with the test DNA in the form of either metaphase or interphase cells. The single-stranded complementary DNA sequences will bind together and the fluorescent signal from the probe can be visualized using a microscope.
- **Outline the main steps in a typical FISH procedure**
  The basic steps include the denaturation of both slide and labelled probe, hybridization of the probe to the chromosomes on the slide, washing off the excess unbound probe, and detection (with amplification) of the signal.
- **State the types of FISH probes**
  Types of FISH probes include unique sequence probes, repeat sequence probes, or chromosome paints.
- **Give examples of the use of FISH probes**
  Unique sequence probes can be used to identify microdeletions such as cri-du-chat and Williams syndrome. Dual fusion and break-apart probes can be used to identify rearrangements in leukaemia. Chromosome paints can be used on unidentified chromosome segments or marker chromosomes.
- **Describe some advantages and disadvantages of FISH**
  The advantages of FISH include the increase in resolution to 1–2 kilobases rather than the 2–5 megabases of conventional cytogenetics. FISH can be used on interphase cells, avoiding the need for cell culture and allowing testing of archived material. It can also be used to determine the chromosomal composition of rearrangements and extra structurally abnormal chromosomes (markers). Disadvantages include the fact that only the area targeted by the probe can be investigated. There is also a limit to the number of colours that can be visualized in any one experiment and hence how many areas can be analysed in one experiment.
- **Give examples of other techniques used as a supplement to cytogenetic analysis**
  Multiple ligation-dependent probe amplification (MLPA) and quantitative fluorescent PCR (QF–PCR) can be used to detect changes in the numbers

of copies of DNA sequences. MLPA is used to detect sub-telomeric deletions and duplications. QF–PCR is used for the rapid detection of the common aneuploidies, trisomies 13, 18 and 21.

- **Explain the technique of microarray comparative genomic hybridization (CGH)**
  Microarray CGH follows the same principle as FISH, such that the probes are attached to a fixed substrate such as a glass slide. Many thousands of tests can be undertaken in one experiment with a resolution of 100 kb or less. This generates enormous amounts of complex data which requires the use of bioinformatics for analysis. Many novel syndromes are being discovered along with numerous copy number variants of unknown significance or without apparent phenotypic effect.

6.1 INTRODUCTION

In the last chapter it was seen how the detection of chromosomal abnormalities depends on the limits of resolution of the light microscope. This depends on the quality of the preparations and is generally considered to be between 3 and 5 Mb of DNA for deletions and 2 Mb for amplifications. This represents large numbers of base pairs and can include a number of genes. New techniques have been developed based on molecular technologies and our better understanding of DNA structure; these have improved this resolution and can be used to supplement the findings from routine cytogenetic analysis.

The main technique used in routine diagnostic laboratories is fluorescent *in situ* hybridization (FISH) which is also sometimes called molecular cytogenetics. FISH can increase the resolution to 10 kb. However, the disadvantage is that it only examines the area targeted by the probe itself so you need to have some idea as to the abnormality you suspect may be present.

Other moleular techniques that are used to detect abnormalities include MLPA (multiple ligation-dependent probe amplification) and PCR (polymerase chain reaction) and these are described further in *Chapter 7*. More recently the resolution has been increased even further using the technique of microarray comparative genome hybridization (CGH) or molecular karyotyping.

In this chapter these techniques will be explained, along with some of the uses and limitations for routine clinical practice.

6.2 FLUORESCENCE *IN SITU* HYBRIDIZATION

Fluorescence *in situ* hybridization (FISH) involves the hybridization of a target DNA sequence or probe to metaphase or interphase preparations from the patient.

Probes

It is possible to design a short piece of DNA complementary to a region of interest on a particular chromosome. This piece of DNA is called a **probe.** The DNA probes used in FISH are usually 2–55 kb in size. The length of DNA which represents the human probe is often called an **insert**, because it cannot be copied in isolation, but must be inserted into another (often circular) piece of DNA called a **vector** (or cloning vehicle). This carries the human insert through all of its subsequent manipulations. If it is used for reproducing the DNA fragment, it is called a '**cloning vector**'.

Vectors are in turn usually found in bacteria or yeast (the **host**), which can easily be grown up in the laboratory; as the bacterium divides, so does the vector and its insert, and thus produces many more copies. This is referred to as **cloning.** Commonly used vectors include **plasmids, cosmids** and **artificial chromosomes inserted into bacteria** (**BACs**) or **yeast** (**YACs**). **Plasmids** are circular, double-stranded DNA molecules that exist in bacteria and in the nuclei of some eukaryotic cells. They can replicate independently of the host cell.

A **cosmid** is a type of hybrid plasmid that contains *cos sequences*, DNA sequences originally from the Lambda phage, and the DNA sequence to be cloned. The choice of vector will depend to some extent on the size of the insert required (see *Box 6.1*).

A large number of ready to use probes are now available from commercial companies for the most commonly needed areas of interest. In addition, knowledge of the human genome sequence means that it is also possible to grow and label probes for other specific DNA sequences as required. However, it is essential that these probes are validated to check the region of the hybridization before use; this is carried out by hybridizing to chromosome preparations.

Box 6.1 Types of vectors

- Plasmids – can carry small inserts <10 kb
- Cosmids – can carry medium inserts 30–40 kb
- BACs (bacterial artificial chromosomes) can carry large inserts of up to 250 kb
- PACs (P1 phage-derived artificial chromosomes) can carry large inserts up to 250 kb
- YACs (yeast artificial chromosomes) can carry very large inserts >1 Mb

Basic principles of FISH

Figure 6.1 shows the general principles, and technical details of the FISH method which is routinely used in cytogenetics laboratories.

Chromosome preparations are made in the routine way from an appropriate tissue (see *Section 5.2*). The patient's chromosomal DNA on the slide is denatured by heating in a solution of formamide to make it single-stranded. The labelled probe is also denatured so it is also single-stranded. The

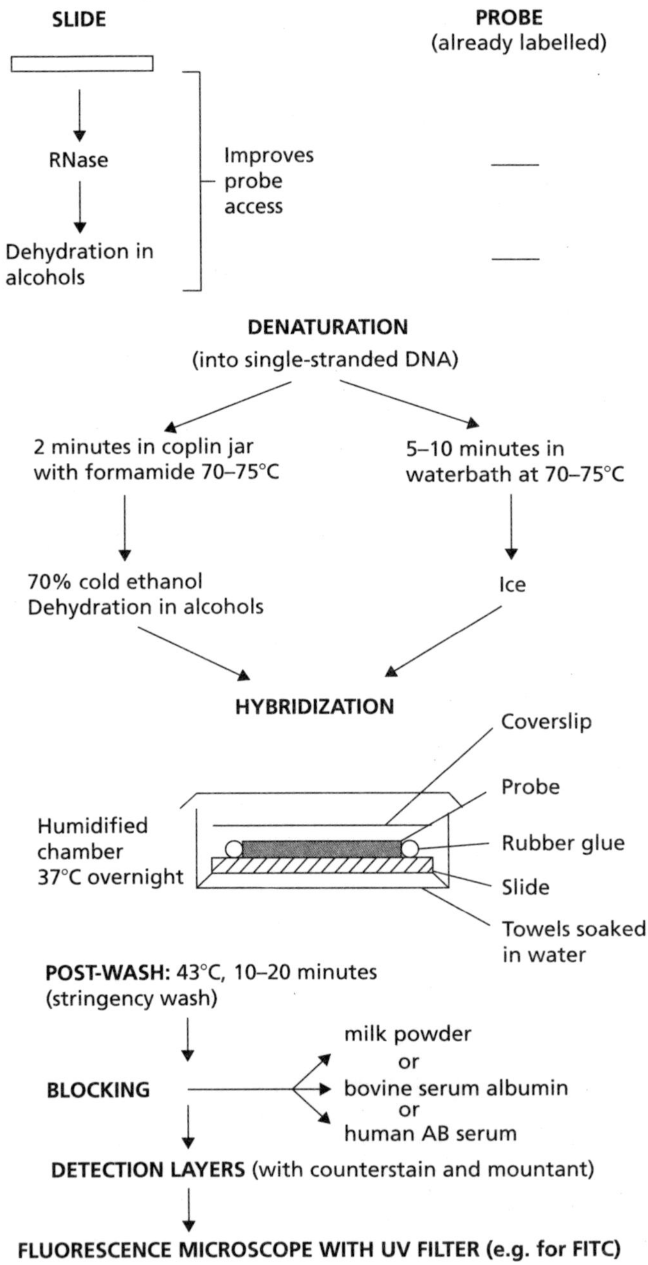

Figure 6.1
General principles of *in situ* hybridization.

labelling of the probe for use in FISH procedures is usually done by a process known as **nick translation** (*Figure 6.2* and *Box 6.2*), which is capable of labelling the large amounts of probe needed for FISH.

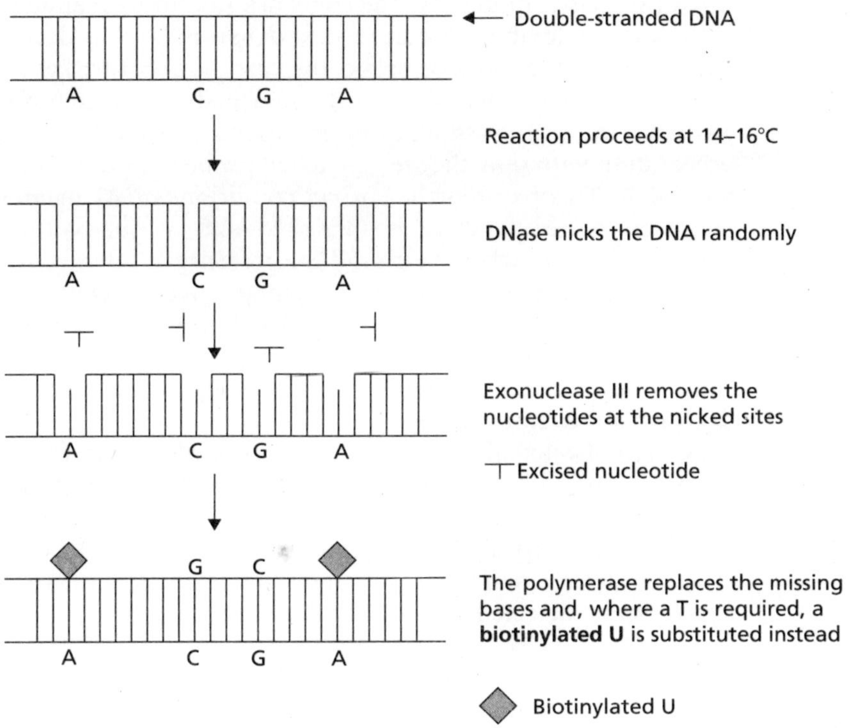

Figure 6.2
Labelling a probe by nick translation.

Box 6.2 Nick translation

Around 10 ml of probe is added to the three normal dNTPs (dATP, dCTP and dGTP), together with the biotinylated dUTP and enzymes called deoxyribonuclease 1 (DNase 1) and DNA polymerase 1. This reaction works on double-stranded DNA. The basic method for nick translating a probe is as follows:

1. The tube containing the reaction mixture is placed at a temperature of 14–16°C.
2. The DNAse 1 proceeds to **nick** the DNA randomly.
3. The DNA polymerase 1 comprises two enzymes: an exonuclease III and a polymerase. The exonuclease III chops out the nicked nucleotides and thus leaves gaps in the DNA structure. The DNA polymerase has a 'copying' facility and it recognizes the missing complementary bases (from the opposite strand), and replaces them with an A, C, G or biotinylated U, as appropriate.
4. The reaction is stopped after about 90 minutes and the DNA is purified to remove the unincorporated nucleotides (either by precipitation using ethanol or down a column of Sephadex beads).
5. The labelled probe DNA is then dried and re-dissolved in buffer ready for use or storage.

The probe solution is then added onto the surface of the slide. As the probe DNA has a very specific complementary sequence, it will find that matching sequence on the patient's DNA and anneal (hybridize) to it. Formamide is used in the hybridization step as it alters the melting temperature of the DNA duplex. As the concentration of formamide in a solution is increased, the temperature needed to denature DNA is decreased. Instead of having to boil the slides or probe, they can be denatured at about 70°C.

Probe that is not hybridized is then washed away. It is important to achieve a balance between washing away any unattached probe, and leaving a clean preparation with only the well-matched probe hybridized to its complementary DNA. This requires the correct conditions of **stringency**. Stringency is increased as temperature is increased and as the salt concentration is decreased. Too much background fluorescence is often caused by the probe remaining attached to non-complementary areas of DNA and this effect can be reduced by increasing the stringency. Likewise, if there is no fluorescent signal because all of the probe has been washed away then the stringency should be decreased (usually by increasing the salt concentration, or by decreasing the temperature).

A counterstain such as DAPI is then applied to stain the chromosomes. A counterstain is a second fluorescent stain of contrasting colour to that used for the probe. DAPI stains the chromosomes fairly evenly and so it is not possible to identify all of the chromosomes with confidence.

The slide can then be viewed under a fluorescence microscope so that the signal can be detected.

Filters and microscopy

The fluorescence microscope is specially adapted and has a UV (ultraviolet) light source in order to see the signals. Fluorescent molecules absorb light at one wavelength and emit light at a longer wavelength. For example, fluorescein isothiocyanate (FITC) when excited at a wavelength of 495 nm emits light at 523 nm (see *Box 6.3*). Short high-energy **UV light** is absorbed by the fluorescent dye and re-emitted at a longer wavelength which is within the visible range. Modern filter combinations include an excitation filter which provides maximum absorption at an appropriate wavelength for the fluorochrome, and a barrier (or suppression) filter, which maximizes emission of light at a longer (visible) wavelength.

The microscope comprises a mercury lamp as the light source, an excitation filter to transmit light of the desired wavelength, a dichromatic beam splitter which reflects the light onto the specimen and transmits light of longer wavelength, and an emission filter which transmits light at specific emission wavelengths. Filters are available which allow the excitation and emission of more than one fluorochrome simultaneously. For example, a **dual band pass** can be used for both spectrum red and spectrum green, or both FITC and Texas red. A triple band pass can be used for the combination of three fluorochromes such as spectrum red, spectrum green, and DAPI blue.

Box 6.3 Examples of commonly used fluorochromes

	Max. excitation wavelength (nm)	Max. emission wavelength (nm)
Fluorochrome		
Spectrum aqua	433	480
FITC	495	523
Spectrum green	497	524
CY3	548	562
Texas red	596	615
CY5	650	670
DNA stains		
Hoechst 33258	352	461
DAPI	359	461
Propidium iodide	535	617

Interphase FISH

As well as producing signals on metaphase chromosomes of **cultured cells**, it is also possible to detect signals on the interphase nuclei of **uncultured cells** such as buccal cells or tumour cells. Using specifically designed probes, this allows identification of the presence or absence of the DNA regions targeted by the specific probe.

6.3 TYPES OF FISH PROBES

Probes can be divided into three main types:

- single copy, unique sequence probes, or locus-specific probes
- centromere/repeat sequence probes
- chromosome paints

These probe types allow a variety of applications in clinical cytogenetics.

Single copy probes

These map to specific well-defined regions of usually 50–200 kb in size. They produce small discrete signals. A large probe is usually desirable as it results in a brighter signal if the sequence is present. Many are available commercially and non-commercial probes are relatively easy to generate. Single copy probes can be used for detection of a wide variety of clinical conditions.

Detection of microdeletion syndromes

Probes hybridizing to unique sequences include the **microdeletion probes**, whose function is to hybridize to either a unique sequence of DNA representing a particular gene or to a larger stretch of DNA with several genes which is known to be involved in syndromes having overlapping symptoms (known as a **contiguous gene syndrome**). The name **microdeletion** refers to a very small deletion of DNA, usually at the limits of detection by the light microscope. There are a group of well-characterized syndromes on various chromosomes which are typified by such borderline deletions, and are therefore well suited to detection by FISH; these are the **microdeletion syndromes** (see *Box 6.4*).

Some syndromes are caused by faulty genes, many of which have yet to be identified. Although we may not know the unique gene sequence for such a candidate gene, a probe can still be made for the general area, or shortest region of overlap (the SRO), which is known to contain the gene. The SRO is found in patients with a similar phenotype but having differing lengths of deletion which share a common region of overlap.

The probe is usually mixed with a control probe from another appropriate unrelated locus on the same chromosome as the region of interest. The control probe is used simultaneously with the locus-specific probe to identify the pair of homologous chromosomes under investigation.

There is always a compromise between the size of the probe (which maximizes visibility of the signal) and the size of the microdeletion detected (the resolution of the probe). A microdeletion probe is usually used to determine the presence or absence of a specific area on a metaphase chromosome and is therefore not usually used on interphase preparations.

When the probe is hybridized to the slide of a patient who may have a microdeletion syndrome, the normal chromosome in the pair will still have its gene sequence present. The probe will therefore find that complementary sequence, and a coloured signal will result. If the other chromosome has a small deletion including the gene sequence, the probe will be unable to hybridize, and no signal will be visible.

Technical details of microdeletions commonly detected by FISH are given below, whereas descriptions of clinical features of the associated syndromes are listed in *Appendix: Glossary of disorders.*

Box 6.4 Examples of microdeletion syndromes detectable by FISH

Wolf–Hirschhorn	4p16
Cri-du-chat	5p15
Angelman	15q12
Miller–Deiker	17p13
Smith–Magensis	17p11.2
Kallman	Xp22

Deletions of 22q11

There are a group of apparently unrelated syndromes which are characterized by a deletion of the long arm of chromosome 22 (see *Box 6.5*). These

Box 6.5 The 22q11 syndrome

Individuals with the 22q11 syndrome have a very broad phenotypic spectrum and it occurs with an incidence of 1 in 400 live births. The most frequent reason for referral is for heart defects which occur in 75% of patients. The genes implicated in these syndromes are contiguous. In 10–15% of cases the deletion is inherited from a mildly affected or phenotypically normal parent, so it is important to check, once a person is identified as having a deletion, to see if one parent is a carrier. Carriers have a 50% chance of passing on the deletion to their offspring.

include DiGeorge syndrome, velo-cardio-facial syndrome (VCFS) and Shprintzen syndrome. The commercial probe TUPLE contains one 120 kb probe complementary to the region on chromosome 22 which is deleted in these patients, together with an extra distal probe to identify the two chromosome 22s. In a normal person both probes will be visible on both chromosome 22s; the probe signal for the locus will be absent if there is a deletion (*Figure 6.3*).

Angelman syndrome

Angelman syndrome occurs with a frequency of 1 in 25 000 live births (see *Appendix: Glossary of disorders*). It is characterized by a deletion of 15q11–q13, although it can also arise from UPD and imprinting centre mutations (see *Chapter 3*). The size of the deletion may range from 5 Mb, when it is visible by light microscopy, through to several kilobases, when it is detectable by FISH, down to the molecular level, where it is detectable by Southern blotting or PCR.

The most commonly used FISH probe for the Angelman syndrome locus at 15q11–q13 is D15S10. There is also a control probe which maps distal to D15S10, called PML. Patients with Angelman syndrome will have the D15S10 signal missing on one 15.

Williams syndrome

Williams syndrome occurs with a frequency of 1 in 10 000 live births (see *Appendix: Glossary of disorders*). It is characterized by a deletion at the elastin locus at 7q11.23. The children have a characteristic elfin face with bulbous nasal tip, wide mouth, long philtrum, full lips and small widely spaced teeth. They have developmental delay with variable mental retardation ranging from severe to low average. They have an overfriendly personality and short attention span and show hypercalcaemia. They also have a particular type of heart defect called supravalvular aortic stenosis (SVAS).

Probes for small chromosome rearrangements

Unique sequence probes can also be used to detect rearrangements such as translocations, some of which may not always be visible down the light microscope, by finding the location of the probe(s) in different chromosomes from that expected. There are two types of probes used, **dual fusion** and **break-apart** probes.

Dual fusion probes

Translocations are commonly found in cancer cells (see *Chapter 8*) including leukaemia. Dual fusion probes have been developed which span the breakpoint of the two chromosomes involved in the translocation.

An example of this would be the BCR–ABL probe used in the detection of a very specific translocation found in chronic myeloid leukaemia (CML). The usual form of the translocation occurs when a small segment of chromosome 9 (breakpoint 9q34) carrying the *ABL* oncogene (see *Chapter 8* for more on oncogenes) is exchanged with a piece of chromosome 22 (breakpoint 22q11). A novel fusion gene is formed on the remains of chromosome 22, at the breakpoint cluster region (*BCR*).

The dual fusion probes are a red probe mapping to *ABL* on chromosome 9 and a green probe mapping to *BCR* sequence on chromosome 22. In a normal cell there will be two red (R) signals, one on each of the normal chromosome 9s, and two green (G) signals, one on each of the normal chromosome 22s. This result can be written as 2R2G. In abnormal cells where the BCR–ABL translocation has occurred, the red signal and the green signal from the translocated chromosomes are brought together to form a yellow fusion signal (F) on the derivative chromosomes 9 and 22. There will also be one normal red and one normal green signal (2F1R1G; see *Figure 6.4*).

It is very easy to score large numbers of cells, including interphase cells, to determine if the rearrangement is present even at low levels amongst the normal cells.

Break-apart probes

Break-apart probes consist of a red probe one side of the breakpoint and a green probe on the other side of the breakpoint. The result is a fusion signal or a close red/green in normal cells (2F). In abnormal cells there is a break between the probes and one signal is moved to another area of the chromosome such as in an inversion or to another chromosome.

One example is the translocation involving breaks within the immunoglobulin heavy chain gene on chromosome 14 and chromosome 18 which occur in mantle cell lymphoma (see *Figure 6.5*).

Not all genetic changes can be seen using chromosome analysis or FISH, and so other techniques based on the polymerase chain reaction (PCR) described in *Chapter 7* are used. These include real-time PCR, reverse transcriptase PCR, and standard PCR.

These methods are extremely sensitive and can detect very low levels of abnormal cells. This is very important when following a leukaemia patient as they have their treatment or bone marrow transplant to remove the disease. Levels down to 1 abnormal cell in 10 000 can be detected. This detection of very low levels of disease, or minimal residual disease, is described further in *Chapter 8*.

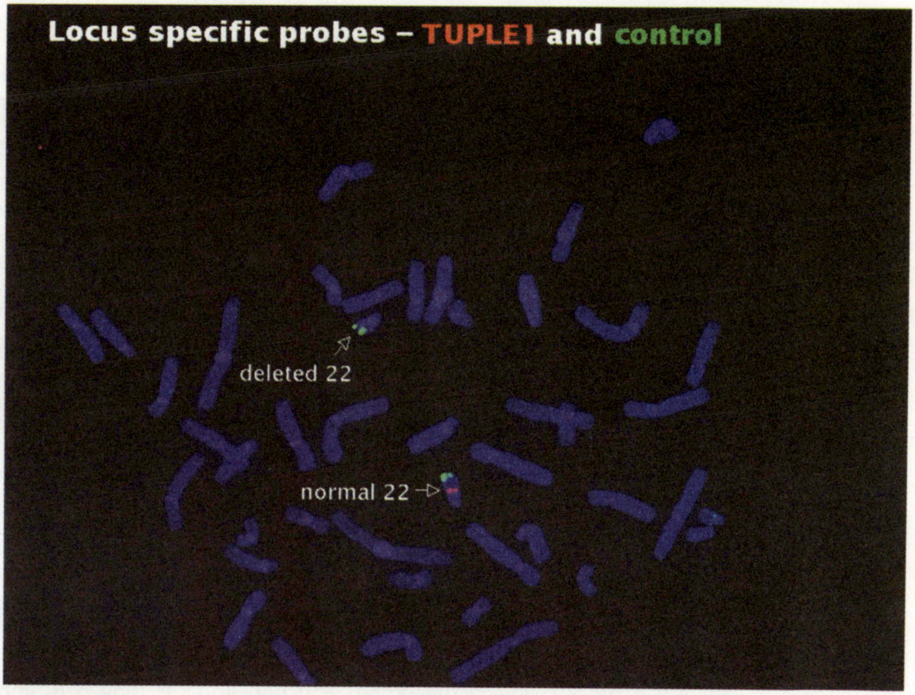

Figure 6.3
The detection of deletion of 22q11 by FISH. 22q11.2 TUPLE1 (spectrum orange) probe and 22q13 control probe ARSA Spectrum green hybridized to metaphase cell; the absence of the orange/red signal on one chromosome 22 indicates the deletion of the TUPLE1 locus.

Normal pattern 2R2G

Typical abnormal pattern:

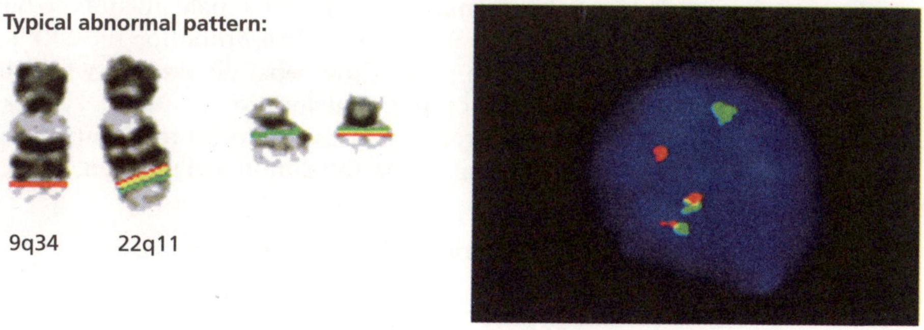

Results of hybridization: 2 fusions 1 red 1 green

Figure 6.4
Dual fusion translocation probes in the detection of the BCR–ABL rearrangement. Normal signal is two green signals on chromosome 9 and two red signals on chromosome 22 (2R2G). Abnormal dual fusion occurs when the probe spans the breakpoint of the two chromosomes involved in the translocation; if the BCR–ABL rearrangement is present, only part of the signal will move to the other chromosome, combining to form a yellow fusion on both abnormal derivative chromosomes (2F1R1G).

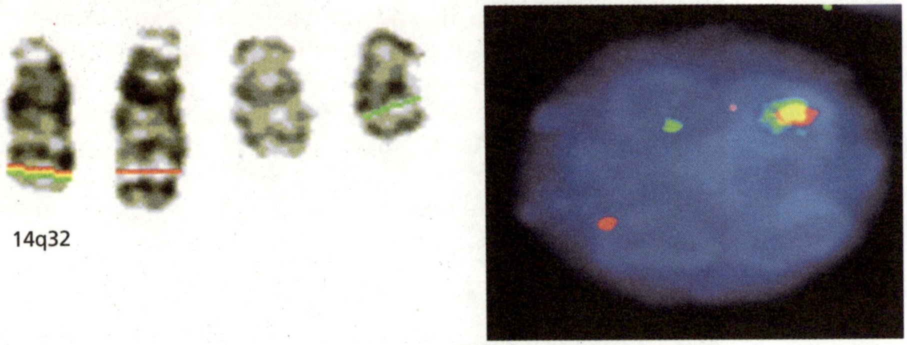

Figure 6.5
Break-apart probes in the detection of rearrangements involving the immunoglobulin genes. These probes consist of a red probe one side of the breakpoint and a green probe the other side of the breakpoint, which result in a fusion signal (or red and green close together) in normal cells (2F) on chromosome 14. In abnormal cells there is a break between the probes and one signal is moved to another area of the chromosome; this results in a 1F1R1G signal.

Sex chromosome rearrangements

Sometimes there is reason to suspect that male gene sequences are present in an individual, but these are not detectable by cytogenetic techniques. One example would be in an individual with a male phenotype but with a female karyotype, 46,XX (see *Section 5.5*). The probe for the *SRY* locus at Yp11 is commonly used. It comprises three separate cosmids which provide a signal large enough to detect cryptic translocations.

Sex chromosome probes can also be used to investigate patients after a bone marrow transplant where the donor and recipient are of opposite sex.

Centromere and repeat sequence probes

Alpha satellite probes are derived from areas of high repetitiveness, such as the **centromere**. Most human chromosomes have repetitive sequences specific for each centromere of each chromosome, but there are exceptions. The pair of acrocentric chromosomes 13 and 21 share 99% homology at the centromere, such that the use of this probe results in four chromosomes being highlighted instead of two. Chromosomes 14 and 22 interact in the same manner, so they do not have unique centromeric probes.

Beta satellite probes are derived from repeat sequences usually found in the centromeric heterochromatin of the acrocentric chromosomes, together with chromosomes 1, 9 and Y.

Chromosome enumeration probes (CEP) are a group of chromosome-specific FISH probes that hybridize to highly repetitive DNA sequences such as alpha and beta satellite DNA. They produce very bright signals and are used to allow the identification and enumeration of human chromosomes in interphase and metaphase cells from fresh and archived samples. They can also be used to investigate marker, dicentric and ring chromosomes.

Both centromeric and unique sequence probes can be used in interphase cells to look for trisomy/monosomy of whole chromosomes or certain chromosomal regions. For example, the Williams elastin probe can be used to look for monosomy 7 in leukaemia or for mosaicism in tissues.

Interphase FISH for prenatal diagnosis of common aneuploidies

Prenatal diagnosis is used to detect chromosome aberrations during pregnancy (this is discussed in more detail in *Chapter 9*). The most common reason for prenatal testing is when a pregnancy is identified as being at a high risk of trisomy 21. In order to produce chromosomes for analysis, cell culture is required, and this means an anxious wait of 7–14 days for the parents to get the result. In response to this, FISH on uncultured fetal cells from amniotic fluid samples was developed which produces a limited result in only 24–48 hours.

Probes specific for the common trisomies 13, 18 and 21, and also X and Y, are hybridized to the cells and the number of signals counted. Three signals indicate three copies of the chromosome. Although one might expect each Down interphase nucleus to have three clear signals in every cell, in practice some cells may lie in orientations which mask one or two signals. Also, if the stringency of the washes is not accurate, cross-hybridization may result in more than the correct number of signals.

Chromosome analysis is still required to determine if the three signals are due to free trisomy or to a translocation. If it is due to a translocation then one of the parents may be a carrier and will then have an increased risk of recurrence in a future pregnancy.

Low level mosaicism and other abnormalities affecting other chromosomes also cannot be detected. Increasingly, molecular techniques have replaced interphase FISH. **Quantitative fluorescent PCR (QF–PCR)** is now more widely used as multiple sites can be examined and it is cheaper to perform.

QF–PCR is based on the polymerase chain reaction (as described in *Chapter 7*). It uses short tandem repeat DNAs (STRs) or microsatellite sequences repeated in tandem. These are scattered throughout the genome with more than 10 000 known. They are also used to determine genetic profiles in forensic science. STRs are known to be highly polymorphic in the number of repeats. Any individual is likely to have different numbers of repeat units on each chromosome homologue. Different allele markers are chosen in the region of interest with the highest heterozygosity in the population and therefore most likely to be different in different individuals.

After PCR amplification, the products (which are labelled with fluorescent dye) are separated by size and a trace produced. As the reaction is quantitative, the number of copies of an allele (and by inference the number of each chromosome) can be detected. Multiple pairs of primers are used in a single reaction in order to investigate multiple regions along a chromosome; this is called a multiplex **PCR,** and avoids problems with markers being non-informative.

Normal individuals are expected to show two alleles of different sizes. A trisomic fetus will have three alleles representing the three copies of the appropriate chromosome (see *Figure 6.6*). A typical trace may show three peaks or two peaks of which one is twice as high as the other. The large peak is due to the double dose of one allele.

As the reaction is very sensitive, contamination with maternal cells can lead to problems with interpretation. Comparison of the maternal genotype with that from the amniotic fluid will usually determine if the results are fetal or maternal in origin (see *Figure 6.7*).

The multiplex can also include markers for the sex chromosomes to allow rapid sex determination. However, results need to be treated cautiously as structural rearrangements of the sex chromosomes are not uncommon.

As with interphase FISH, low-level mosaicism and other chromosome abnormalities will not be detected. For this a full chromosome analysis is still required.

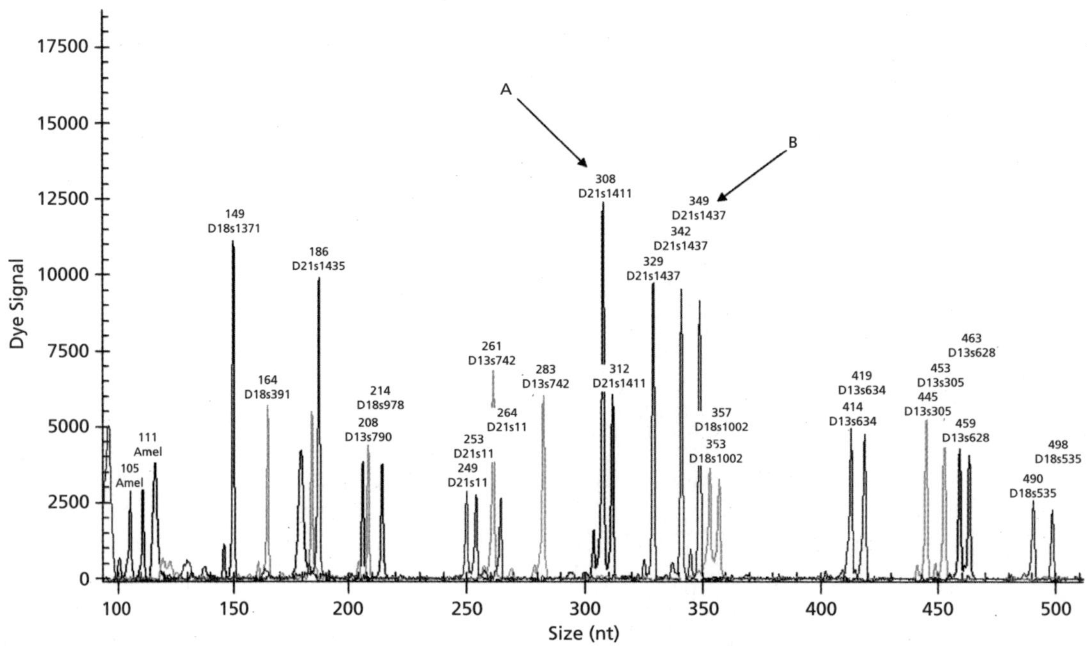

Figure 6.6
Trisomy 21 in an amniotic fluid sample detected by QF–PCR.
A shows peak ratio of 2:1 for marker D21s1411 of 308 and 312 base pairs in size.
B shows three peaks, one from each chromosome for marker DS21s1437 with 329, 342 and 349 base pairs.

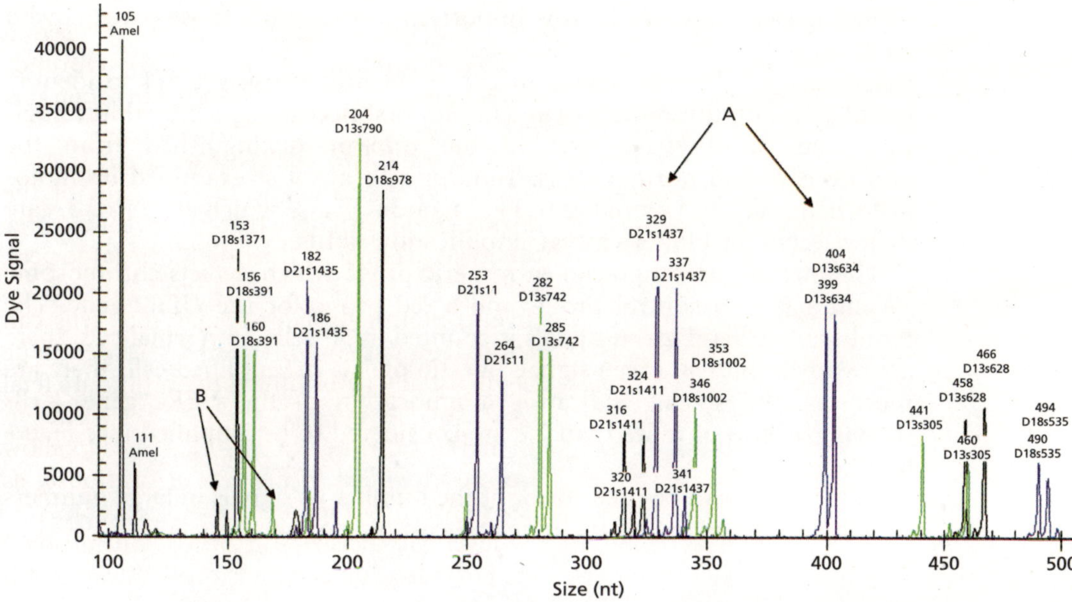

Figure 6. 7
Amniotic fluid sample showing contamination with maternal DNA detected by QF–PCR. There are two clear peaks **A** for all markers, but in addition there are extra small peaks **B** representing contamination with a smaller amount of DNA from another sample. The maternal and fetal samples will have one peak in common for each marker. Comparison of the peaks with a sample of maternal blood shows that in this sample the majority of the DNA is from the maternal genotype.

QF–PCR can be used on amniotic fluid and chorionic villus samples for rapid prenatal testing and on blood samples from newborns suspected of having Down syndrome or with ambiguous genitalia.

FISH detection of HER2 amplification in breast cancer

Gene amplification is frequently detected in human tumour cells and is thought to make an important contribution to tumourigenesis (see *Chapter 8*). The *HER2* gene is a member of the type 1 tyrosine kinase growth factor receptor family. It is overexpressed in 20–30% of human breast cancers. In 90–95% of these cases, overexpression is a direct result of gene amplification. *HER2* overexpression has been shown to correlate with poorly differentiated high grade tumours, relative resistance to some types of chemotherapy, and poor clinical prognosis.

Trastuzumab (Herceptin) is a humanized monoclonal antibody that specifically targets the *HER2* receptor. Clinical trials have shown it to be effective in cases of breast cancer in which *HER2* is overexpressed or amplified, but it does not provide any benefits to those cases of breast cancer with normal expression levels of *HER2*. The drug is expensive and has side-effects, including cardiotoxicity, so it is important to treat only those patients who are likely to benefit.

Patients suitable for treatment can be identified using FISH to identify *HER2* gene amplification. Tissue samples first need to be assessed histologically and the invasive part of the tumour distinguished from the non-invasive and normal areas. Tumour cells are first examined immunohistochemically and scored as 0, 1+, 2+ or 3+. Those which are 2+ or 3+ are then selected for FISH to assess amplification status.

The method uses a green centromeric probe which detects chromosome 17 and acts as a control probe, and a red probe for the *HER2* gene. The number of red and green signals is counted in 30 cells. In normal cells there are two red and two green signals, a ratio of 1:1. In amplified cells there are more cell red signals indicating amplification of the *HER2* gene. Cells showing a ratio of more than 2:1 are considered to be amplified (see *Figure 6.8*).

Automation can be used to count the number of signals in large numbers of cells and hence reduce the time and cost of analysis.

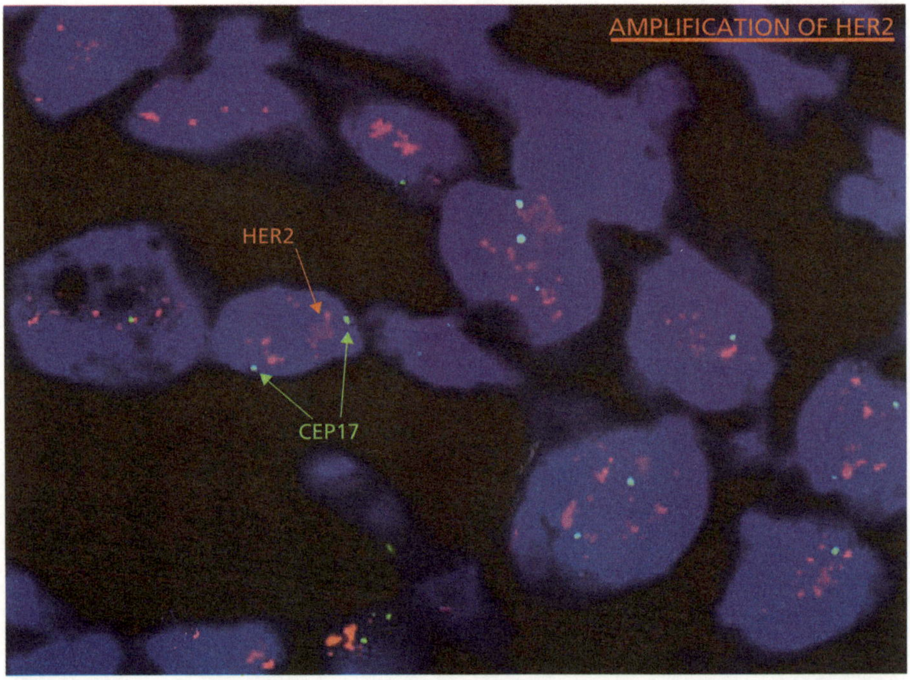

Figure 6.8
Amplification of the oncogene *HER2* in tissue from a patient with breast cancer. Note the multiple copies of the *HER2* signal (red) compared to the control probes for the chromosome 17 centromere CEP17 in green.

Multi-telomere FISH

The tip of a chromosome or telomere in humans consists of a tandem repeat of the sequence TTAGGG and ends in a 3′ extension which may be bent over like a hairpin (see *Chapter 1*). All ends of a chromosome must have a telomere cap to be stable. The region of the chromosome just next to the telomere (the sub-telomeric region) is frequently involved in rearrangements. Deletions and duplications of these can be the basis of abnormal phenotypes and are found in 5–10% of children with abnormal development and dysmorphism.

Unique sequence probes can only target one region of interest at a time and it is necessary for the clinical phenotype to give clues as to which area of chromosome may be affected and hence identify which probe is most likely to be useful.

Multi-telomeric FISH kits have been devised to look at several regions together in one experiment. They comprise 15 mixes of both p and q arm unique sequence probes that map to the sub-telomeric regions of the chromosomes. The probes for the short and long arms are labelled with different colours (one colour for each end). The process involves using three slides, each with five areas of cell suspension, and hence is very time-consuming and expensive. It is for these reasons that multi-telomere FISH is often replaced with a more cost-effective technique called multiple ligation-dependent probe amplification (MLPA; described in *Chapter 7*). This method allows for the relative quantification of up to 40 different DNA sequences and can identify deletions and duplications of any unique sequence used. Probes specifically designed for the sub-telomeric regions are combined into a commercially available kit which is now widely used in diagnostic cytogenetics laboratories.

Chromosome painting of entire chromosomes

FISH using unique sequence probes is limited to analysis of one or a small number of chromosome regions in one experiment. By combining enough unique sequence probes at many different loci on one chromosome, the effect will be to hybridize with and 'paint' the whole of that chromosome. These are called **painting probes**, and the resulting FISH technique is called **chromosome painting** (see *Figure 6. 9*).

Chromosome painting can be used to investigate the composition of marker chromosomes and chromosome rearrangements and be used to improve the interpretation of the complex chromosome changes observed for example in metaphases from tumours.

The usefulness of chromosome painting is limited by the resolution that can be achieved which is only approximately 5 Mb, so it can not be used to investigate small chromosomal segments. In addition, although in principle chromosome 'paint' should cover the entire chromosome, in practice there sometimes appear to be unpainted areas, or gaps, which remain uncoloured. These are usually the regions containing certain repetitive or satellite DNAs such as the centromere, heterochromatic regions (e.g.

9q,1q), or the telomeres. The unique sequence pooled probes are not designed to anneal with common repetitive sequences such as these.

Multicolour FISH

By 1996 it became possible to 'paint' the entire human genome simultaneously so that each chromosome fluoresces a unique and different colour. The combined labelling of five fluorophores makes it possible to assign a specific fluorophore combination and thus a unique spectral signature to each human chromosome.

Multicolour FISH (M-FISH, see *Figure 6.10*) utilizes a series of filters with defined emission spectra, which when coupled with imaging software generate a unique and distinct fluorescent image for each chromosome. This technique has been used for the identification of markers and derivative chromosomes and also for the identification of multiple chromosomal abnormalities in complex karyotypes, particularly in cancers.

The high cost of the probes, the limit of resolution and complexity of the method mean that it is rarely used in routine diagnostic laboratories, and is now being superseded by other techniques to look at the whole genome.

Cross-species FISH

This technique (also known as R_xFISH) uses paints from gibbon chromosomes labelled with different coloured fluorochromes, hybridized to human chromosomes. This results in a colour-coded banding pattern resembling the barcode in a supermarket.

Although human probes should be unique to human DNA, many genes have been so important in evolution that the sequences have remained unchanged for millions of years. In this case human probes would be able to detect analogous **conserved** sequences in other species such as the primates. The number of conserved sequences we share with other animals also gives us an idea of how long ago we diverged from a particular line, because the number of accumulated mutations will increase with time.

Even just looking at the karyotype of a chimpanzee makes it easier to appreciate that humans share 98.5% homology with this species. Many of their chromosomes are recognizable, and many others are derived from inversions or translocations.

If preparations of chromosomes are made from a non-human species (e.g. chimpanzee or orang-utan), and human probe DNA is applied, then the extent of cross-hybridization is readily seen.

6.4 COMPARATIVE GENOMIC HYBRIDIZATION

The FISH techniques described so far are limited in that they only examine the specific areas detected by the probes used and do not examine all other areas of the genome. Combining multiple probes together is expensive and time-consuming to analyse with limited resolution. Chromosomal comparative genomic hybridization (CGH) is a technique based on the same principles as

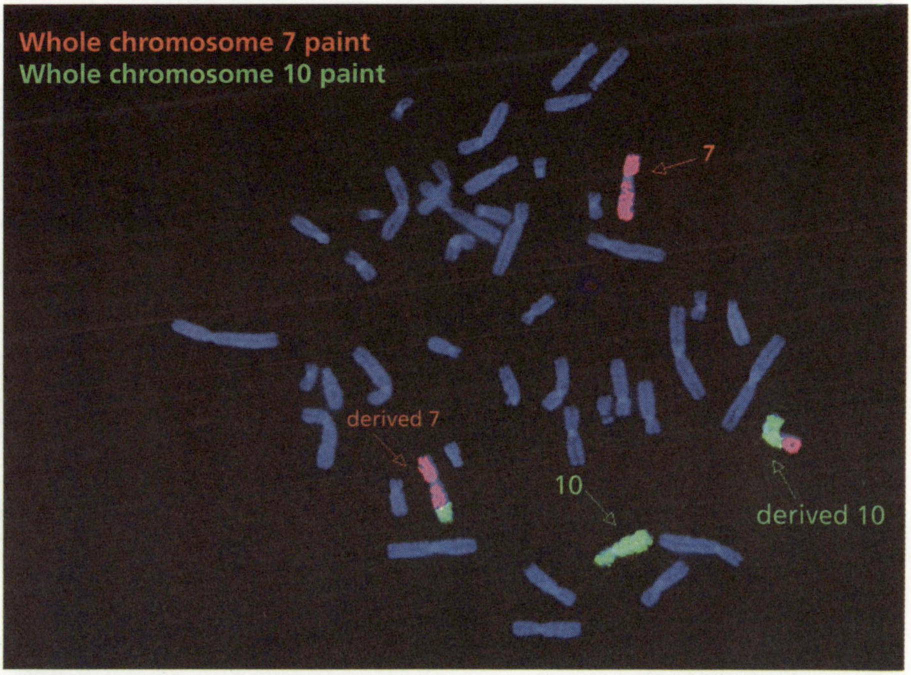

Figure 6.9
A metaphase spread showing a reciprocal translocation between chromosomes 7 and chromosomes 10.

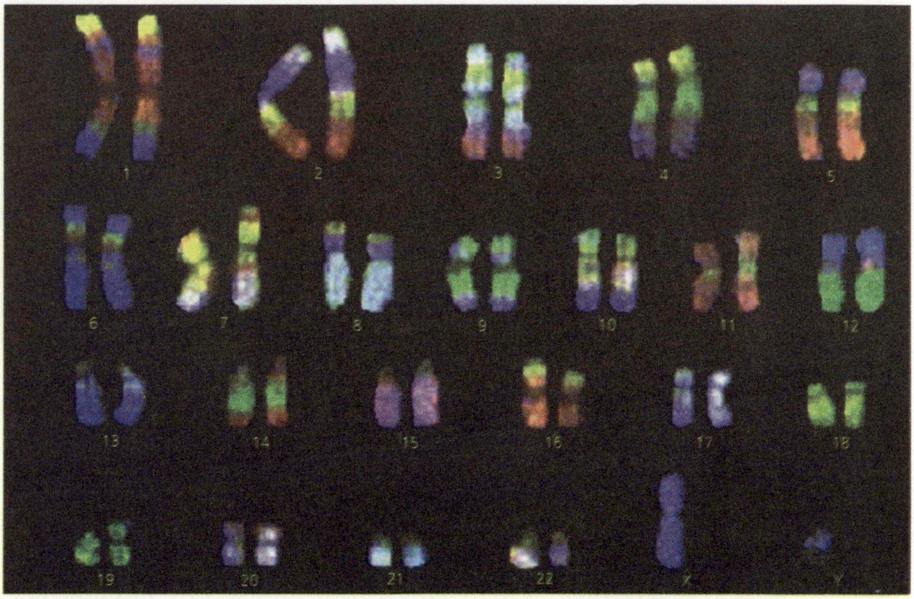

Figure 6.10
Multicolour FISH.

routine FISH and can be applied to situations where a karyotype may be particularly difficult to elucidate due to multiple or unusual abnormalities.

Chromosomal comparative genomic hybridization

In this technique the test and control are labelled in two different colours, usually green and red, respectively. These are mixed together in a 1:1 ratio and applied to a target of normal metaphase chromosomes. Sophisticated computer software is used to analyse the chromosome images from a number of cells. Excess red (control) colour indicates loss of test chromosomal material. Excess green (test) colour indicates gain of test material. For example, if a tumour karyotype included trisomy 1, then there would be one and a half times as much chromosome 1 DNA arising from the tumour compared to a normal karyotype.

One advantage of this test is that it examines the whole genome in one experiment. If more than one unbalanced abnormality is present in the same metaphase (which is often the case in tumours), then CGH can display all the imbalances at once in a single test. In addition, as only DNA is required for the test sample rather than metaphases, cell cultures can be avoided along with the need for living tissue. The test can therefore be applied to archived material.

One disadvantage is that this technique cannot detect balanced rearrangements such as translocations, as the total amounts of DNA would be the same. It will also not identify ploidy levels (such as triploidy) and mosaicism. The process is also time-consuming and has a low resolution of only 2–5 Mb, which is similar to conventional cytogenetics.

Array comparative genomic hybridization

Array CGH allows the rapid detection of copy number changes across the whole genome at a very high resolution. It is based on the same principles as chromosomal CGH, except that the targets are mapped genomic clones located on a microarray, instead of whole metaphase chromosomes.

A microarray (sometimes referred to as a chip) is an ordered array of genetic elements or probes on a solid substrate. New DNA technology has provided large capacity for miniaturization and automation and so it is now possible to 'spot' thousands of DNA fragments onto a microscopic glass slide or silicone matrix in an orderly fashion.

The array CGH procedure relies on the labelling of reference (normal) DNA with genomic DNA test samples with different fluorochromes, most usually red and green (see *Figure 6.11*). The labelled single-stranded DNA samples are mixed and applied to the normal single-stranded DNA immobilized on the microarray slide, and the fluorescence detected using laser excitation.

Where there is no change in sequence copy number there will be equal binding of test and reference sample DNA, equal amounts of green and red fluorescence, and a net emission of yellow light. For sequences where there has been amplification in the test DNA there will be more green than red fluorescence and an overall green emission. Conversely, deletions will result in a reduced level of green fluorescence relative to red from the reference sample, and a net red emission.

Fluorescence ratio imaging uses automated digital analysis of the images produced on laser excitation of the hybridized arrays to generate fluorescence ratio profiles; these are outputs with peaks and troughs representing areas of DNA amplification and deletion respectively. Their relative levels of fluorescence can also be used to determine the degree of amplification (see *Figure 6.12*).

Although cell culture is not required, array CGH does require good quality DNA. The technology is increasingly amenable to automation and array platforms are becoming cheaper and easier to use. However, there are limitations in the use of array CGH, most notably the inability to detect certain forms of chromosomal abnormality, including balanced rearrangements and low level mosaicism.

The resolution of array CGH depends on the number of probe sequences used, how long they are, and how widely spaced throughout the genome. Initially microarrays were made using large insert genomic clones (BACs – bacterial artificial chromosomes). These give a typical resolution of 1 Mb, which is 3–5 times greater than a karyotype. Other sources of clones include cDNA (single-stranded DNA produced from RNA templates by the process of reverse transcription) or oligonucleotide (a short fragment of a single-stranded DNA that is typically 5–50 nucleotides long) probe sequences.

Higher resolution can be achieved using **tiling path genomic arrays**. These tiling path arrays span chromosomes with overlapping reporter sequences, making coverage of the genome even more extensive. The downside is that these require more complex data analysis to produce results. This problem has been overcome to some extent by using probes that avoid sequences known to hybridize to multiple genome locations. Also, arrays can be prepared that comprise a series of reporter probes, each of which is specific to a particular chromosomal region or known genetic abnormality. Coverage is typically smaller than whole genome CGH arrays, but resolution is maximized and they are a useful tool for confirming the presence or absence of selected genetic factors such as known genes or syndromes.

Analysis and interpretation of array CGH results

One of the first questions to be answered on detecting a change in copy number in one or more areas of the array is whether or not the abnormality is a **polymorphic variant** of no significance, or a genuine **pathogenic** finding that causes the phenotype. We have already seen that cytogenetically visible copy number variations (CNVs), the **euchromatic variants**, can occur (see *Chapter 5*).

The human genome is structurally dynamic with thousands of heritable CNVs recently discovered among genomes of clinically normal individuals. It is now known that up to 12% of the genome can vary in copy number without having an effect on the human phenotype. DNA copy number polymorphism represents a major source of genome variation between individuals. The high frequency of CNVs in the human genome from an individual reveals a personal signature. A study published in *Nature* in 2008 (Kidd *et al.*, 2008) mapped the genome of eight individuals and revealed even larger regions of genetic variability than had previously been suspected. The

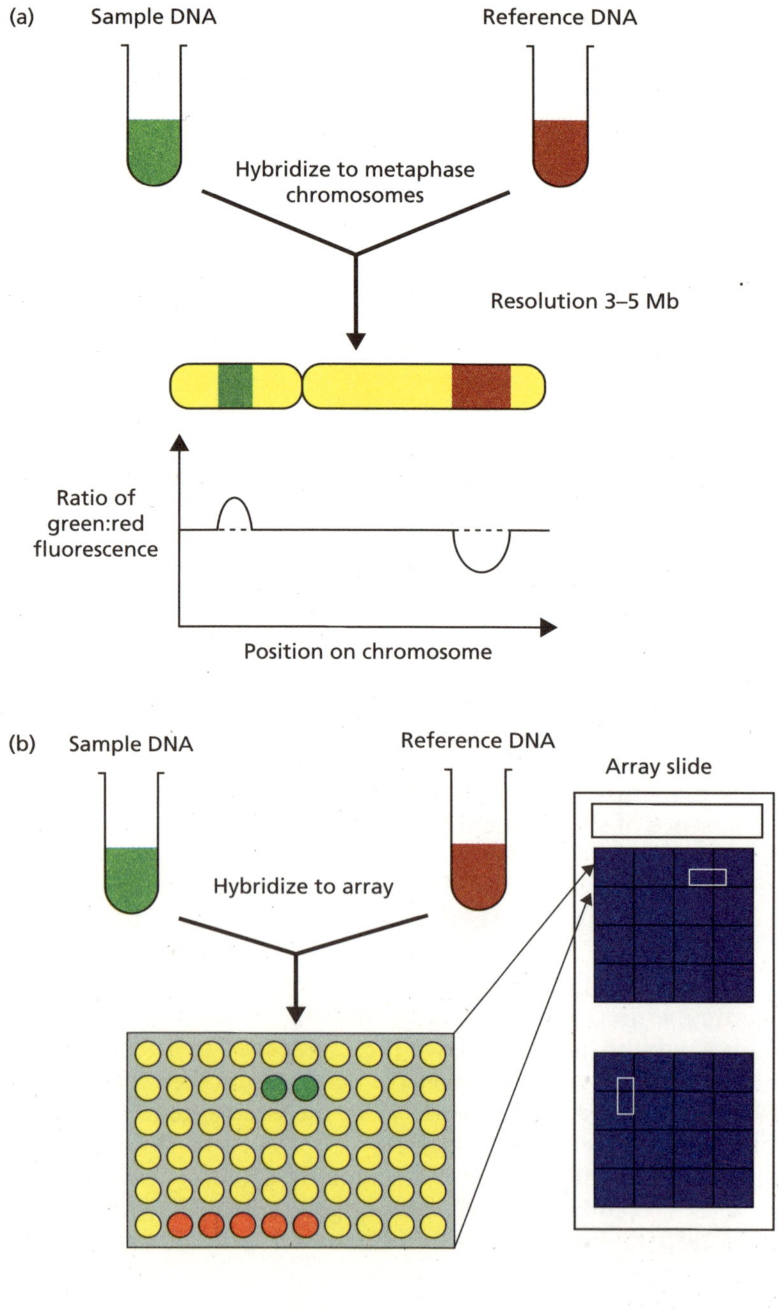

researchers identified nearly 1700 sites of structural variability in the genome, including deletion, insertion, inversion or multiple copies of large segments of DNA or genes. Fifty per cent of the DNA segments were previously unrecognized as areas of large variability in the genome. This also suggests that the human genome sequence identified in the Human Genome Project is still incomplete. Our understanding of benign and pathogenic genomic variation is clearly still in its infancy.

The higher the resolution of the arrays and the greater the coverage of the array probes, then the more likely it is that these changes will be detected. Thus there is significant potential for the detection of novel CNVs that can be difficult to interpret. It is not uncommon for 4–6 'calls' or deviations from the 1:1 ratio to be found in the testing of any one patient.

Any imbalance identified must be fully validated. It may be useful to review the cytogenetic karyotype, if this is available, as it is possible the finding is actually visible but had been missed or was close to the limit of cytogenetic resolution. It is important to investigate the parents in order to determine if this is a *de novo* imbalance or inherited. Of course, both parents are not always available for study, and there is the potential to reveal non-paternity (see *Chapter 10*), so such information needs to be handled sensitively.

Even if a variant is present in an affected individual, but absent from the 'normal' parents, it does not necessarily follow that it is a pathogenic change and it may still represent a polymorphism without effect that has arisen *de novo*.

Figure 6.11

(a) Chromosomal CGH. Patient DNA is labelled in green and reference DNA in red. The mix is simultaneously hybridized to normal metaphase spreads. Fluorescence is detected and the changes in intensities of the two fluorochromes are measured along the chromosomes – yellow means equal amounts of red and green; green means gain of patient DNA; and red means there is loss of patient DNA. However, as chromosomes are used as the template, the resolution achieved remains low at 3–5 Mb, and the procedure is labour-intensive and time-consuming.

(b) Array CGH. This technique is based on the same principles as chromosomal CGH, except that the targets are mapped genomic targets immobilized on a microarray instead of metaphase chromosomes. Each microarray is made up of many bits of single-stranded DNA fragments arranged in a grid pattern on the glass or plastic surface. When sample DNA is applied to the array, any sequences in the sample that find a match will bind to a specific spot on the array. The two genomes, a test (patient) and a reference (control), are differentially labelled and co-hybridized to a selected set of pre-spotted genomic fragments. There are multiple copies of each clone on each array to improve robustness of results. Measurement of spot intensities is carried out on a laser scanner. As with chromosomal CGH – green indicates increased patient DNA and red indicates less patient DNA compared to reference, i.e. deletion. Array CGH can detect multiple DNA changes across multiple loci simultaneously (provided they are represented on the array) – it is equivalent to one enormous FISH test.

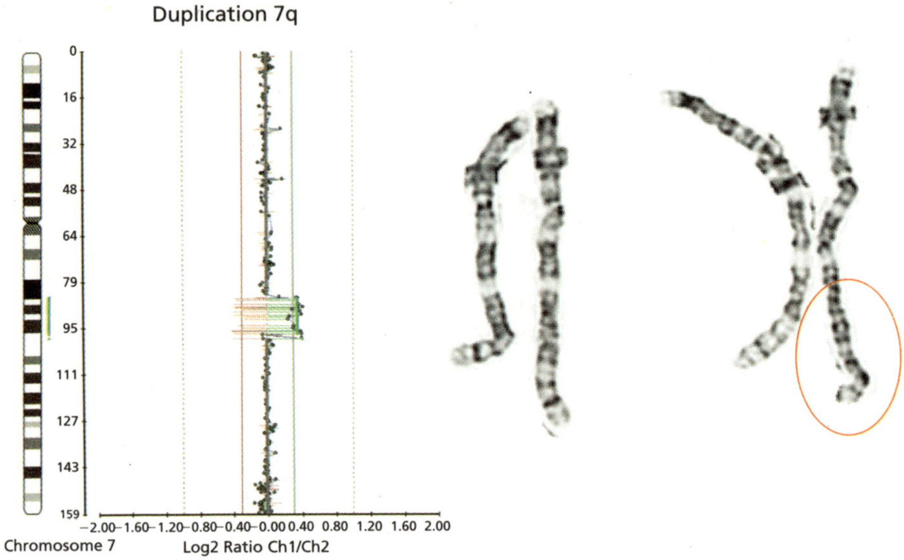

Figure 6.12
Duplication of chromosome 7q identified by BAC array CGH. The region that is duplicated has a log2 ratio of 0.4. Each spot is produced from a single BAC. The duplication spans 20 BAC clones and is estimated to be between 14.3 and 15.1 Mb in size.

The array finding can be confirmed using a variety of techniques including FISH, by using higher resolution arrays, or by MLPA. FISH has the advantage of being a well-proven method which is widely available. It does, however, require the availability of suitable material in the form of metaphase chromosome preparations and the availability of the required probes. The probes can be produced in-house from sequence information available from the Human Genome Project or, increasingly, can be purchased from commercial suppliers. MLPA is a rapid, cheap and relatively simple method that will detect deletions smaller than those detectable by FISH, and it will also detect more duplications than can be detected by FISH. It can be difficult with FISH to see differences in size of probe signals, but with MLPA many regions can be examined simultaneously (e.g. as in sub-telomeres described earlier in this chapter). MLPA does require the availability of good quality DNA. A higher resolution array can be used and this will confirm and extend the original array finding, help to size the region, define breakpoints more accurately, and add information about the exact genes involved.

Bioinformatics

As our knowledge increases, more 'targeted' arrays are being developed to overcome these problems. Detection down to the level of 100 kb is now commonplace. The higher the resolution and the more probes used the greater the problems with noise and the identification of CNVs, and the greater the amount of data produced.

Subtle chromosome changes can occur anywhere in the genome but particular changes are individually very rare. Bringing information on these rare changes together so that it can be shared by clinicians and scientists is helping to accelerate progress towards understanding such rare conditions and their related gene functions. Such collaborative databases are increasing medical and scientific knowledge about chromosomal microdeletions and duplications which in turn is leading to improved care and advice for affected families, and facilitating research into genes that affect human development and health.

Bioinformatics (see also *Box 6.6*) is the branch of life sciences that deals with the study and application of information technology to the vast amounts of data produced using molecular biology. A number of international databases have been established to share information obtained from array CGH. One such database that is widely used is called DECIPHER.

DECIPHER (DatabasE of Chromosomal Imbalance and Phenotype in Humans using Ensembl Resources) is a web-based resource for recording clinical information about chromosomal microdeletions/duplications, insertions, translocations and inversions linked to phenotypic descriptions and genome mapping. DECIPHER shows immediately whether a similar chromosomal imbalance has been reported previously and also the phenotypes of the individuals concerned. It lists all of the known and predicted genes that are implicated in altered regions of established high clinical significance, and so, for example, tumour suppressor genes can be identified immediately. It has a gene prioritization tool where genes within the affected area are identified and prioritized according to their relevance to the phenotype(s), using advanced text mining tools to search the scientific literature such as PubMed. DECIPHER also facilitates contact between clinicians in contributing centres, thereby accelerating the recognition and publication of novel syndromes.

Box 6.6 Bioinformatics

Over the past few decades rapid developments in genomic and other molecular research technologies, and development of information technologies, have combined to produce a tremendous amount of information related to molecular biology. Bioinformatics is the application of information technology to data produced in the field of molecular biology. The term bioinformatics was coined by Paulien Hogeweg in 1978 for the study of informatic processes in biotic systems. Nowadays it entails the creation and advancement of databases, algorithms, computational and statistical techniques and theory to solve formal and practical problems arising from the management and analysis of biological data. Common activities in bioinformatics include mapping and analysing DNA and protein sequences, aligning different DNA and protein sequences to compare them, and creating and viewing 3-D models of protein structures. Assembly of the human genome sequence is one of the greatest achievements of bioinformatics.

Another database is called ECARUCA (European Cytogeneticists Association Register of Unbalanced Chromosome Aberrations), and this is a European database that covers both the common and rare chromosome aberrations. It has over 4000 entries and brings together cytogenetic, molecular and clinical data.

The use of array CGH in clinical cytogenetics

Array-based whole genome investigation or 'molecular karyotyping' has a number of potential applications in human clinical genetics, but careful consideration needs to be given as to which disorders the test would be most applicable to. In addition, molecular karyotyping can be performed in different ways using different array platforms and software, so the outcome of an experiment on patient DNA may be very different in different laboratories; defined quality standards and guidelines are required and these are gradually being produced.

The challenge for diagnostic labs has been the selection of the most suitable platform for this analysis. Factors to consider include the resolution, ease of interpreting the results, sensitivity, specificity and cost. Array CGH can be used as an adjunct to cytogenetic investigations in situations where the karyotype is normal and second-stage testing by FISH or MLPA in clinically selected cases is also normal. As it has the ability to diagnose hitherto unrecognized syndromes, array CGH is particularly useful for individuals and families suspected of having chromosome disorders but which remain undiagnosed.

Enhanced resolution has led to the identification of many new microdeletion and duplication syndromes. Detection rates for chromosome abnormalities with array CGH in constitutional cytogenetics ranges from 5 to 20% in individuals with normal karyotypes (as determined by prior routine cytogenetic testing).

The use of arrays in the investigation of children with learning difficulties

Several studies have demonstrated the potential value of genomic arrays in the investigation of children with unexplained learning difficulties. Learning disability (see *Box 6.7*) is a serious and lifelong condition characterized by the impairment of cognitive and adaptive skills. It is one of the few clinically

Box 6.7 Coding of learning disability using IQ score

- Mild: 50–55 to approximately 70
- Moderate: 35–50 to 50–55
- Severe: 20–25 to 35–40
- Profound: <20–25

important disorders with an aetiology that is poorly understood. It has a prevalence of about 3%, i.e. 3 in every 100 living individuals have a learning disability. The occurrence of learning disability is influenced by genetics, environmental infections and perinatal factors, and a definitive cause cannot be identified for up to half of all cases.

The proportion of cases for which a definitive cause is identified varies according to the severity of the learning disability (see *Box 6.8*). Approximately 30% of causes remain unidentified in children with severe learning disability and 70% in children with mild to moderate learning disability. Genetic factors have been estimated to be the main cause of learning disability in around 50% of all patients with severe learning disability and around 15% in patients with mild to moderate learning disability. Chromosome abnormalities account for 4–28% of learning disability and may involve whole chromosomes such as in Down syndrome (trisomy 21), parts of chromosomes such as in cri-du-chat syndrome (deletion 5p), or submicroscopic deletions only detectable by FISH such as in Angelman syndrome (deletion 15q). Learning disability can also be caused by defects in specific genes such as in Fragile X or Rett syndrome (see *Chapter 3*). Other genetic causes can be uniparental disomy and mosaicism.

Box 6.8 Causes of learning disability identified in the literature

Cause	% of learning disability
Chromosomal abnormalities	4–28
Recognizable syndromes	3–7
Known monogenic conditions	3–9
Structural central nervous system abnormalities	7–17
Complications of prematurity	2–10
Environmental or teratogenic causes	5–13
Familial multifactor LD	3–13
Unique monogenic syndromes	3–12
Metabolic/endocrine causes	1–5
Unexplained	30–50

Data adapted from Curry *et al.* (1997).

Identification of a cause for the learning disability brings many advantages to the patient and their families including:

- having a name for the diagnosis
- allowing better counselling (see *Chapter 10*)
- being able to find out more about the condition
- allowing parents to find others affected by the condition
- helping to understand the future needs of the child
- helping to plan for future children by making informed reproductive choices
- helping to find child services and support

Microarray CGH has increased the detection of genetic imbalances in children with learning disability by an average of twofold. This technique is now helping many more families both in diagnosing the cause of learning disability in a child who previously had none and also offering the potential for prenatal testing where the genetic imbalance is inherited. The detection rate of imbalances appears to increase slightly by using higher resolution arrays. Overall, approximately 20% of all detected imbalances are smaller than 1 Mb.

Investigation of apparently balanced rearrangements

The finding of an apparently balanced rearrangement in an individual with an abnormal phenotype raises the question 'is the abnormality the cause of the problems?'. Possible explanations for balanced rearrangements being causative include disruption of a gene(s) or a position effect altering gene expression (see *Chapter 5*). It is also possible that what appears to be a balanced rearrangement may hide unsuspected degrees of additional complexity that could explain the phenotype.

Use of array CGH has demonstrated that up to 40% of apparently balanced *de novo* rearrangements are in fact unbalanced.

The use of array CGH in prenatal diagnosis

It could be tempting to introduce array CGH into a prenatal setting and so avoid the birth of many children with genetic abnormalities. Some would argue that as a pregnancy has been put at risk through an invasive procedure it is unethical not to undertake a detailed genome-wide screen.

Despite the ability of array CGH to detect disorders that often have a more severe post-natal presentation than Down syndrome, its use for prenatal diagnosis remains controversial. Array CGH technology is still evolving and the current knowledge of the extent of genomic CNV and phenotypes associated with specific CNVs is also still incomplete. Moral objections against its use have been voiced, primarily because a detected CNV, even of uncertain significance, may result in a decision to terminate a pregnancy. Concerns that array CGH will produce overwhelming information, which may mislead and/or cause unnecessary anxiety in parents have also been raised.

Certainly more extensive pre- and post-test counselling will be needed to explain the test and its limitations and the potential difficulties in interpretation. Detailed counselling for each disorder represented would be too lengthy to be practicable and may produce an overload of information that may be confusing to prospective parents. In addition to the detection of imbalances in regions resulting in well-established disorders, there will also be the detection of imbalances in regions for which we do not know the phenotypic consequences, or where very little information is available due to the extreme rarity or limited family data and genotype–phenotype correlations.

Some of the concerns about using array CGH in prenatal testing could be addressed using targeted arrays designed to:

- interrogate only specific regions known to be associated with mental retardation and malformation syndromes
- exclude loci which are associated with adult onset conditions
- minimize coverage of regions of unknown clinical significance

Most prenatal testing is currently carried out in the context of increased risk of Down syndrome identified from screening (see *Chapter 9*), and rapid testing is already in place for this using FISH or QF–PCR. The use of arrays targeted to detect genomic imbalances associated with aneuploidy and known microdeletion/duplication syndromes may be of benefit. In addition, it could be used to investigate further apparent *de novo* balanced rearrangements which carry a 5–10% risk of congenital anomaly (some of which will be related to a submicroscopic imbalance), or to elucidate the content of extra structurally abnormal chromosomes (see *Chapter 5*).

It has been suggested that array CGH could be limited to use in pregnancies with anomalies identified on ultrasound scan. This also has the potential to greatly accelerate our insights into the genetic aetiology of abnormal fetal development. However, it is important to bear in mind that many known microdeletion and microduplication syndromes cannot be detected by ultrasound. In addition, the emerging data on the role of CNVs in learning difficulties, as described above, suggest that reserving array CGH for pregnancies with an abnormal prenatal ultrasound may not be the optimal strategy.

The technology has the potential to decrease the time taken for a result to be obtained as no cell culturing is required. With reduced costs of arrays and greater use of automation it is predicted that array CGH on fetal DNA obtained through amniocentesis or CVS (or in the future non-invasive collection from maternal blood) will replace conventional prenatal karyotyping and transform the practice of prenatal diagnosis. It may also be possible in the future to extend offering array CGH at the stage of pre-implantation genetic diagnosis, allowing only normal embryos to be returned for implantation and avoiding having to consider termination of abnormal embryos (see *Chapter 9*).

Infertility and recurrent abortions

Chromosomal translocations are a significant cause of infertility and recurrent miscarriages are often caused by unbalanced transmission of a chromosomal translocation from one parent (see *Chapter 5*). As balanced rearrangements cannot be detected by array CGH, conventional cytogenetic analysis will remain the key diagnostic test for infertility and recurrent abortions caused by chromosomal translocation. However, it could be used to investigate the actual fetal tissue from abortions and potentially highlight the need to investigate the parents.

Cancer

Analysis of chromosomal aberrations is particularly important in cancer, where amplification of oncogenes or deletions of tumour suppressor genes are involved in the multistep process of cancer development. Rapid and accurate identification of such genetic imbalances could improve diagnostics which might lead to better decision-making regarding the choice of available treatments.

Array CGH allows examination of tumour genomes in unprecedented detail and identifies novel genetic alterations not seen before. In recent years there have been numerous reports of high resolution array CGH studies on CNVs in tumour genomes. Identification of common gains and losses has led to the discovery of genes involved in tumourigenesis, has helped in disease classification, and in drug response.

Genetic alterations such as duplications and deletions frequently contribute to tumourigenesis (see *Chapter 8*). These alterations change the level of gene expression which can modify normal growth control and survival pathways. Characterization of these DNA copy number changes is important for both the basic understanding of cancer and its diagnosis. Identifying and understanding the genes involved in cancer will help the design of therapeutic drugs that target the dysfunctional genes and/or avoid therapies that cause tumour resistance.

6.5 SUMMARY

The development of FISH techniques has permitted identification of small deletions not previously visible, identification of unknown markers and rearrangements, and even provided information from uncultured (interphase) cells at a resolution not previously possible.

There is no doubt that the number of probes and the number of colours will continue to increase, enabling the laboratory to provide more information to the clinicians and ultimately improve both prenatal and post-natal diagnosis, and hence patient care.

The development of more sensitive molecular techniques such as MLPA and PCR and more recently array CGH is increasing the number of genetic changes we can identify whilst increasing speed and reducing cost. However, conventional cytogenetic analysis remains important for the foreseeable future and is the only method suitable for the detection of balanced rearrangements and changes in ploidy.

SUGGESTED FURTHER READING

Curry, C.J., Stevenson, R.E., Aughton, D., *et al.* (1997) Evaluation of mental retardation: recommendations of a Consensus Conference: American College of Medical Genetics. *Am. J. Med. Genet.* **72**: 468–477.

Fan, Y.-S. (2003) *Methods in Molecular Biology, Vol. 204: Molecular Cytogenetics Protocols and Applications*. Totowa: Humana Press.

Kidd, J.M., Cooper, G.M., Donahue, W.F., *et al.* (2008) Mapping and sequencing of structural variation from eight human genomes. *Nature,* **434**: 55–64.

Lichter, P. and Cremer, T. (1992) Chromosome analysis by non-isotopic *in situ* hybridization. In: *Human Cytogenetics A Practical Approach Volume 1,* Rooney, D.E. and Czepulkowski, B.H. (eds). Oxford: IRL Press.

Lockwood, W.W., Chari, R., Chi, B. and Lam, W.L. (2006) Recent advances in array comparative genomic hybridization technologies and their applications in human genetics. *Eur. J. Hum. Genetics,* **14:** 139–148.

Speicher, M.R. and Carter, N.P. (2005) The new cytogenetics: blurring the boundaries with molecular biology. *Nature Rev. Genet.* **6:** 147–151.

Vermeesch, J.R., Fiegler, H., de Leeuw, N., *et al.* (2007) Guidelines for molecular karyotyping in constitutional genetic diagnosis. *Eur. J. Hum. Genetics,* **15:** 1105–1114.

Ward, D.C., Boyle, A. and Haaf, T. (1995) Fluorescence *in situ* hybridization techniques. In: *Human Chromosomes: Principles and Techniques,* Verma, R.S. and Babu, A. (eds). New York: McGraw-Hill.

DECIPHER – www.sanger.ac.uk/PostGenomics/decipher
ECARUCA – www.ecaruca.net

SELF-ASSESSMENT QUESTIONS

1. What is a probe?
2. Name three uses of FISH in clinical cytogenetics.
3. Name three clinical syndromes detected by FISH, their chromosome location, and two clinical features of each (use the *Appendix: Glossary of disorders*).
4. After performing FISH, your slide has a lot of background interference. Define stringency, and explain what the result would be of increasing the stringency.
5. What are the limits of resolution of FISH?
6. Explain the role of filters in fluorescence microscopy. What is the difference between a dual pass and a triple pass filter?
7. Describe the steps you might take to determine whether or not a sample contained a copy polymorphism?

CHAPTER 7

Molecular genetics

Learning objectives

After studying this chapter you should confidently be able to:

- **Define molecular genetics**
  Molecular genetics uses technology which works at the DNA level and therefore makes it possible to detect point mutations and other small changes undetectable by conventional cytogenetics and FISH.
- **Define the term probe and give examples of nomenclature and use in clinical work**
  Probes used in routine clinical work are usually pieces of double-stranded DNA, either complementary or linked to a region of interest. Probes are inserted into vectors for cloning.
- **Define polymorphisms**
  Changes in DNA sequence may either lead to a clinical defect or a harmless non-clinical change termed a polymorphism. These can be revealed as different lengths of DNA using restriction enzymes or sequencing.
- **Describe the principles of Southern blotting and give examples of its use**
  Southern blots are used to detect large pieces of DNA, or when the gene locus is not known. An appropriately labelled probe is hybridized to the patient's denatured DNA and the result detected as a band pattern.
- **Explain gene tracking and 'informativeness'**
  This is an indirect method of predicting disease or carrier outcome where the gene locus is not known. A closely linked probe detects polymorphisms, one of which is inherited (and therefore tracks) along with the disease status in a family.
- **Describe the principles of PCR and give examples of its use**
  The polymerase chain reaction is a method of amplifying up to a million copies of DNA from very little starting material. It is quicker and safer than Southern blotting, although the conserved sequences flanking the exons of a gene usually have to be known, and there is a contamination risk. PCR is used in forensic science, archaeology, or studies of evolution, as well as for clinical diagnosis.

- **Outline some methods of mutation screening, using examples of disorders**
  As some diseases are caused by different types of mutations, a pre-screen may be necessary, followed by automated sequencing in order to determine the exact nature and position of the specific mutation.
- **Outline the nature and validation of a pathological variant**
  If an altered DNA sequence is found, it is important to distinguish between a true pathological mutation and a polymorphism.

7.1 INTRODUCTION TO MOLECULAR GENETICS

Genetics has always been regarded as a comparatively modern science; DNA structure was elucidated in 1953, and cytogenetics grew from the discovery of the correct number of human chromosomes by Tjio and Levan in 1956.

In the 1970s a whole new technology called molecular biology became available, which enabled the analysis of human DNA down to the single base pair level. By the 1980s this led to a new approach to screening for previously uncharacterized clinical disorders. In the 1990s the techniques became automated such that the complete DNA sequence of the 46 human chromosomes was elucidated, a huge endeavour entitled the 'Human Genome Project'. In the 2000s, automation reached the routine diagnostic laboratory; robots and DNA analysers are now used to manipulate and sequence DNA.

Molecular biology has now been incorporated into the study of inherited disorders and is known as **molecular genetics**. Molecular genetic analysis is carried out for two principal reasons:

- molecular techniques may provide a more specific or more accurate result; some molecular genetic tests provide a less expensive alternative to cytogenetic analysis, e.g. for Fragile X syndrome and Prader–Willi/Angelman syndrome.
- because molecular genetics examines the DNA at the base pair level, certain syndromes which cannot be detected cytogenetically can be analysed for mutations or deletions using molecular techniques (e.g. cystic fibrosis and Duchenne muscular dystrophy), which has led to an understanding of how pathogenic mutations relate to genetic disease.

The two techniques upon which most molecular analysis is based are **Southern blotting** and the **polymerase chain reaction (PCR)**.

7.2 SOUTHERN BLOTTING

This is the older of the two techniques which, although superseded by PCR in the late 1980s, still plays a valuable role in the molecular analysis of particular clinical disorders. Southern blotting detects specific sequences of DNA amongst the entire human genome using unique DNA probes. It is so sensitive that it can detect less than 0.1 pg of DNA, i.e. a single copy of a gene.

The following four elements provide the key to understanding Southern blotting:

- probes
- restriction enzymes
- gel electrophoresis
- polymorphisms

Southern blots are named after their inventor, Professor E. Southern working in Edinburgh. One of the most useful techniques in molecular biology was achieved using plastic trays and paper towels!

DNA probes

As already briefly described in *Chapter 6*, these double-stranded pieces of DNA are complementary to the gene or region of interest (usually the site of a genetic disorder). The probes used in Southern blotting are generally 0.3–5 kb (see *Box 7.1*).

Box 7.1 Probes in the diagnostic laboratory

A clinical laboratory rarely receives a probe as an independent solution of DNA fragments. The probe (or **insert**) is usually inserted into a double-stranded circular piece of DNA called a **plasmid** by enzyme cutting and pasting. As it carries the probe the plasmid is termed a **vector**.

The vector (which generally carries a gene for antibiotic resistance) and insert are exposed to bacteria which are **competent** to take up the plasmid. The bacteria are grown in broth containing an antibiotic to which the plasmid is resistant. The plasmids give the bacterium a selective advantage, and enable the plasmids and their inserts to divide; this process enables large quantities of probe to be generated, and is called **cloning**.

Nomenclature

Probe names. Names may include the laboratory of origin or the probe size. Ox1.9 is a fragile X probe; it originated in Oxford and has a length of 1.9 kb. The name may indicate the number of attempts at production, the initials of their inventor, or have a prefix, e.g. 'p' for plasmid.

Locus name. The area on the human chromosome to which the probe is complementary is called the **locus**. The locus may therefore have a separate name which (unfortunately for the student), is often different from the probe name. Locus names often take the form given in the example below.

D7S8: D=**D**NA
7=chromosome **7**
S=unique **s**egment
8=unique segment number **8**

For example, the old cystic fibrosis probe pJ311 is complementary to the locus D7S8.

Molecular geneticists generally use the probe name, as several different probes may be synthesized for the same locus (e.g. LS6-1 and LS6-2 for D15S113). However, a large commercial FISH microdeletion probe such as TUPLE1 may cover a range of loci, so nomenclature may not be consistent. Older probes, or those hybridizing to an important critical region may have the same locus name as probe name (see *Table 7.1*). Probes may also be described generically from their position with regard to the gene of interest (*Figure 7.1*).

Table 7.1 Examples of probe nomenclature

Locus name	Probe name	Genetic disorder where used
D22S609	TUPLE1	22q11.2 deletion syndrome
D22S942	TUPLE1	22q11.2 deletion syndrome
SNRPN	KB17	Prader–Willi and Angelman syndromes
MET	METD	Cystic fibrosis
D15S113	LS6-1	Prader–Willi and Angelman syndromes
D15S113	LS6-2	Prader–Willi and Angelman syndromes

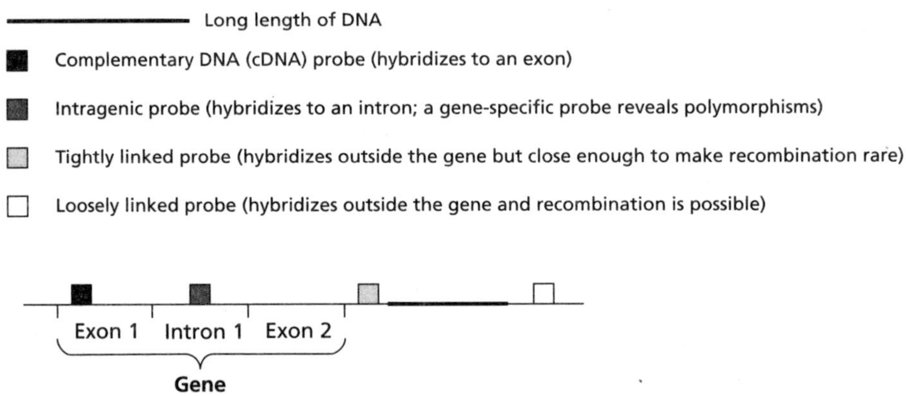

Figure 7.1
Probe nomenclature and positions with respect to a gene.

Restriction enzymes

In the early 1970s it was discovered that certain bacteria produce enzymes which cleave foreign DNA. These are known as **restriction endonucleases** (see *Box 7.2* for a definition of this and other molecular genetic terms); the enzyme names reflect the bacterium of origin:

*Taq*1: *Thermus* ***aq****uaticus*
*Eco*R1: ***E. co****li* type **R1**

Box 7.2 Definitions of some molecular genetics terms

Allele (general genetics) – the alternative form(s) of a gene at the same locus on homologous chromosomes.

Allele (molecular genetics) – the alternative length(s) of DNA (i.e. RFLPs) generated due to the presence or absence of a restriction site at the same locus on homologous chromosomes. Alleles may also be defined as alternative lengths of DNA (at the same locus on homologous chromosomes) detected by a probe or a pair of PCR primers.

DNA probe – a short length of DNA, usually radiolabelled and made single-stranded, which hybridizes with complementary DNA sequences linked to, or containing, the gene of interest.

Haplotype – a combination of linked alleles inherited together as a unit (from one parent on one chromosome).

Linkage disequilibrium – the association of two linked alleles more frequently than would be expected by chance.

Polymorphisms – in molecular genetics: a naturally occurring variation in a DNA sequence, usually in an intron.

Restriction enzymes – naturally occurring endonucleases from bacteria, which cleave double-stranded DNA at specific base sequences; they usually recognize a 4–6 base sequence. For example:

*Eco*R1	——G*AATTC——		*Taq*1	——T*CGA——
	——CTTAA*G——	← staggered		——AGC*T——
*Hae*III	——GG*CC——	← blunt ended	*restriction site	
	——CC*GG——			

Restriction fragments – the different lengths of DNA produced by a restriction enzyme.

RFLP (restriction fragment length polymorphism) – a length variation (i.e. polymorphism) in a DNA fragment, revealed by a restriction enzyme, and due to the presence or absence of a restriction site.

*Hind*III:	***Haemophilus influenzae*** type **dIII**
*Msp*1:	***Moxarella sp.***

Restriction enzymes cut double-stranded DNA at very specific base sequences (see *Box 7.3*) called **restriction sites** (see *Figure 7.2*), which occur many times in the genome. The resultant **digest** contains a mixture of differently sized **fragments** which can be separated by gel electrophoresis.

Box 7.3 Sequence differences revealed by restriction enzymes

The pair of homologous chromosomes shown in *Figure 7.2* display a sequence difference in an intron. On the first chromosome a 3 kb length of DNA is produced between two restriction sites. On the second chromosome a mutation has occurred such that an extra restriction site has been created. The enzyme now produces two differently sized pieces of DNA – one of 1 kb, and one of 2 kb. These will run to different levels on the gel.

The probe detects a 3 kb fragment from one chromosome, and a 2 kb fragment on the other homologue (the probe has not hybridized to the 1 kb fragment, so although the band is present on the gel it will not be visualized).

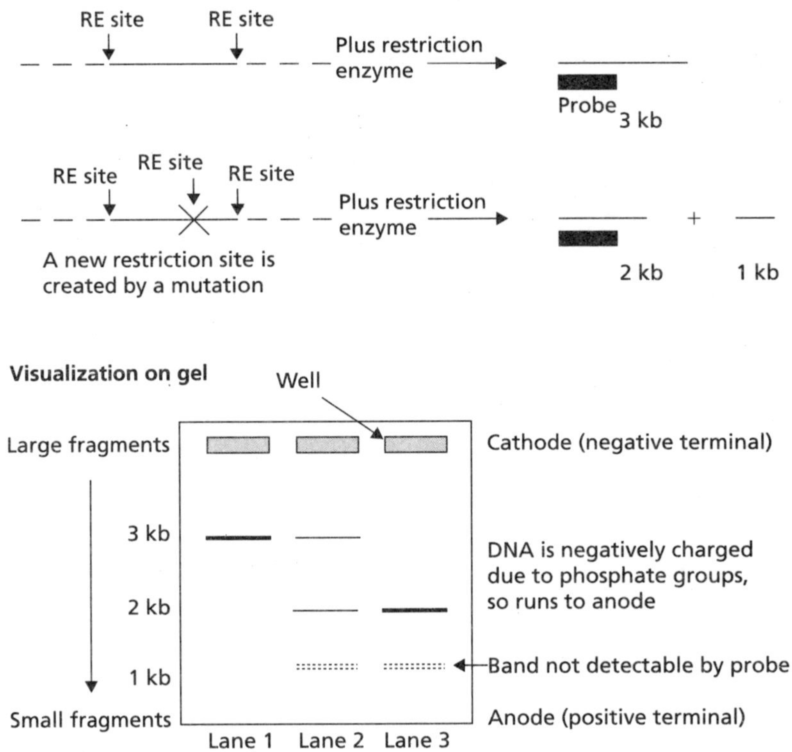

Figure 7.2
Exploitation of RFLPs using gel electrophoresis.
Lane 1 is homozygous for the 3 kb polymorphism, as there is no mutation (i.e. has 2 alleles 3 kb in length, running the same distance).
Lane 2 is heterozygous for the polymorphism, as one chromosome has no mutation and the other has a mutation creating a restriction site (i.e. has 1 allele of 3 kb and 1 allele of 2 kb).
Lane 3 is homozygous for the 2 kb allele as both chromosomes each have the mutation creating the restriction site (i.e. has 2 alleles 2 kb in length, running the same distance).

General principles of gel electrophoresis

During gel electrophoresis, the patient's restricted DNA fragments are subjected to an electrical current. As **DNA is negatively charged** due to the phosphate groups, it will **migrate to the positive anode** of the gel tank (*Figure 7.2*). Small DNA fragments run faster and therefore further than the larger fragments in a set time, and appear as bands nearer the lower end (anode) of the gel.

Gels comprise a matrix of differently sized pores. They are either made from **agarose** powder which is dissolved by boiling in buffer and then sets 'mechanically' like jelly, or an **acrylamide** monomer is used which polymerizes by chemical means, using a catalyst. This is known as a **polyacrylamide** gel.

Gels are made and run in **buffer** which contain **ions** for **electrical conduction.** They are either set in plastic trays (to run horizontally) or between glass plates (to run vertically). The DNA is loaded into **wells** which are created at one end of the gel using plastic combs with teeth of suitable sizes. The comb is removed and the tray and running buffer is placed in a gel tank.

Different percentages of gels can be made depending on the size of the DNA fragments to be separated. A 0.8% agarose gel will separate fragments around 1–30 kb, a 2% agarose gel around 100–500 bp, while a 13% polyacrylamide gel will separate fragments as little as 3 bp apart.

Polymorphisms

Approximately every 100–200 bp in the introns of human DNA, a natural variation in DNA sequence may occur (see *Box 7.3*). These variations may be within the introns of genes, or within the intervening sequences between genes. If variations occur in the conserved exons, the consequences may be serious, resulting in a pathogenic mutation. Over time, however, introns tend to accumulate sequence differences which may vary between individuals, or even between the two chromosome homologues in one individual. These sequence differences are called **polymorphisms**, which can be detected by Southern blotting using a probe complementary to the restriction site(s) of interest, or by PCR.

A single nucleotide polymorphism (SNP) is any polymorphic variation at a single nucleotide. Although less informative than short tandem repeats, SNPs are very abundant and therefore useful as genetic markers.

Restriction fragments of different sizes derived from introns are called **length polymorphisms**. The correct term for these digested fragments is therefore **restriction fragment length polymorphisms (RFLPs)**. RFLPs are used in particular circumstances, often when the exact location of a gene is not known. Linked probes are then used, and the polymorphisms can be associated with a gene disorder and tracked through a family (see section on *Gene tracking* later on in this section).

Probes which are some distance away from the gene have disadvantages, as there is a chance of **recombination** between the locus at which the probe hybridizes, and the gene of interest (see *Figure 7.1*).

Principles of Southern blotting

The patient's double-stranded DNA is denatured by chemical treatment to make it single-stranded. The double-stranded probe DNA is heat-denatured and labelled. The label may be non-radioactive (e.g. digoxygenin, detected using chemiluminescence) or radioactive (the isotope ^{32}P is generally used in Southern blotting).

When the denatured probe is added to the single-stranded human DNA, it will find its complementary sequence and hybridize, producing a length of labelled double-stranded DNA. If it is non-radioactive the result may be detected visually using dyes as with FISH, or chemiluminescent technologies

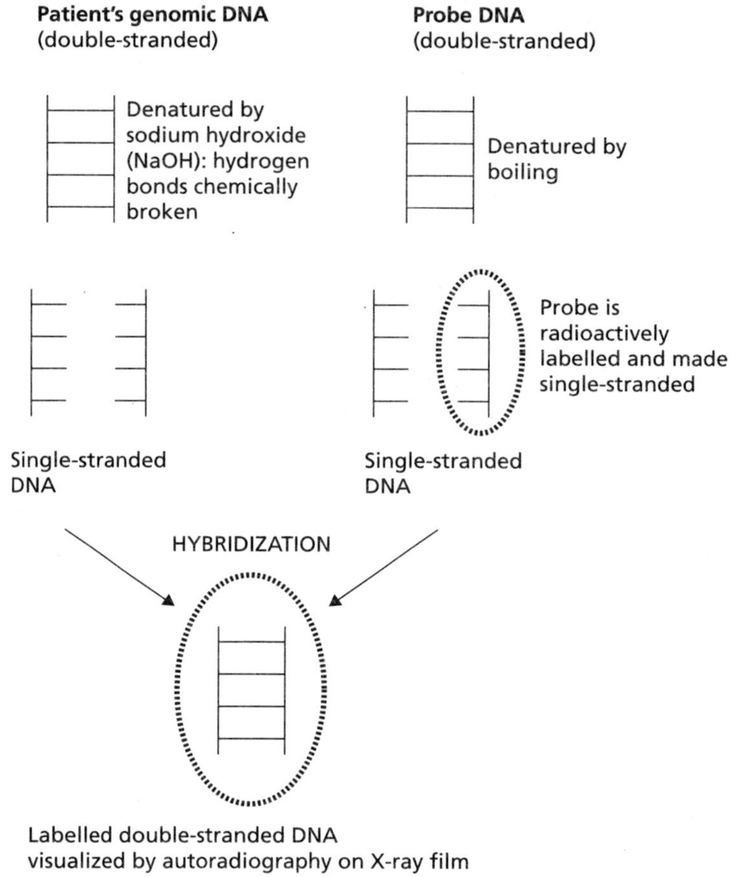

Figure 7.3
Principles of Southern blotting.

in Southern blotting. Radioactive methods use autoradiography; this relies on the radioactivity fogging a piece of X-ray film to produce black bands (*Figure 7.3*).

Steps involved in Southern blotting

The steps involved in Southern blotting are illustrated in *Figure 7.4* and are described below. Very small weights and volumes are used in DNA technology and the various terms are clarified in *Box 7.4*.

DNA extraction

DNA can theoretically be extracted from any **nucleated cells**. In medical genetics the usual source is a blood sample containing nucleated white cells. Other sources include chorionic villi (derived from the placenta, hence the tissue is fetal in origin), amniotic fluid cells, tumours and histopathological paraffin wax preparations. A 10 ml whole blood sample usually yields about

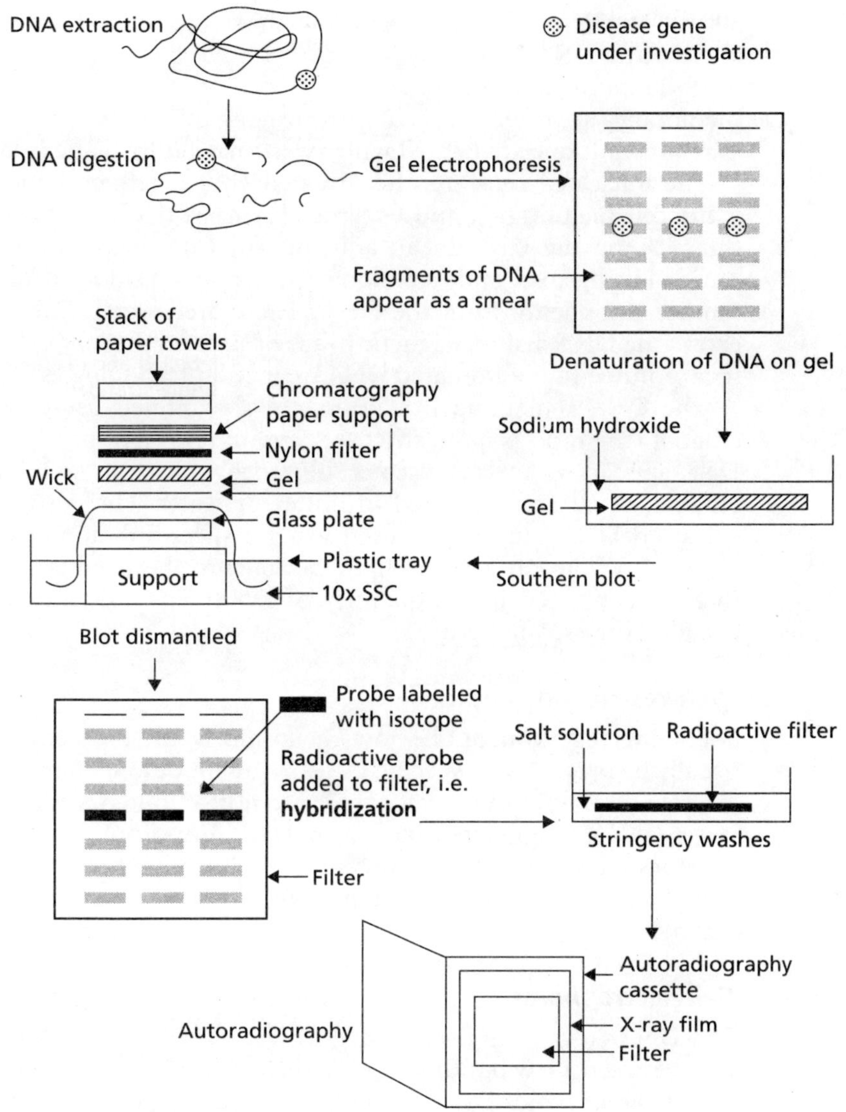

Figure 7.4
Steps involved in Southern blotting.

Box 7.4 Definition of weights and volumes

Very small weights and volumes are used in DNA technology. The units are as follows:

Weights:	grams (g)		**Volumes**:	litre (l)	
	milligrams	(1 mg = 10^{-3} g)		millilitre	(1 ml = 10^{-3} l)
	micrograms	(1 μg = 10^{-6} g)		microlitre	(1 μl = 10^{-6} l)
	nanograms	(1 ng = 10^{-9} g)			
	picograms	(1 pg = 10^{-12} g)			

500 μg of DNA, whereas 3–4 mg of chorionic villus sample (CVS) yields about 50 μg DNA.

If contaminants such as red blood cells are present, the red cell membranes are **lysed** (broken open) leaving the white cells to be spun down. Wax preparations are treated with xylene until only tissue is left.

The nucleated cells are often treated with detergent which ruptures the white cell membranes, and enzymes (proteases) which digest the proteins, thus leaving the DNA in an aqueous solution. This 'aqueous' solution is purified using an organic solution such as phenol; chloroform is then used to remove the phenol from the DNA. There are several different methods of extracting DNA, using magnetic beads or protein precipitation, which can be incorporated into automated DNA extraction technologies.

The DNA remaining in solution is precipitated using two volumes of ethanol (the ratio is quite strict – 2.5 volumes might precipitate RNA). The DNA appears as a white viscous substance which can be hooked out of the ethanol, dried and dissolved in buffer or water. The concentration of the extracted DNA can be measured using a spectrophotometer. Nucleic acids absorb UV light at a wavelength of 260 nm; the resulting absorbance can then be converted into a quantity of DNA. The concentration of DNA is usually expressed in μg/μl.

DNA restriction digests

Following digestion of the total genomic DNA, the gene of interest will usually be present on a specific DNA fragment of known size. Sometimes the enzyme does not cut at every available restriction enzyme site. This creates larger DNA fragments due to **partial digestion**. Although restriction enzymes are supposed to be very site-specific, under certain conditions such as high enzyme concentrations they will cut other sites. This is known as **star activity**.

Gel electrophoresis

The DNA fragments produced during a routine Southern blotting digest vary between a few hundred base pairs to 50 kb, depending on the restriction site of the enzyme. The fragments are separated by gel electrophoresis and when each patient's DNA migrates from the well it leaves a streak in that particular track or **lane**.

The lanes of DNA are visualized under UV light by staining the gel with the fluorescent dye **ethidium bromide**, which intercalates into the major grooves of the DNA helix.

Denaturation

The DNA on the gel must be denatured in order to hybridize with the specific probe, which is added later. Sometimes an initial step is performed where the gel is soaked in hydrochloric acid. This **depurination** step nicks the large molecular weight DNA, which will help it to transfer more easily at the Southern blotting stage. The gel is then immersed in a tray of sodium hydroxide and gently shaken; this will break the hydrogen bonds and

denature the patient's DNA. A final **neutralization** step may be performed using 'TRIS' buffer; this will bring the pH back to neutral.

Southern blot

As the denatured gel is too fragile for further manipulations, the DNA must be transferred onto a more durable membrane (sometimes called a filter) by the technique of Southern blotting. A salt solution is drawn by capillary action via the bridge (which acts like a wick), through the gel and into the dry paper towels. As it passes through the gel the DNA is transferred on to the filter. The DNA is permanently fixed or **cross-linked** to the membrane using ultraviolet light.

Most modern types of membrane are made of nylon, which is sometimes positively charged, so there is no need to cross-link the DNA. The older nitrocellulose membranes were very prone to grease marks and, after baking for 2 hours at 80°C, very brittle.

Prehybridization

Prehybridization solution is added to the filter for a few hours or overnight, at a temperature determined by the amount of formamide in solution (see *Box 7.5*).

Box 7.5 Components of prehybridization and hybridization solutions

Prehybridization and hybridization solution may contain some or all of the following ingredients:

Formamide:	this alters the T_m (melting temperature) so the hybridization can proceed at a lower temperature.
Salt (e.g. SSC):	provides ions for the reaction.
Denhardt's solution:	contains the heavy chemical ficoll to weight the solution onto the filter; also bovine serum albumin (BSA) which acts as a protein blocker.
SDS:	a detergent (sodium dodecyl sulphate) which wets the filter evenly and also provides weight.
Dextran sulphate:	speeds up the hybridization.
Denatured salmon sperm:	preblocks human repetitive sequences on the filter.

Alternatively a 'Church–Gilbert' solution may be used, comprising phosphates and sodium dodecyl sulphate but no formamide.

Labelling the probe

Radiolabelled probes are still commonly used as they are very sensitive. The usual method of labelling is **random primed labelling** (*Figure 7.5*), which has several advantages over the nick translation method used in FISH: the probe need not be pure, only 25–50 ng is required, which is labelled at room temperature to a high specificity (i.e. very radioactive), and there is no need to purify the labelled product to eliminate unincorporated nucleotides. The DNA insert (i.e. the probe) is cut from the vector before labelling (see *Box 7.6*).

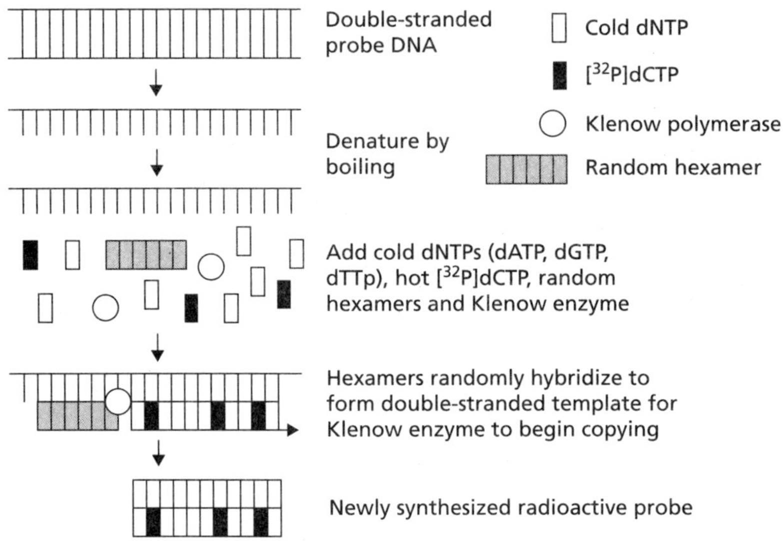

Figure 7.5
Random primed labelling.

Box 7.6 A typical method of random primed labelling

- Double-stranded probe DNA is denatured by boiling.
- It is added to a solution containing buffer, hexadeoxynucleotides, BSA, water, dATP, dGTP, dTTP, [^{32}P]dCTP (the isotope) and the large fragment of the enzyme DNA polymerase 1 (Klenow enzyme).
- The hexadeoxynucleotides or 'hexamers' (which are random sequences of DNA six nucleotides long) find complementary sequences along the single-stranded probe DNA and hybridize to them.
- The Klenow enzyme attaches to the double-stranded template formed by the hexadeoxynucleotides and, using its polymerase function, starts to copy the probe strand, incorporating the appropriate complementary cold (unlabelled) dNTP or, if the probe sequence contains a G, a 'hot' (radiolabelled) C is added.
- The 'hot' Cs essentially ensure that the whole (copied) probe is radioactive. This reaction proceeds very quickly at room temperature; the incorporation of radioactive isotope reaches a maximum after 1–3 h.

Hybridization

The prehybridization solution is replaced with labelled probe, which has been reboiled and added to a small amount of hybridization solution (see *Box 7.7*). Hybridization may last from 30 mins to 24 hours or more, depending on the type of probe.

Stringency washes

These follow the same principles as for FISH; the stringency is empirically determined for each probe, and usually involves a specified **concentration of**

Box 7.7 Melting temperature

The melting temperature (T_m) of a DNA duplex is the temperature at which the two strands separate, or alternatively, the temperature at which they begin to hybridize.

For every 1% of formamide concentration, the T_m (in this case the temperature of hybridization) is reduced by 0.7°C. Hybridization temperature is therefore determined by the proportion of formamide.

salt solution (often SSC) at a particular temperature. **SDS** is usually added to wash the isotope off more thoroughly.

Autoradiography

In a suitable darkroom, the washed filter is wrapped in clingfilm and placed face down on a piece of medical X-ray film. The film is placed in a light-tight autoradiography cassette, and exposed at –80°C for 1–10 days. It is then developed; the X-ray film should display black bands where the hybridized radioactive probe has emitted hard beta particles which fog the X-ray film.

Advantages of Southern blotting

- The exact gene location of a disease need not be known (see sections on *Gene tracking* and *Linked probes* below).
- It can detect large pieces of DNA.

Disadvantages of Southern blotting

- The radioactive part of the process takes 7–14 days (chemiluminescence is slightly faster).
- It needs micrograms of DNA.
- Radioactivity has exposure limits.
- It is not suitable for detecting mutations at the base pair level, unless that mutation creates or destroys a restriction enzyme cutting site.

Interpretation of Southern blots

Deletions

A clinical probe is designed to be complementary to or tightly linked to a disease gene locus. If that gene (or closely associated DNA) is deleted, the probe will have no target with which to anneal, resulting in a missing band on the X-ray film.

Gene deletions occur in X-linked Duchenne muscular dystrophy. Around 60% of affected boys have a deletion which can now be detected in diagnostic laboratories by PCR (see sections on *Multiplex PCR* and *MLPA* later in this chapter). Historically, however, the disease was first screened using cDNA probes and Southern blotting.

Figure 7.6a shows a hypothetical example of some typical band patterns produced by a normal boy, a carrier mother, and several patients with different deletions. The DNA would have been digested previously, and the

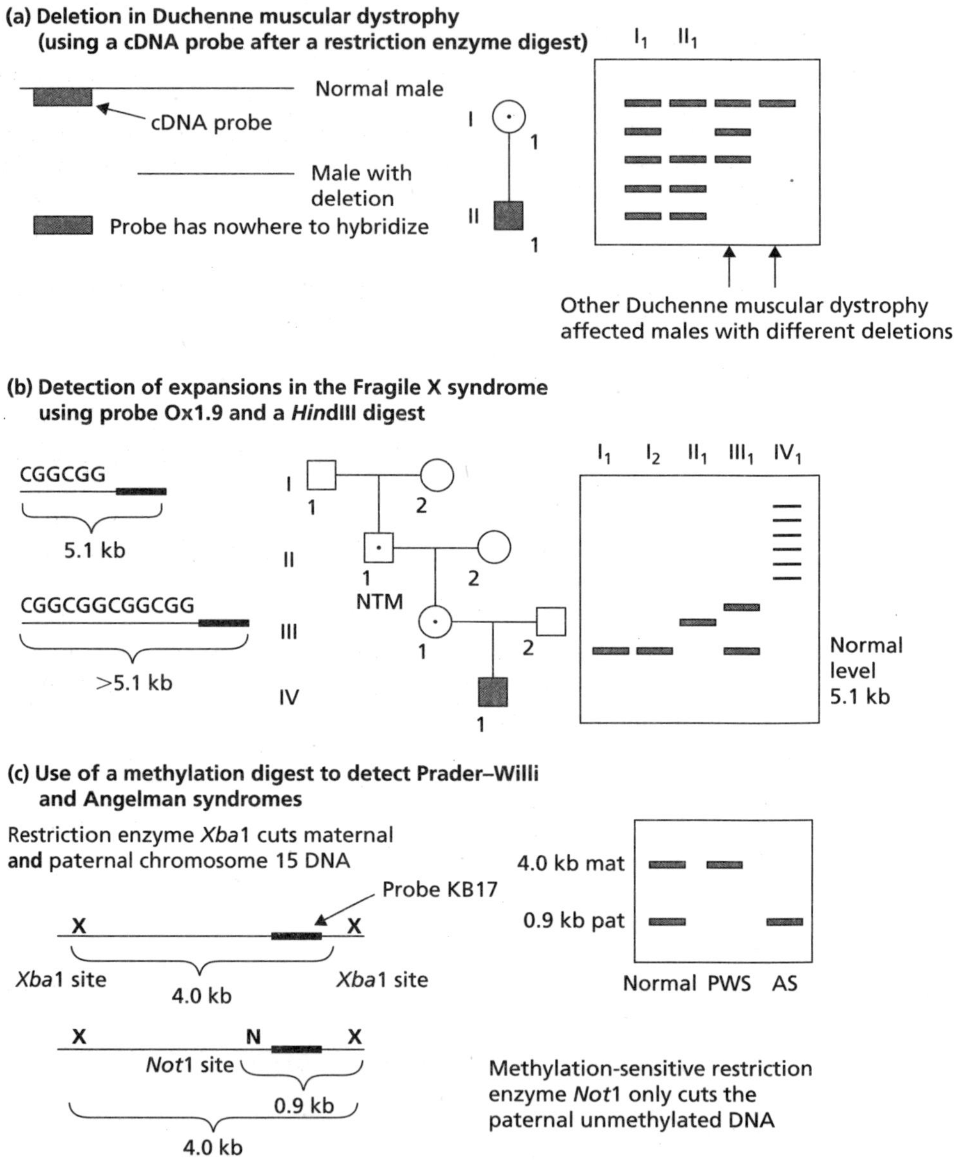

Figure 7.6
Detection of deletions, expansions and methylation by Southern blotting.

absence of various restriction fragments can be seen. As the dystrophin gene is very large this technique needed many probes to cover the large number of exons, and proved too time-consuming.

Additional DNA

Any increase in the length of DNA in or around a gene will also be detectable using a suitable probe. Occasionally DNA duplications arise, which will increase the size of a characterized restriction fragment. Often, however, duplications are too small to detect by Southern blotting, as are small inserted pieces of DNA.

The major application of Southern blotting has been in the detection of a group of disorders known as **triplet repeats** or **dynamic mutations** (see also *Chapter 3*). At various locations in the human genome, individuals have runs of three repeating nucleotides, the sequence of which is consistent for a particular disorder. The number of repeats will vary between individuals, and often between the two homologous chromosomes. As a general rule there will be a numerical range of repeats considered to be 'normal', and another range of repeats considered to be 'abnormal' – usually larger numbers of repeats, which then result in the disease phenotype in affected patients (see *Table 7.2*). Some individuals may have an 'intermediate' number of repeats, which can expand into the disease range when passed on to the next generation. These individuals are not usually affected themselves.

The example shown in *Figure 7.6b* represents a typical pedigree from a fragile X family. The mentally normal grandfather II_1 has more than 50 repeats, such that the band size detected by the fragile X probe is slightly increased from normal, and does not run so far down the gel. As a male has only one X chromosome, he will only produce one band (or allele). This **normal transmitting male** passes on his expanded X to his daughter III_1, and in doing so this allele may expand again by just a few repeats. His daughter inherits two bands of different sizes comprising one normal X from her mother and the slightly expanded X from her father. She is mentally normal but carries a **pre-mutation** (see *Figure 7.7a*, using a methylation double digest). When she has a son (IV_1), there is a 1 in 2 chance of his inheriting the expanded X. If he does, the numbers of repeats may expand again, and this time produce the disease phenotype.

This **full mutation** can be very unstable when many hundreds or thousands of repeats are involved. It is possible for an affected individual to carry different sizes of repeats in every somatic cell, or to have several clones

Table 7.2 Examples of triplet repeat diseases

Name of disorder	Triplet	Normal range	'Intermediate' range	Affected range
Fragile X syndrome	CGG	0–50 repeats	50–200[1]	>200
Myotonic dystrophy	CTG	5–35	36–49	50–2000
Huntington disease	CAG	10–29	36–39[2]	36–121

[1]More accurately, the intermediate range in fragile X is actually 50–59 repeats; 60–200 repeats fall into the pre-mutation category

[2]36 repeats is the lowest number associated with Huntington disease, 36–39 is late onset, 40–55 adult onset, >60 juvenile onset.

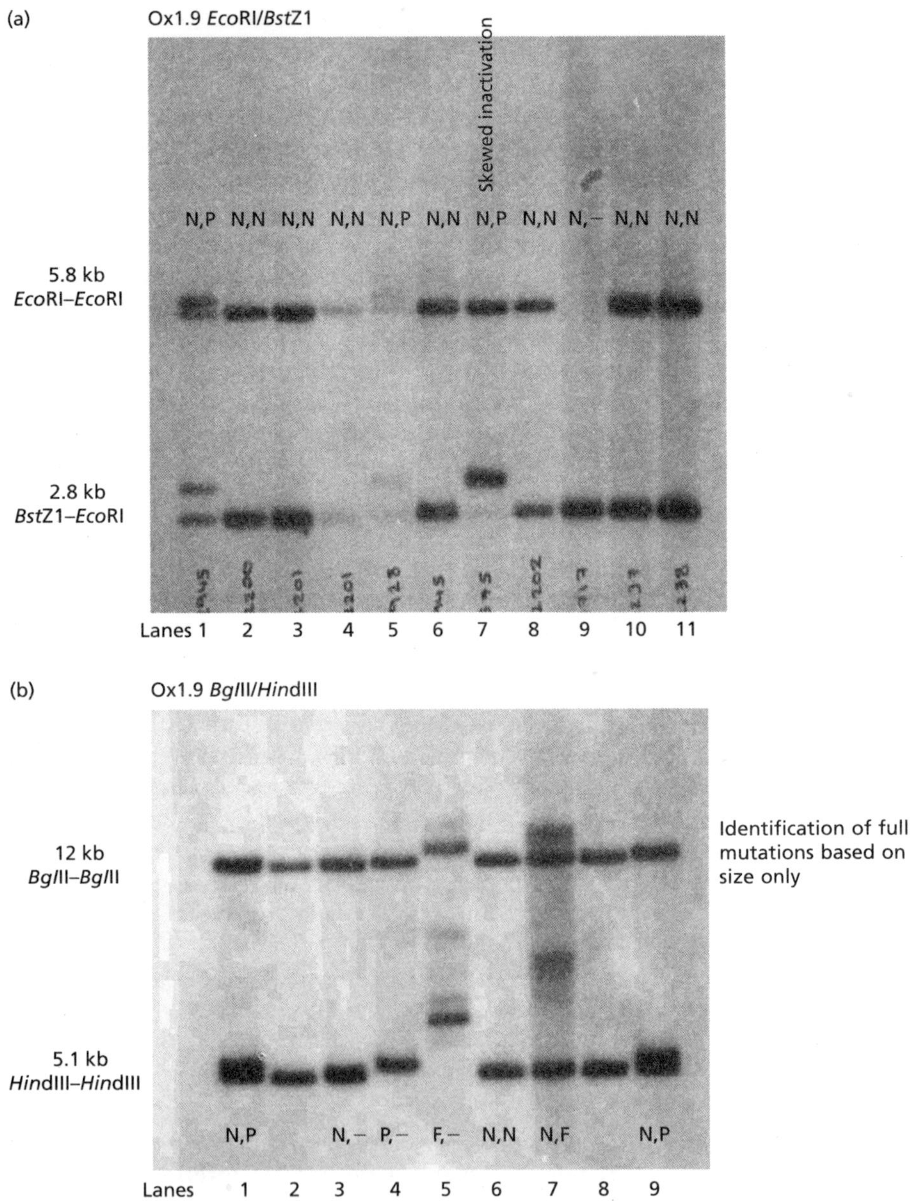

Figure 7.7
Fragile X autoradiographs.
(a) Fragile X Southern blot showing pre-mutation females (lanes 1, 5 and 7).
(b) Fragile X Southern blot showing full mutation male and female (lanes 5 and 7).

(groups of cells) comprising large repeats. The clones produce large bands on X-ray film and the **heterogeneous** (different) populations of cells appear as a smear. This is known as **somatic heterogeneity**.

Females may also carry full mutations (as well as their normal X); they have an increased risk of being affected (see *Figure 7.7b* using single digests).

As the numbers of repeats increase in both males and females, the fragile X gene becomes more likely to be methylated. It cannot be transcribed, and the normal protein (FMR-1) cannot be produced. This leads to neurological problems in the brain, resulting in mental retardation.

Methylation digests

This is a specialized form of double digest, where more than one type of restriction enzyme is added to the same patient's DNA. One of the most common reasons for using this technique is to compare methylated DNA with non-methylated DNA.

Example detecting the parent of origin in Prader–Willi/Angelman syndrome. Prader–Willi and Angelman syndrome may be analysed by cytogenetic and molecular cytogenetic methods. To study these syndromes at the molecular level, a Southern blot can be done using a methylation digest (although this has been superseded by a PCR in most diagnostic laboratories). This exploits the fact that at the PWS and AS region on chromosome 15, the **maternal** DNA is **methylated**. The **paternal** DNA at the locus detected by the probe KB17 is **unmethylated**.

The first enzyme *Xba*1 (which is **not** methylation-sensitive) is added to the patient's DNA, and cuts the DNA from both the maternal and paternal DNA into a 4.2 kb size, detected by the probe KB17. The second enzyme *Not*1 (which **is** methylation-sensitive) is added; this cannot cut the maternal DNA as it is methylated, but can cut the paternal DNA again, to the smaller size of 0.9 kb.

Following Southern blotting a normal person will have two bands derived from the two normal parental chromosome 15s. A person with Prader–Willi syndrome will only have a 4.2 kb maternal band; a person with Angelman syndrome will only have the 0.9 kb paternal band (*Figure 7.6c*).

Although this test does not differentiate between deletion, uniparental disomy, or small mutations at the imprinting centre (responsible for the resetting of imprinting of the nearby PWS and AS genes), it will show an abnormal result at the molecular level.

Gene tracking

Background to gene tracking

Historically this was once the most common application of Southern blots. The approximate location of a mutated gene responsible for a specific disorder could often be deduced from gene disruption or position effects resulting from the translocation of a particular chromosome.

Duchenne muscular dystrophy was known to be X-linked by observing the inheritance pattern. An approximate location was deduced by studying the chromosomes of girls who were phenotypically affected. In each case there was a translocation involving the band Xp21.

The breakpoint would give a starting point from which loosely linked probes could be constructed (see *Box 7.8*).

Box 7.8 Reverse genetics

This method was used on the globin protein isolated from red blood cells. Once the amino acid sequence was determined the mRNA sequence could be elucidated. Subsequently mRNA can be converted into cDNA using the enzyme reverse transcriptase. The globin cDNA can then be used as a probe which is complementary to the normal globin sequence.

Probes such as these were useful in diagnosing particular haemoglobinopathies such as the thalassaemias, where a defective globin protein is produced.

By looking at RFLPs using a variety of restriction enzymes, it could be shown that in some families (with a particular genetic disorder) certain DNA polymorphisms were inherited together with the disease state, therefore the probe was probably linked to the gene locus. Once probes were made which mapped nearer to the gene location (tightly linked), the recombination risk could be reduced and the probe used in clinical diagnosis (see *Box 7.9*).

See *Box 7.10* for some examples of gene tracking using linked probes.

Box 7.9 Shotgun cloning and gene libraries

The gene location of disorders such as cystic fibrosis and Huntington disease were discovered by **shotgun cloning**. The first probes were made from known chromosomes, which had been cut into small (probe-sized) random pieces. These are called **gene libraries**.

Many families were then studied who had extensive pedigrees in which the pattern of inheritance was very clear. One of the 'anonymous' probes from a gene library would be used in conjunction with a restriction enzyme known to produce RFLPs. The inheritance of the differently sized RFLPs was compared to the inheritance of the disease.

While it took several years to find linked cystic fibrosis probes, one of the very first probes to be tested for Huntington disease proved to have a 4% recombination risk, which was low enough to be used in clinical diagnosis.

'Informativeness'

RFLPs may produce one of three types of results when used in tracking.

- **Uninformative** – for example, if every member of the family has the same homozygous result. In this instance another probe/enzyme combination has to be tried.
- **Partially informative** – only limited information may be deduced (for example, if one parent is heterozygous and one parent is homozygous). It may be possible to combine one partially informative probe/enzyme combination with another, such that two partially informative results can be combined as **haplotypes** to give a fully informative pedigree. A haplotype represents all the alleles inherited from one parent on an ideogram of that parental chromosome (*Figure 7.9*).
- **Fully informative** – any combination of bands inherited by an individual can be interpreted as either a normal result, a carrier (if recessive or

Box 7.10 Examples of gene tracking using linked probes

Using the family pedigrees in *Figure 7.8*, decide which allele or alleles track with the three diseases, and predict the results of the prenatal diagnoses (to help you, band sizes have already been transferred to the pedigree). Follow rules 1–6 (answers are given below):

1. The **index case** (the first affected member of a family) must have a **definitive** (clinically definite) **diagnosis**.
2. The **pattern of inheritance** must be identified (autosomal recessive, autosomal dominant, or X-linked). With an autosomal disorder there will be a possibility of two sizes of allele corresponding to the two homologues; with X-linked diseases, the female will have a choice of two alleles, but the male has only one X chromosome and will only have one allele.
3. The band sizes on the autoradiograph must be transferred to the pedigree for each family member (see *Box 7.11*). If more than one probe/enzyme result is used, they are each considered separately to begin with.
4. The band sizes inherited by the **affected index case** are studied **first**. This establishes which size (or sizes) of all the possible choices of alleles are associated with the inheritance of the disease.
5. The band size(s) associated with the disease are tracked back to the parents, one or both of whom will be carriers depending on the inheritance pattern.
6. The inherited pattern of the sizes is examined to see how **informative** the probe/enzyme result is. Tracking using RFLPs must be applied each time to a new unrelated family. It cannot be assumed that an allele tracking with the disease state in one family will do so in another (it may track with the normal state the next time). Remember that linked **RFLPs are indicators** or **markers** of the disease state, not the disease itself.

Answers

- Late onset autosomal dominant disorder = Normal. The disease tracks with the B allele. Notice that the status of the parent II_3 need not be known.
- Autosomal recessive disorder = Affected. The disease tracks with the 6.6 kb allele from each parent.
- X-linked recessive disorder = Normal. The disease tracks with allele 1.

X-linked) or affected. This is especially desirable when predicting fetal genotype in a prenatal diagnosis, when ideally both parents are heterozygous.

7.3 THE POLYMERASE CHAIN REACTION

Principles of PCR

PCR is a method of amplifying very small amounts of DNA (although theoretically as little as one cell nucleus can be amplified in a PCR, in practice usually 25–500 ng of genomic DNA is used). Total genomic DNA is denatured; the length of DNA to be amplified is delineated by two short pieces of complementary DNA called **primers** which anneal to the target DNA.

A heat-stable polymerase attaches to this double-stranded template and copies the existing DNA strands. There is now twice the initial amount of

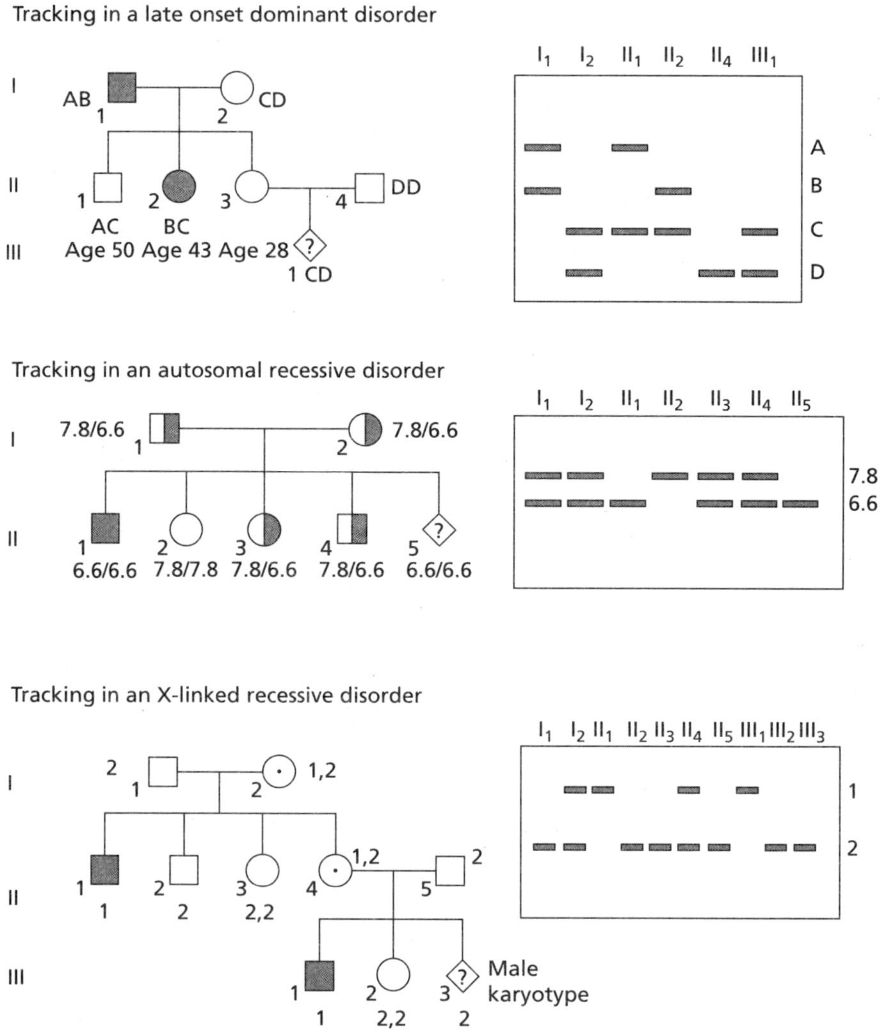

Figure 7.8
Gene tracking.

> **Box 7.11 Size nomenclature of alleles**
>
> There are several ways of writing band sizes (the largest allele of the pair is usually written first):
>
> - the **size** of the bands in kb, e.g. 5.5/4.3, or 2.1/1.4
> - **letters** of the alphabet, e.g. AB or CD (A>B>C>D)
> - **numbers,** e.g. 1,1 or 1,2 or 2,2 (1>2>3>4)
> - + or – where + represents the presence of a restriction site, – absence of a restriction site, e.g. +/+ or +/– or –/–. Here, a '+' will be the smaller allele.
>
> Numbers are recommended by the Clinical Molecular Genetics Society (CMGS).

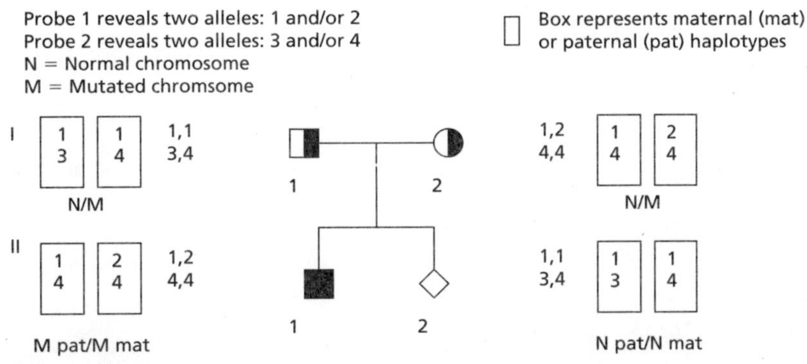

Figure 7.9
Using haplotypes to produce a fully informative pedigree in cystic fibrosis.

DNA (four strands instead of two). These heat cycles are repeated, ensuring exponential amplification of the target DNA (*Figure 7.10*).

Details of a typical PCR method

The standard polymerase enzyme used in PCR limits the size of fragment to be amplified; usually the optimal size is up to a few hundred base pairs. In practice therefore, exons of genes are suitable targets for amplification, although longer lengths can be amplified using modern polymerases specific for long range PCR.

Primers are designed complementary to **conserved** DNA sequences flanking the chosen exon. The size of the PCR product is determined by the distance between the annealed primers. PCR is therefore most often applied to diseases where the gene location is known. Primers are short, single-stranded pieces of DNA around 18–30 bp in length. If the primer is shorter than 18 bp, the sequence of the target may not be unique in the genome; if the primer is longer than 30 bp, there may be problems with the length of time the primer takes to anneal in a PCR reaction.

A typical PCR is set up using an eppendorf tube for each patient in a contamination-free area. The reaction usually comprises a suitable buffer, the four dNTPs (dATP, dCTP, dGTP and dTTP), the pair of primers, the patient's DNA and the heat-stable enzyme Taq polymerase. The eppendorf tubes (containing the above reaction mixture) are placed in an automated programmable heating block, which proceeds through several heating cycles as follows:

- **Denaturation** (94°C) – the patient's DNA becomes single-stranded.
- **Annealing** (55–66°C) – the single-stranded primers anneal to the complementary sequences on the two strands of the patient's DNA.
- **Extension** (72°C) – the Taq polymerase attaches to the double-stranded template and copies the appropriate strand.

Usually 28–35 cycles are sufficient to produce 10^5–10^6 copies of DNA.

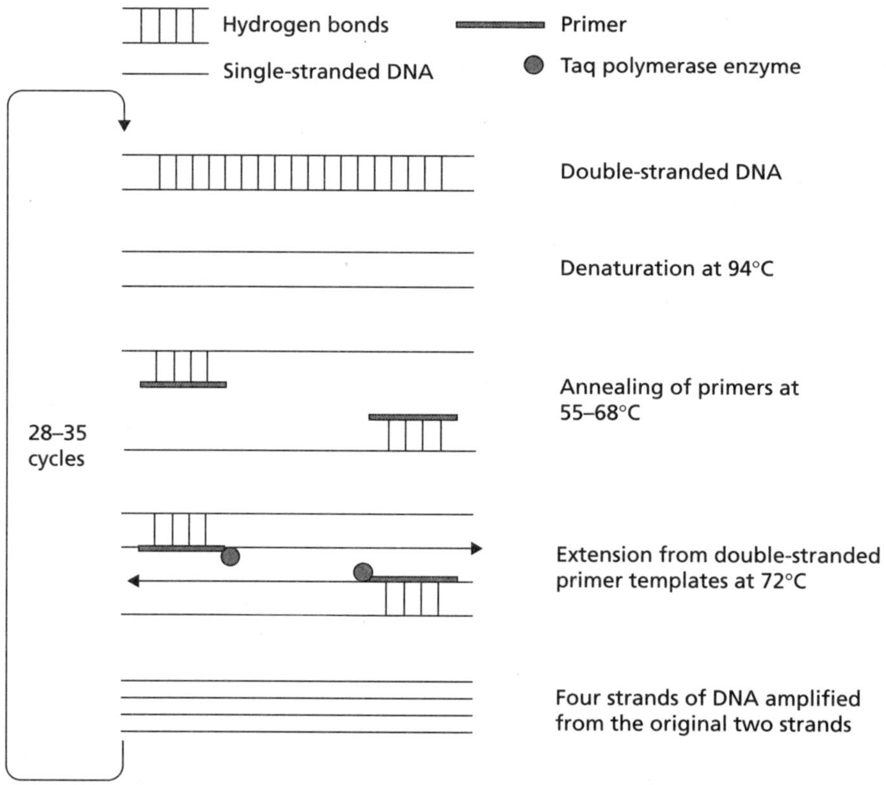

Figure 7.10
Principles of the polymerase chain reaction.

Advantages of PCR

- Very small amounts of DNA are required.
- The procedure is fast; results are obtainable in hours (30 min to 48 hours), depending on the steps following PCR.
- PCR is safe; no radioactivity need be used.
- Mutations can be detected at the molecular level.

Disadvantages of PCR

- Large DNA sequences cannot be amplified.
- Some of the DNA sequence usually has to be known to design the primers.
- The reaction is easily contaminated.
- Dosage is difficult to estimate.

Applications of PCR

Apart from clinical diagnostic use, PCR may be applied to areas such as forensic science or archaeology and anthropology (and hence evolution),

Box 7.12 Hypothetical use of PCR

It is hypothesized that Abraham Lincoln had Marfan syndrome, typified by tall thin stature, long fingers and weakness of the aorta (the main artery of the heart). It has been suggested that a PCR could be done on the dried blood from the jacket he was wearing when he was assassinated in order to study the fibrillin gene which is mutated in Marfan syndrome.

where only minute traces of DNA may be found (see *Box 7.12*). If primers are used which can detect numerous polymorphisms, PCR can be used in paternity testing and also criminal pathology, as the polymorphic band patterns will be almost unique to an individual, and constitute a DNA fingerprint. The more closely two people are related, however, the likelihood of inheriting the same polymorphisms increases. Identical twins will share identical genetic fingerprints.

Interpretation of PCRs

Direct visualization of PCR products

If a pair of primers flank a region which displays size variations, then the products themselves will also vary in size. This is useful where a known deletion, insertion or expansion is characteristic of a particular disorder.

The cystic fibrosis mutation p.Phe508del. The most common cystic fibrosis mutation in the Caucasian population is p.Phe508del (phe=phenylalanine at amino acid position 508, del = deletion), comprising 70–80% of north-western European *CFTR* mutations. A deletion is found in exon 10 of the *CFTR* gene (see also *Box 7.13*). This 3 bp deletion (CTT) can be detected by running a high percentage polyacrylamide gel to separate the differently sized PCR products produced by normal individuals, carriers and affected patients (*Figure 7.11*).

A normal product using a specific pair of primers is 98 bp in size; the deleted product will only be 95 bp. A carrier will therefore display one 98 bp (normal) product from one chromosome 7, and a 95 bp (deleted) product from the other chromosome 7. When a carrier produces two different sizes of PCR product, they attempt to pair, but due to the size discrepancy a unique band pattern is produced which runs at a different level to that expected on

Box 7.13 The biochemical basis of cystic fibrosis

As cystic fibrosis is an autosomal recessive disease, the affected children have two mutations (which may be different), one on each chromosome 7. The mutations may result in an altered or truncated protein. The normal protein is called CFTR (cystic fibrosis transmembrane conductance regulator); its normal components form a chloride channel which is essentially a pore in the epithelial cell membrane. In the pathological state, chloride ions and hence also sodium ions are not reabsorbed in the appropriate area of the sweat gland due to the absence of a chloride channel. This leads to an increased salt concentration in the sweat.

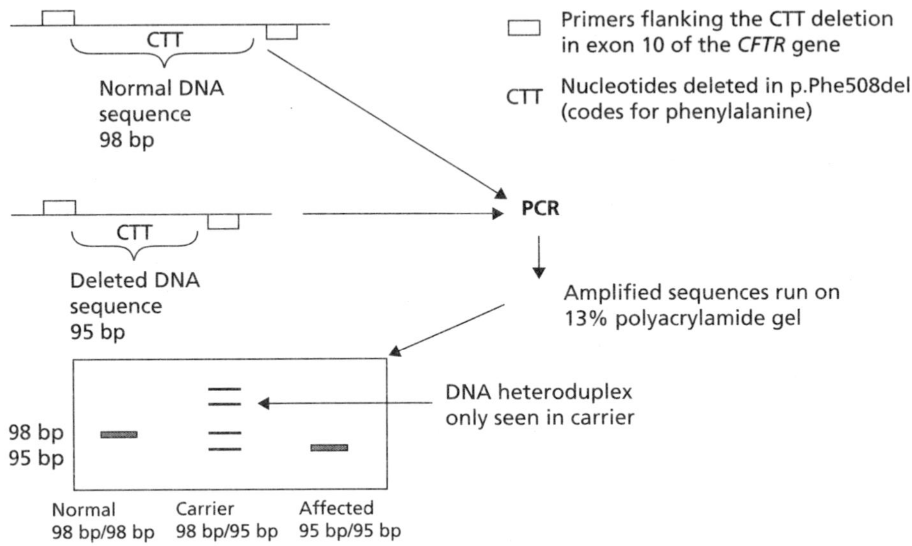

Figure 7.11
Direct visualization of a PCR product using the cystic fibrosis mutation p.Phe508del.

the gel. This is a **heteroduplex**, which has a different **conformation** to the correctly paired 98 bp or 95 bp products.

An affected child will have two faulty chromosomes; the gel will display two 95 bp products (indistinguishable on the gel). The gel is stained with ethidium bromide for visualization (*Figure 7.12*). However, as over 1600 *CFTR* mutations have been found, this method has now been superseded by kits detecting multiple *CFTR* mutations which commonly occur in the local population.

The triplet repeat of Huntington disease. Primers have been designed to flank the region of CAG repeats in Huntington disease which may expand in some families, resulting in affected individuals. When run out on a gel with suitable molecular weight markers giving bands of known sizes, an estimate of the numbers of repeats can be obtained. It can then be determined whether a patient is unaffected (up to 34 repeats), affected (over 36 repeats), or borderline (34–36 repeats). In the last category it is difficult to assess whether these numbers of repeats may expand in future generations.

The duplication in hereditary motor sensory neuropathy. The hereditary motor sensory neuropathies are a heterogeneous group of neurological disorders also known as Charcot–Marie–Tooth disease. The most common form is inherited in an autosomal dominant manner, and is caused by a duplication of part of the short arm of chromosome 17.

The duplication may be detected in the laboratory using different pairs of PCR primers (or markers) which map to the duplicated region. In the normal state two alleles will be detected. If there is a duplication, either three alleles are detected, or there is a **dosage** effect such that two alleles are detected, one of which may be visible as a heavier band. One of the

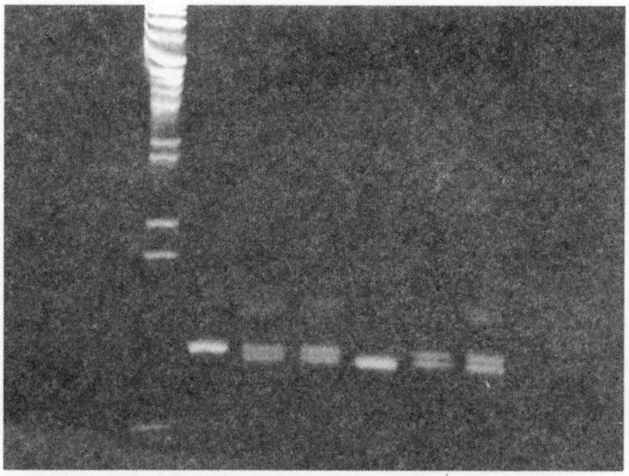

Figure 7.12
A polaroid of a gel showing detection of the p.Phe508del *CFTR* mutation. The bands of DNA are stained with ethidium bromide. Lane 1 is a molecular weight marker, lane 2 is a normal child (N/N), lanes 3 and 4 are the heterozygous carrier parents (N/p.Phe508del), lane 5 is the affected index case (p.Phe508del/p.Phe508del), and lane 6 is a prenatal diagnosis revealing the fetus to be a phenotypically normal carrier of p.Phe508del (like the parents). Lane 7 is a control sample from a known carrier.

disadvantages of PCR becomes apparent when looking at 'dosage'; it takes experience to know when more than one copy of an identical allele is present, and often there is preferential amplification of the smaller allele. Automated techniques such as MLPA (see the *Mutation screening* section below) are now used to standardize dosage analysis.

Restriction enzyme digests and PCR

A restriction enzyme digest may be performed on the PCR product from a patient, exploiting the fact that a mutation may create or destroy a restriction site (see *Boxes 7.14* and *7.15*).

Amplification refractory mutation system (ARMS)

A pair of primers is designed such that one primer is complementary to a common conserved sequence in both normal individuals and patients carrying a particular mutation. The other primer is only complementary to the

Box 7.14 HGVS mutation nomenclature

Mutation nomenclature in molecular genetics is HGVS (Human Genome Variation Society) compliant. A typical example is the *CFTR* mutation, p.Gly542X:

Gly is the amino acid glycine coded for by the **normal DNA sequence**
542 is the **amino acid position** in the protein
X is a stop codon coded for by the **mutated DNA sequence**

Box 7.15 Creation of a restriction site by the exon 11 *CFTR* mutation p.Gly551Asp

During the PCR of the exon 11 *CFTR* mutation p.Gly551Asp, primers which flank exon 11 amplify a product of 425 bp in every individual, so a further step is necessary to detect that mutation.

The mutation p.Gly551Asp is a G>A base substitution such that glycine (Gly), at amino acid position **551**, is changed to aspartic acid (Asp). Fortuitously, a restriction site for the restriction enzyme *Mbo*1 is created in individuals carrying the mutation when the G>A transition occurs, and this cuts the 425 bp product into two fragments; one of 242 bp and one of 183 bp. In normal individuals the restriction site is not present and *Mbo*1 can no longer cut, resulting in one fragment of 425 bp.

Carriers will have one normal chromosome (which does not cut) and one chromosome with the mutation (which does cut), so they will have three bands.

mutated sequence. Only in DNA carrying the mutated sequence will PCR amplification succeed, as both primers anneal, enabling visualization of a band. In normal individuals, only **one** common primer will anneal; the other primer is not complementary to the normal sequence. No band will be produced.

In ARMS reactions, in order to differentiate the carriers of a specific mutation from homozygous affected patients, a primer complementary to the non-mutated DNA sequence can be run with the common primer in a separate reaction.

Multiplex PCR

Rather than amplify one exon of a gene individually using one pair of primers, it is more efficient to use more than one pair of primers (for multiple exons of a gene) in a reaction tube. Several bands will result, each representing different exons. Primers are designed to ensure that the bands are of different lengths and are well separated on the gel.

Problems may arise if too many sets of primers are used which do not respond equally to a common PCR buffer – some exons may 'drop out'.

ARMS multiplex

If a particular disorder can be caused by several different mutations, each present at a reasonably high frequency in the local population (say >1%), one of a pair of primers can be designed complementary to one particular mutation. This is done for a number of common mutations (**ARMS**), and all the pairs of primers are put into one PCR reaction tube (**multiplex**), together with a pair of control primers which detect PCR failure. This has proved to be an efficient method of screening cystic fibrosis mutations, where frequencies of the common European mutations are well characterized.

Nested PCRs

Although primer pairs are designed to be sequence-specific, occasionally the primer match is not perfect, and spurious bands may result. A few microlitres of the first PCR product can be taken and used as the DNA source in a second PCR. **Nested primers** may be designed, both of which are located internally (on the DNA sequence) to the initial pair of primers.

Oligonucleotide ligation assay (OLA)

Specific sequence changes can be detected by using two different allele-specific oligonucleotide probes – one complementary to the normal DNA sequence and one complementary to the mutated DNA sequence. Each allele-specific probe may either be of differing sizes (see *Figure 7.13*), or labelled with different fluorescent dyes.

Each oligonucleotide probe can ligate with a common reporter probe, which is fluorescently labelled. However, only if the match is exact upon hybridizing with the DNA does ligation occur. The product may then be subjected to electrophoresis and analysed by computer software.

Real-time PCR

Real-time PCR (qPCR; see *Figures 7.14* and *7.15*) simultaneously **amplifies and quantifies** a specific DNA sequence. It can therefore determine if a specific sequence is present in a sample, and also the number of copies. If combined with reverse transcriptase PCR (known as RT–PCR), measurement of gene expression is possible.

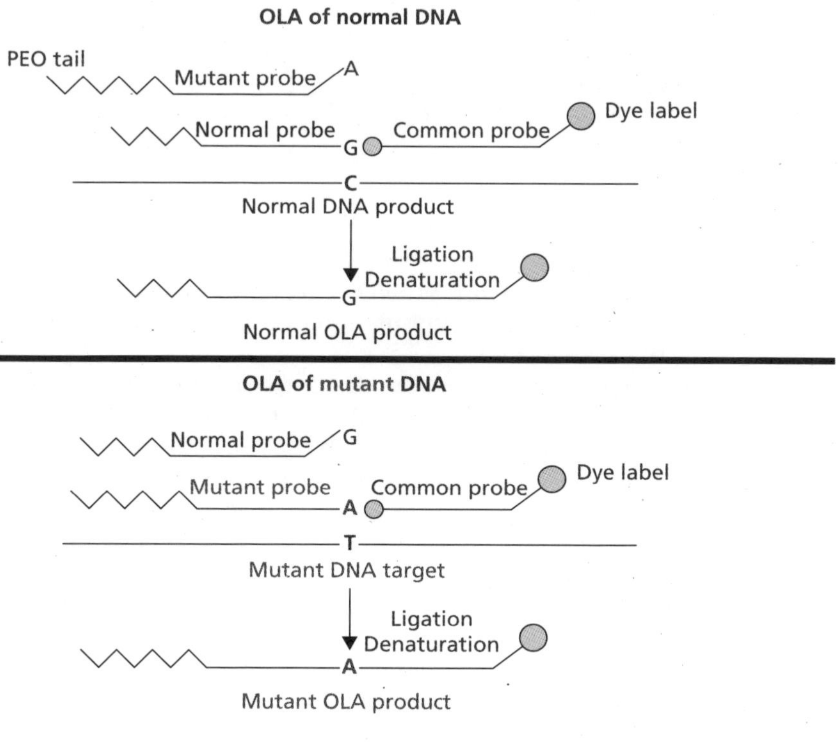

Figure 7.13
Oligonucleotide ligation assay (OLA). In this example, each allele-specific probe has a different length of pentaethylene oxide (PEO) tail. This enables multiplexing, and hence detection of several mutations simultaneously.

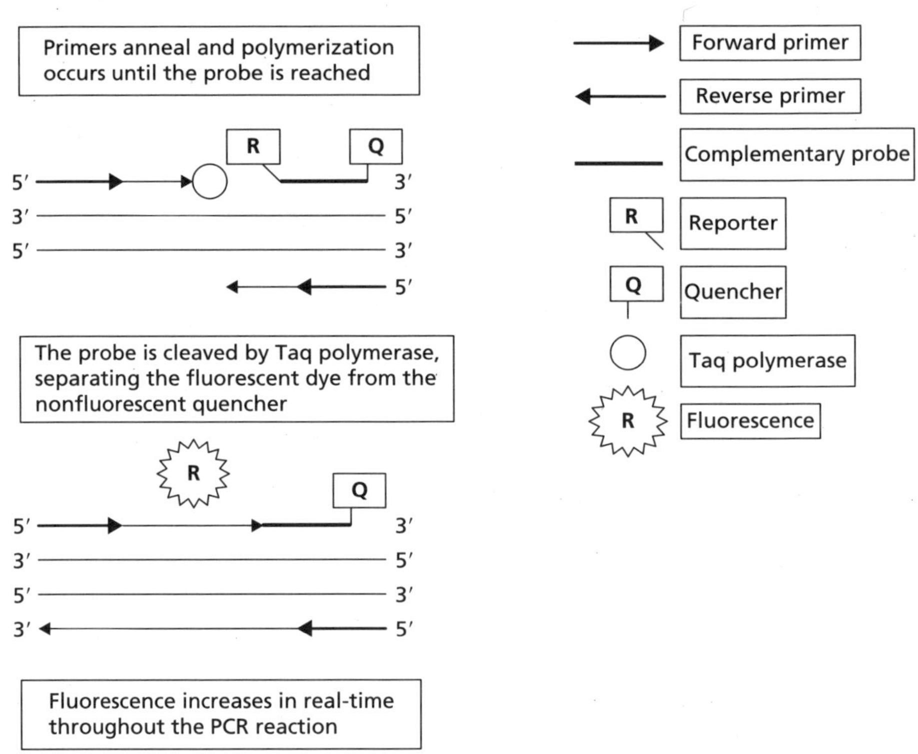

Figure 7.14
Principles of real-time PCR.

Real-time PCR labels the PCR product with a fluorescent label so that accumulation of product can be followed in real time. This is more reliable as it measures the accumulation of product rather than total amount at the end of cycles of PCR.

First, two primers are designed to amplify the region of interest. A fluorescent oligonucleotide probe is then designed complementary to a DNA sequence of interest which has a fluorescent dye attached at one end, and a non-fluorescent quencher at the other. If two allele-specific probes are used to distinguish the normal from the mutant sequence, these can be labelled with different fluorescent dyes at the 5′ end. As the target DNA is amplified, any hybridized probe is cleaved by the Taq polymerase enzyme. This separates the reporter fluorescence dye from the quencher, such that fluorescence increases during the PCR reaction in an allele-specific manner.

The use of fluorescent reporter probes allows multiplexing, so several gene mutations can be assayed for simultaneously, using specific probes with differently coloured fluorescent labels.

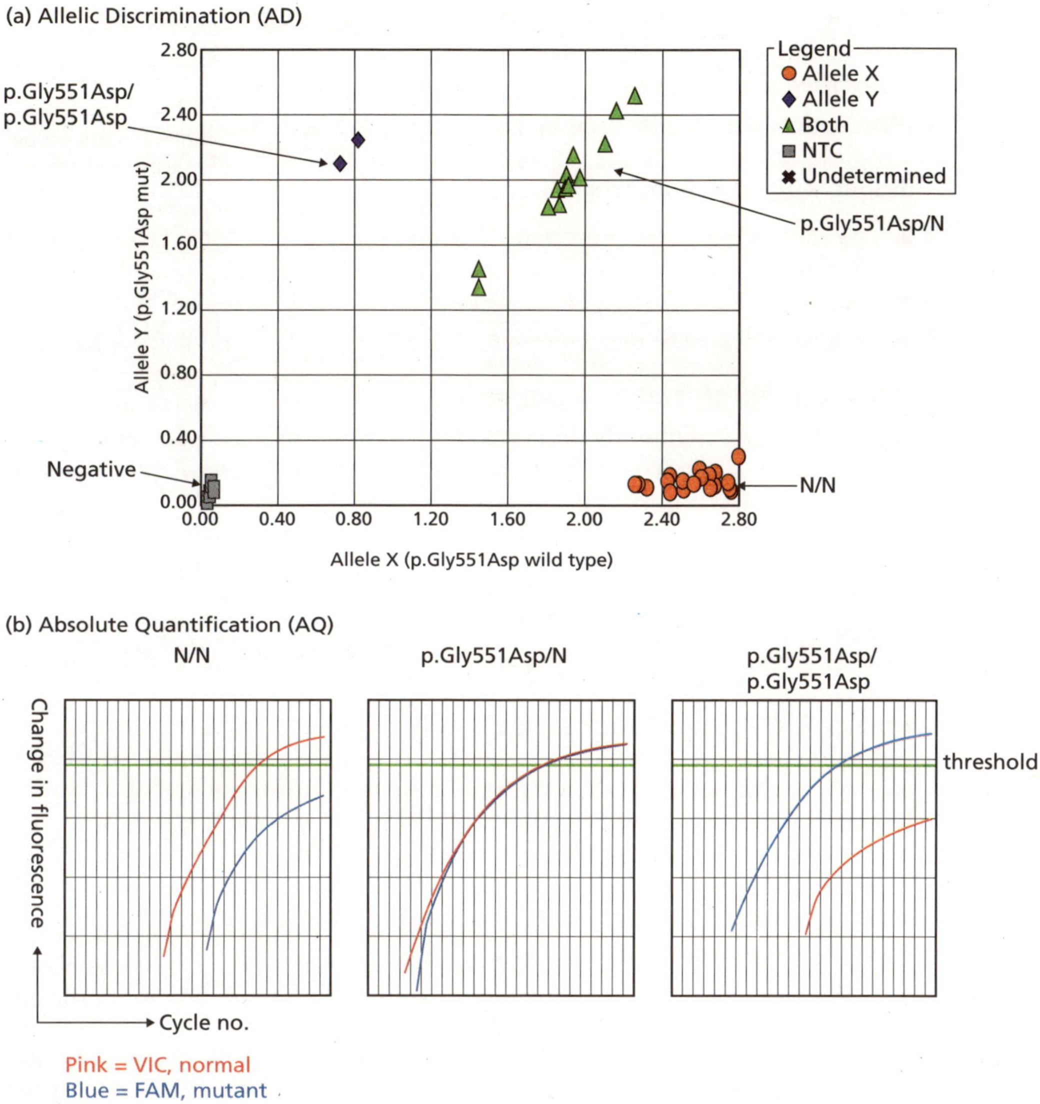

Figure 7.15
(a) RT-PCR Allele Discrimination (AD). A real-time PCR run, showing results for all patients tested for the *CFTR* mutation p.Gly551Asp. Homozygous normal patients are indicated by a red circle, heterozygous carriers are indicated by a green triangle, and homozygous mutant (affected) patients are indicated by a blue diamond.
(b) RT-PCR Absolute Quantification (AQ). A real-time PCR run showing results for three patients tested for the *CFTR* mutation p.Gly551Asp. The X axis represents the number of PCR cycles, the Y axis the amplification due to increased fluorescence during the run. True results have to pass the threshold. In graph 1 the normal (pink) trace is above the threshold; this patient is homozygous normal (N/N). In graph 2 both traces (pink and blue) are above the threshold: this patient is a heterozygous carrier (N/p.Gly551Asp).
In graph 3 the mutant (blue) trace is above the threshold: this patient is homozygous affected (p.Gly551Asp/p.Gly551Asp).

Automated screening strategies using PCR

Pre-screening samples

Automation is generally used in order to achieve high throughput of clinical samples with effective use of staff. Pre-screening may be appropriate for dealing with large numbers of samples:

- measurement of the numbers of triplet repeats (e.g. the CGG repeats of Fragile X)
- heteroduplex-based technologies such as SSCP, DGGE or CSCE
- duplications/deletions using quantitative methods such as MLPA

Pre-screening of fragile X patients

One of the most frequently requested tests in molecular genetics laboratories is for Fragile X syndrome. As previously described, fragile X patients may exhibit increased numbers of the triplet repeat CGG. Patients' DNA samples are amplified by PCR then run on a genetic analyser, which will automatically produce graphs showing one or more peaks (depending on the sex of the patient) which are interpreted by the accompanying software as numbers of repeats (see *Figure 7.16*).

Male patients with one allele in the normal range, and female patients with two alleles in the normal range are reported as normal. Male failures, or females with only one allele go on to Southern blotting. This is because the numbers of repeats are too great to PCR (e.g. an affected male), or there is selective amplification of the smaller allele in a pre-mutation or full mutation female. Some females have two normal alleles of the same size – these appear as one peak, and also have to go to Southern blotting.

Pre-mutation carriers may be detectable if the numbers of repeats are not too large to PCR (under 100 repeats); these can then be reported without recourse to the more expensive and time-consuming Southern blotting.

Heteroduplex-based technologies

DGGE and SSCP may be used to detect differences in band positions on gels (representing mutated DNA sequences) with high sensitivity. However, the exact nature of the mutation is not usually known, and the position of the mutation may only be traceable to a particular exon determined by the PCR. As the above technologies cannot easily be automated, CSCE and automated bidirectional sequencing are becoming the preferred options.

Single-stranded conformational polymorphism (SSCP)

PCR products are denatured and run as single DNA strands such that each strand assumes a particular single-stranded conformation. A normal sequence strand will differ from a mutant strand and produce a different band pattern.

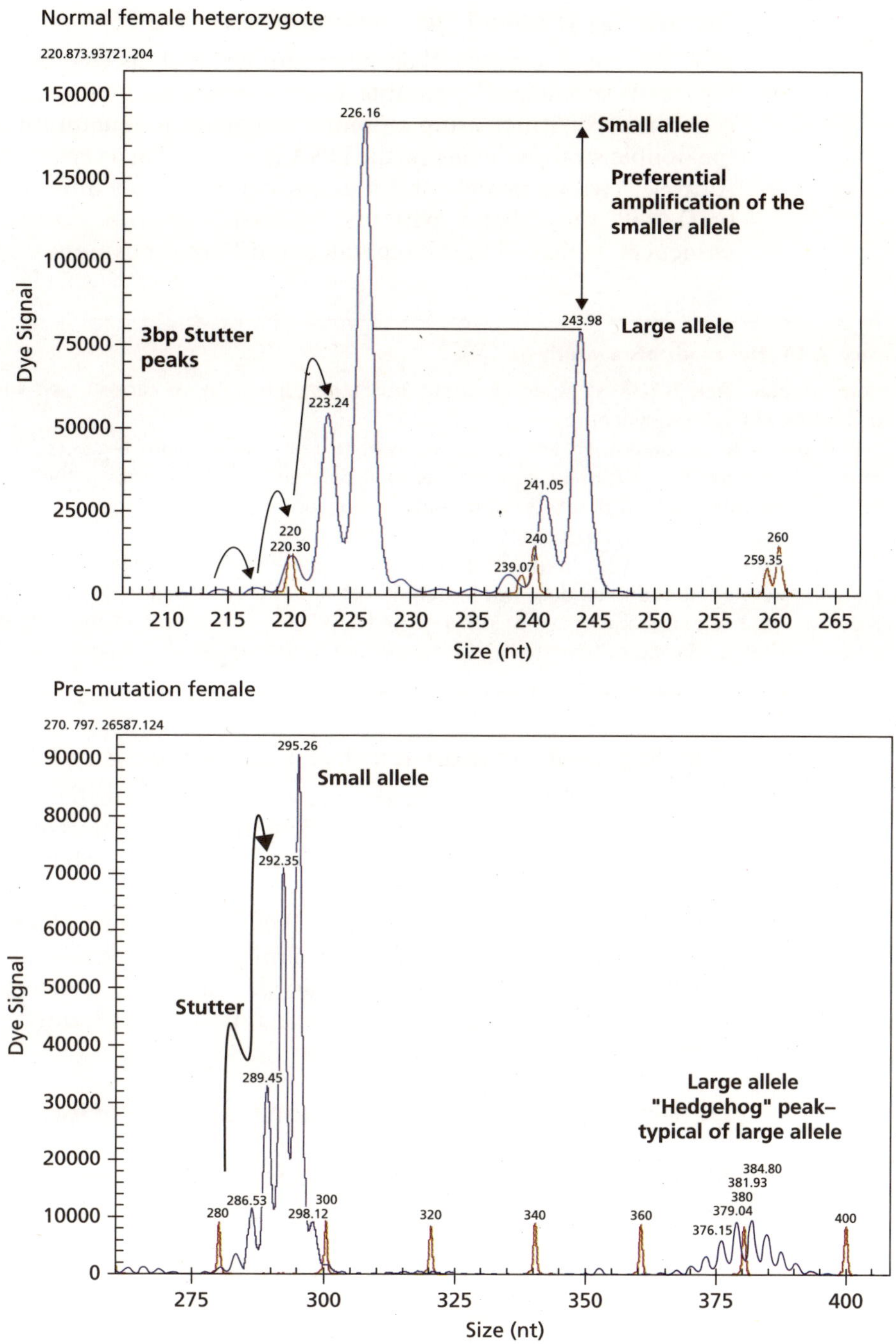

Figure 7.16
Chromatograms from FRAXA PCRs using a gene analyser.

Denaturing gradient gel electrophoresis (DGGE)

DGGE is more sensitive than heteroduplex analysis (see *Box 7.16*) or SSCP, but needs specialized apparatus. Double-stranded PCR products are run on gels comprising increasing amounts of chemical denaturant at 65°C. When the double strands denature, the DNA is retarded at a certain point in the gel. A carrier (whose normal and mutant sequences may differ by only one base pair) will show band patterns representing both normal and mutant sequences, which will have separated at different positions on the gel.

Box 7.16 Heteroduplex analysis

Heteroduplex analysis (HA) is especially useful for carrier detection, as carriers have one normal and one mutated DNA sequence.

PCR products are denatured and slowly allowed to reanneal at room temperature such that the mismatched normal and mutant strands display a different band pattern on the gel compared to homozygous sequences. If detection of mutant homozygotes is required, the suspected mutant DNA must be mixed with an equal quantity of normal DNA.

Modifications of the HA technique include chemical and enzymatic cleavage of mismatches. At the point where the heteroduplex does not match due to a mutation, either a chemical, such as osmium tetroxide, or bacteriophage enzymes can be used to cleave one strand of the heteroduplex at the mutation site, producing differently sized fragments on a gel.

Capillary conformation sensitive electrophoresis (CSCE)

As heteroduplexes form in the presence of a mismatch, they will display abnormal mobility during non-denaturing gel electrophoresis. Mildly denaturing solvents in the polymer may be used to enhance the conformational changes, leading to increased differential migration between the heteroduplexes and homoduplexes. Following the introduction of capillary analysers and fluorescence-based techniques, greater resolution and higher specificity and sensitivity were obtained. If a change is detected, sequencing can be undertaken to elucidate the exact nature of the mutation.

Detection of large duplications and deletions

Multiplex ligation-dependent probe amplification (MLPA)

MLPA (see *Figure 7.17*) is a quantitative high-throughput method which can detect large deletions or duplications of one or more exons. It is therefore useful for large genes comprising many exons, and as a robust quantitative technique enables elucidation of the copy number. As MLPA is a multiplex procedure, it can test for up to 40–45 DNA sequences at once. Therefore up to 40–45 pairs of MLPA probes are required, designed to hybridize to the areas or exons of interest.

Each probe comprises a pair of oligonucleotides which, if they match and bind to adjacent DNA sequences, can be ligated. One of the pair of probes has a stuffer sequence; these will be of different lengths for each probe, so the

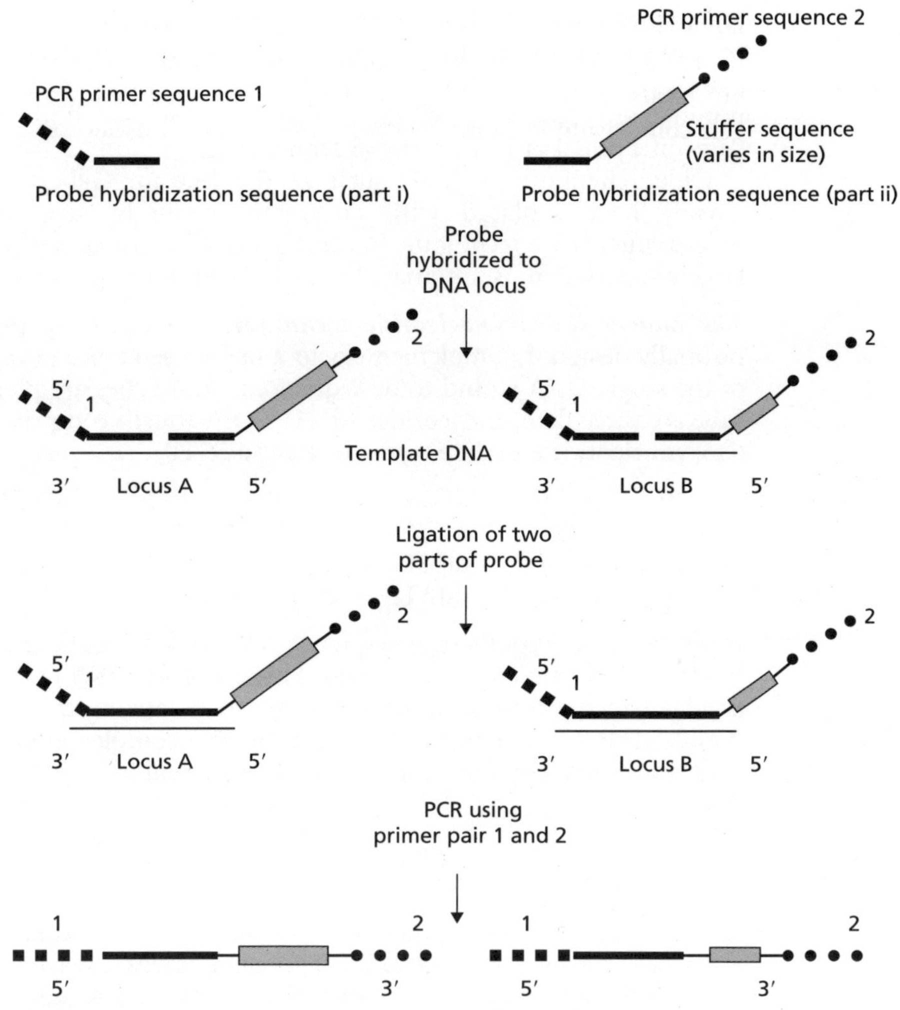

Figure 7.17
Principles of MLPA.

PCR products can be separated and identified. Only one primer pair is needed (one of which is fluorescently labelled), because the same primer binding sites are on the ends of every probe pair.

Capillary electrophoresis results in a graph of the peak height of each amplification product. The height reflects the relative copy number of the target sequence, which is usually displayed as a ratio compared to a normal control sample. Although MLPA tests are usually sold as commercial probe sets, these can also be laboratory-designed for a disorder of interest.

Sequencing

Sequencing can detect very small alterations in genomic DNA, and although this appears to be the ideal mutation detection method, there are constraints due to time and cost. Other screening strategies are therefore often employed to 'pre-screen' for mutations. Once a mutation has been detected by a primary screen, its nature and position have to be characterized by determining the sequence of nucleotides on one or both DNA strands (uni-directional or bi-directional sequencing).

Older chemical methods such as the Maxam–Gilbert technique had already been replaced with enzymatic methods such as the Sanger dideoxynucleotide technique. In turn, the manual Sanger technique has been largely superseded by automated high-throughput sequencing.

The Sanger dideoxynucleotide technique. A sequencing primer (yyy) is normally designed complementary to a known sequence (xxx) at the 3′ end of the single DNA strand to be sequenced. Four tubes are prepared; in each tube are three deoxynucleotides (dNTPs), the fourth comprises a mixture of deoxynucleotide and dideoxynucleotide (ddNTP):

Tube 1:	dATP	dCTP	dGTP	dTTP	**ddTTP**
Tube 2:	dATP	dCTP	dGTP	**ddGTP**	dTTP
Tube 3:	dATP	dCTP	**ddCTP**	dGTP	dTTP
Tube 4:	dATP	**ddATP**	dCTP	dGTP	dTTP

If we take the hypothetical sequence 5′GATCCATxxx3′, the primer (yyy) binds to the complementary 3′ end. This enables a DNA polymerase (which is added to the four separate tubes) to start copying the sequence at the double-stranded template, incorporating the complementary nucleotides. This results in a complementary strand which will be sequenced in a 5′ to 3′ direction:

DNA strand to be sequenced:	**5′GATCCATxxx3′**
Complementary strand:	**3′CTAGGTAyyy5′**

However, because of the ratio of deoxynucleotides to dideoxynucleotides, every time the enzyme tries to incorporate a dideoxynucleotide the reaction is stopped, and a shorter piece of DNA results. In a dideoxynucleotide, the OH^- group on the 3′ carbon of the sugar ring has been replaced by a hydrogen. A further nucleotide cannot be added as no phosphodiester bond can be formed. In the above example, the first complementary nucleotide should be an 'A' (the corresponding dideoxynucleotide is in tube 4). By chance, the reaction may truncate at position 1 if the dideoxynucleotide version of 'A' is incorporated. If not, the reaction proceeds to nucleotide 'A' at position 5 instead, producing a longer piece of DNA.

We now know at which positions 'A' appears in the sequence – position 1 and position 5. Taking all four tubes into account, all possible lengths of DNA are generated (*Figure 7.18*). If these fragments are now run manually on a long 6% denaturing gel (using urea as the denaturing agent), hundreds of base pairs can be sequenced. The four reactions are run in separate wells, and the sequence of bands will appear as shown in *Figure 7.18*.

High-throughput automation. Sequencing has now become an automated procedure, using computer software to reveal any changes in bases from those seen in a normal DNA fragment (see *Figures 7.19* and *7.20*).

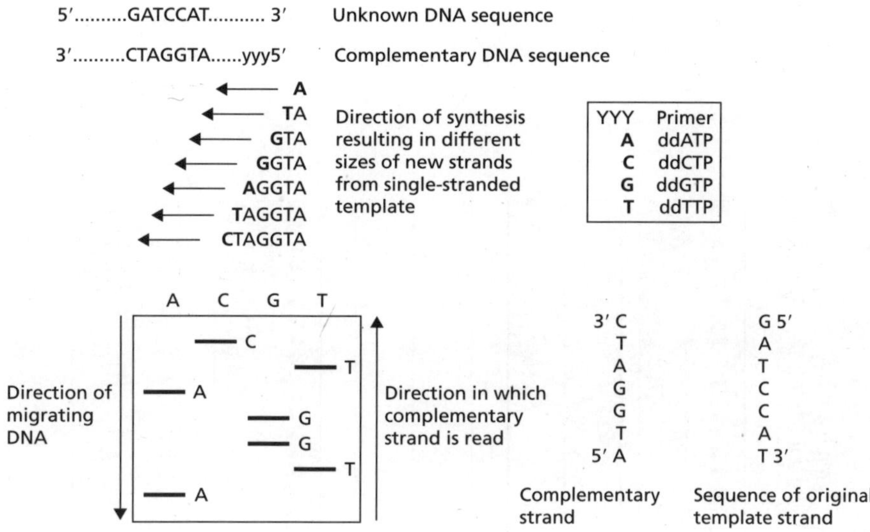

Figure 7.18
DNA sequencing.

While this may be seen as the ultimate detection system, it must be remembered that sequencing generates a vast amount of information, and is not often used as a primary screen for large genes. Sequencing tends to be used for small genes, particular exons, which contain hot-spots for mutations, or to characterize the exact nature of a change in a DNA sequence which has been detected in a primary screen by one of the methods described previously.

However, it is important to confirm that a newly discovered mutation is pathological, rather than a harmless polymorphism.

Unclassified variants: validating a pathological mutation. When setting up testing for a new disorder, or in a routine diagnostic situation, problems may arise when previously unreported changes are found in the gene of interest. These changes may be common polymorphic variants (i.e. non-pathogenic) or true pathogenic mutations. An unclassified variant is therefore a genetic variant of unknown significance. There are various ways to prove that a mutation is the cause of a disease.

- The gene should be checked for the presence of another mutation. If the variant is the only change, nonsense mutations, or mis-sense mutations with highly conserved amino acids, are not likely to be common polymorphisms. Animal models are often helpful to show evolutionary conservation.
- mRNA studies may be undertaken: does the mutation affect mRNA splicing; is the change at a functionally relevant site?
- Pedigree studies may show whether the variant tracks with the disorder.
- A control panel of unrelated normal individuals should be screened for the variant. A suspected pathogenic mutation would not be found in this group.

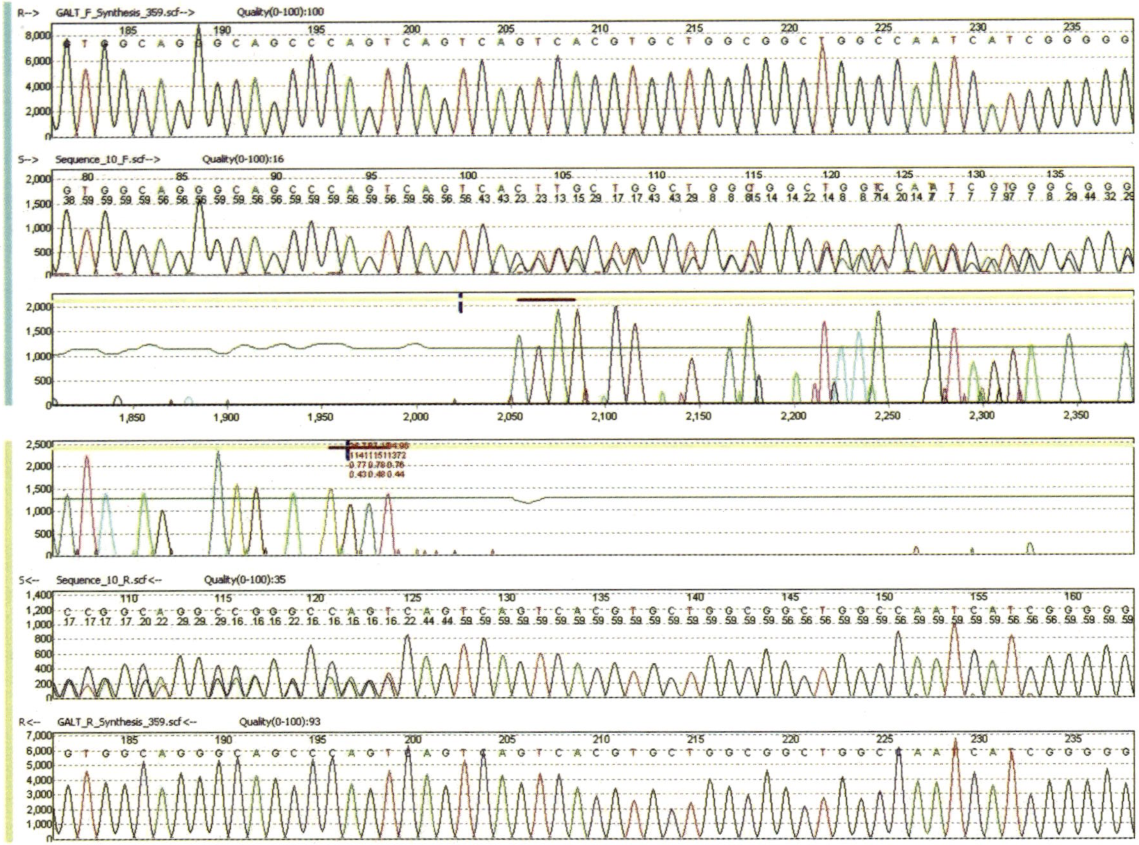

Figure 7.19

Heterozygous deletion in galactosaemia. This sequencing trace (MUTATION SURVEYOR software, Biogene) shows a heterozygous deletion. A = green, C = blue, G = black, T = red. The first (upper) and sixth (lowest) boxes are 'normal' reference sequences (forward and reverse directions respectively). The second and fifth boxes are the patient's DNA sample (forward and reverse directions respectively). The third and fourth boxes are the difference between the two (forward and reverse directions respectively).

In the second (forward) box, reading left to right, a deletion has occurred about half way along, producing a double trace due to a frameshift mutation on one DNA strand. This is mirrored in the fifth (reverse) box.

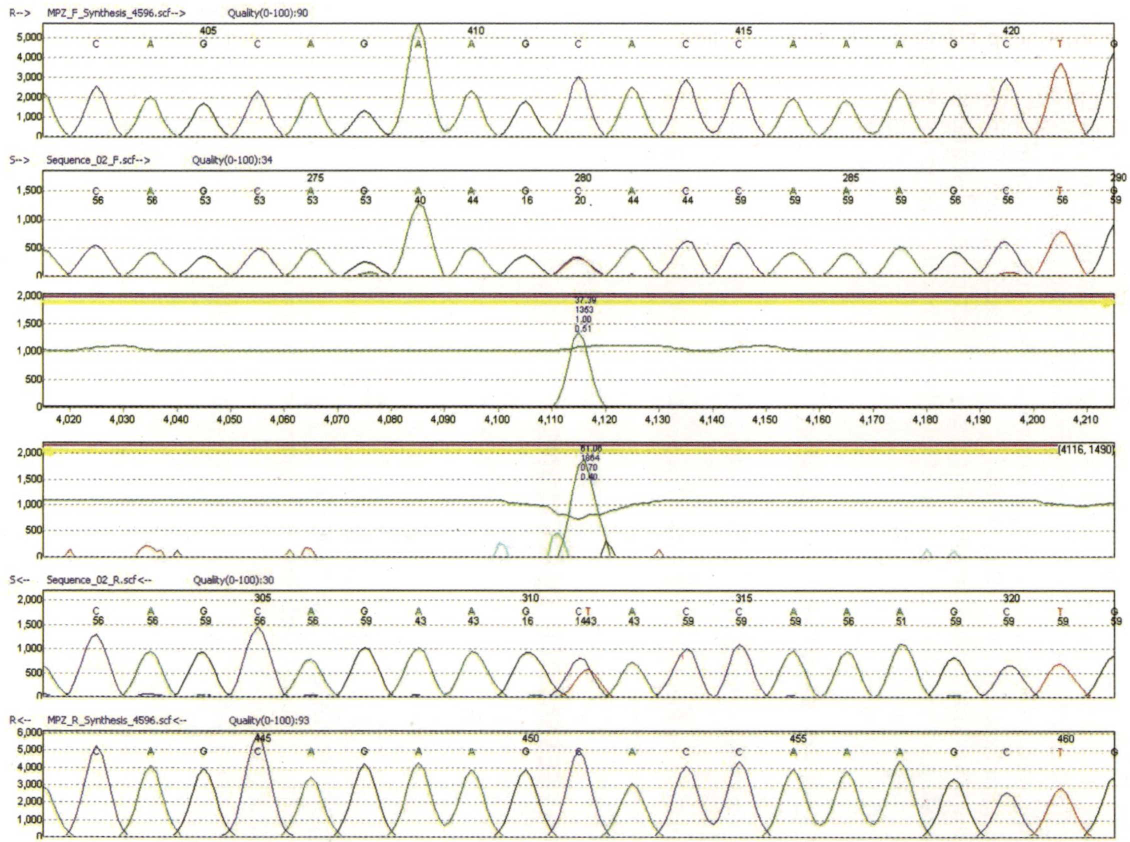

Figure 7.20

Heterozygous point mutation in myelin protein zero (MPZ). This sequencing trace reveals a heterozygous point mutation. A = green, C = blue, G = black, T = red. The first (upper) and sixth (lowest) boxes are 'normal' reference sequences (forward and reverse directions respectively). The second and fifth boxes are the patient's DNA sample (forward and reverse directions respectively). The third and fourth boxes are the difference between the two (forward and reverse directions respectively).

In the second (forward) box a C>T base substitution has occurred on one strand of DNA. Hence there are two peaks: the blue 'C' peak on the normal strand, together with a second red 'T' peak where the mutation has occurred on the other strand. This can also be seen in the fifth (reverse) box.

- Has the mutation been previously reported? There are many useful databases on the internet, often locus-specific.

7.4 SUMMARY

Since the 1970s, molecular genetics has become the tool of choice for the geneticist studying human disease at the level of the gene. Molecular analysis of genetic disorders began with Southern blots using loosely linked probes, and improved following the development of PCR for mutation analysis in clinical genetics, and for fingerprinting in forensic science. Sophisticated screening and automated sequencing technologies have subsequently been developed, and are now the method of choice in most large laboratories today.

SUGGESTED FURTHER READING

Read, A. and Donnai, D. (2007) *New Clinical Genetics.* Bloxham: Scion Publishing.

Strachan, T. and Read, A.P. (2004) *Human Molecular Genetics,* 3rd Edition. Oxford: Garland Science.

SELF-ASSESSMENT QUESTIONS

1. Name three tissues from which DNA may be extracted. What is the defining factor with respect to the cells which ensures the presence of DNA?
2. Two fragments of DNA must be separated by agarose gel electrophoresis. Their sizes are 5.2 kb and 2.8 kb. What percentage gel might you use and why?
3. Part of an intronic double-stranded DNA sequence runs as follows 5′ AGCCTCGAATTCATCGAG 3′.
   A mutation occurs such that the sequence now runs as follows 5′ AGCCACGAATTCATCGAG 3′.
   How many pieces of DNA would you get if you digested with the restriction enzyme *Taq*1 for each sequence? Would the mutation have a clinical effect on the person? What name is given to these differences in sequences?
4. Using all the DNA fragments from question 3, in which order would they run on a gel (use bp)? Now using the 1234 system, allocate numbers to each fragment.
5. Which DNA technique detects large pieces of DNA? Give an example of a disorder to which this applies.
6. What is a linked probe? What disadvantages and advantages are there to using a linked probe?
7. What is the major advantage of the PCR technique compared with Southern blotting? Why is it useful to know the sequence of the gene or exon of interest?
8. Which technical methods are suitable for the detection of large deletions?

CHAPTER

8

Cancer genetics

Learning objectives

After studying this chapter you should confidently be able to:

- **Define cancer**
  Cancer is a loss of cell cycle control such that there is inappropriate or increased cell proliferation. There is a variable genetic component which may require additional environmental factors for the development of full malignancy.
- **State the difference between a constitutional and a somatic mutation**
  A constitutional mutation is an abnormality inherited via the germ cells giving an inborn genetic predisposition to cancer; a sporadic mutation is an abnormality acquired during the lifetime of an individual. In examples shown by some leukaemias and solid tumours, the specific acquired chromosome abnormalities are consistent and predictive.
- **List examples of environmental factors in cancer**
  Examples of environmental components may include chemicals, radiation and viruses.
- **State the normal function of proto-oncogenes, tumour suppressor genes and mismatch repair genes, and outline how mutations in these genes lead to loss of cell control**
  - **Oncogenes**
    Control is lost over genes responsible for cell growth or proliferation. Oncogenes behave in a dominant manner and often arise from point mutations, translocations or inversions, and represent a gain of function.
  - **Tumour suppressor genes**
    Loss of suppression of cell growth occurs when both alleles of a tumour suppressor (TS) gene are lost. Knudson's two-hit hypothesis characterizes TS genes such that the first hit is either constitutional or somatic, and the second is somatic. The second hit is predicted to be found only in the tumour as shown by loss of heterozygosity studies. TS genes act in a recessive manner, and often arise from deletions or point mutations, and represent a loss of function.
  - **Mismatch repair genes**
    Mismatch repair genes proof-read replicated DNA for errors; a sign of mutation in one of these genes is microsatellite instability, where new alleles appear that should have been corrected. Abnormal mismatch repair genes increase the risk of other genes mutating, in turn increasing the risk of cancer.

- **Describe the role of *p53* in the cell cycle and apoptosis**
  Programmed cell death is controlled by many genes; one of the best characterized is *p53*. Failure of *p53* in one of its roles (guarding against abnormal replication of damaged cells which would normally self-destruct) may result in cancer.
- **Define Knudson's two-hit hypothesis**
  Knudson suggested that cancer or tumours may occur as the result of two consecutive mutations. In familial cancer, an inherited constitutional mutation is followed by an acquired somatic mutation. In sporadic cancers, both mutations arise somatically.
- **Describe the role of chromosome analysis in the identification and prognosis of cancer**
  Chromosome analysis is often used in identifying an exact type of tumour, leading to a prognosis and appropriate treatment. This is especially important in leukaemias such as CML, AML or ALL.
- **Outline the multistep nature of cancer together with its limitations**
  Although there are examples of cancer as a stepwise linear progression, it is possible that malignancy is the final result of many disparate pathways, requiring the mutation of caretaker and gatekeeper genes, together with those of the cell cycle checkpoints such as the guardian of the genome *p53*.

8.1 INTRODUCTION TO CANCER CYTOGENETICS

In 1890, David Hanseman suggested that there was a connection between the abnormalities he was finding in tumour cell nuclei, and the origin of cancer. Theodor Boveri, in 1914, published his hypothesis which stated that cancer does have a genetic component such that chromosomal aberrations were responsible for the origin and malignant progression of cancer.

Cancer usually develops in a series of steps. There may be an inborn **genetic** predisposition due to an abnormality inherited via the germ cells (this is a **constitutional** mutation) or an abnormality **acquired** during the lifetime of an individual (this is a **sporadic** mutation in a small number of somatic cells). Sometimes an environmental component contributes to, or causes the genetic mutations leading to the development of cancer. Examples include chemicals (for example, benzene or the tars in cigarettes), radiation or viruses. Note that a **mutagen** is an agent which can cause mutations in DNA and a **carcinogen** is any agent which may induce cancer.

There are three main classes of genes which, when mutated, may lead to cancer. These are:

- **proto-oncogenes**
- **tumour suppressor (TS) genes**
- **mismatch repair genes**

8.2 ONCOGENES

There are very few viruses in humans known to produce tumours. Human oncogenes were discovered through the study of animal tumour viruses, in which oncogenes were first identified. Between 1910 and 1914, Professor Peyton Rous showed that the lysate (clear liquid) from the tumour cells of chickens could be injected into other chickens, who subsequently developed tumours. The virus isolated from the lysate was called **R**ous **S**arcoma **V**irus (RSV).

Viruses either have a DNA or an RNA genome; the animal tumour viruses most closely studied have an RNA genome and are called **retroviruses.**

- As part of their life cycle retroviruses use the enzyme reverse transcriptase to make a double-stranded DNA copy of their RNA. This integrates into the host genome and the virus then uses the host cell's metabolism to produce new viral proteins. Unlike retroviruses, DNA viruses can replicate autonomously without being integrated into the host genome.
- Some viruses have a gene which can induce tumours in animals. This is the **viral oncogene** (v-*onc*). Retroviruses that do not carry an oncogene can inappropriately modify a host gene upon integration, or modify the expression of the host gene using a powerful promoter.
- Viral oncogenes may arise following errors in viral replication after integration into its host. When the virus is excised it takes the host gene with it.
- Animal oncogenes were first discovered due to the remarkable similarities of the viral oncogenic sequences to parts of the animal genome. This reflects the animal origin of the viral oncogene.
- There are very few examples of human retroviruses, but animal examples such as RSV enabled the characterization of human oncogenes. Examples of human tumour or cancer viruses include the Epstein–Barr virus and the papilloma virus (both of which have DNA genomes), and the retrovirus HTLV-1, which has an RNA genome and is responsible for adult T-cell leukaemia–lymphoma (ATLL).

Function of proto-oncogenes

Certain human genes code for various proteins controlling growth and differentiation. Activating these at the correct time and in the correct tissue results in normal cell proliferation.

Growth genes displaying their **normal** functions are called **proto-oncogenes.** They have the potential to be activated at the wrong time or in the wrong place; they mainly exhibit a gain of function and become **cellular oncogenes** (see *Box 8.1*). By studying what happens to cells when a proto-oncogene is transformed into a cellular oncogene, we can deduce the normal function of that gene. The activation of an oncogene generally results in a **loss of control of cellular proliferation and differentiation.** This implies that the normal role of proto-oncogenes often involves the cell cycle. The

Box 8.1 Oncogene nomenclature

As viruses can pick up human or animal cellular genes, a nomenclature has developed using a prefix to differentiate between cellular and viral oncogenes. The oncogenes themselves are usually represented by three letters which reflect the tumour origin:

*MYC** = Avian **My**elo**c**ytomatosis
v-*onc* = **v**iral oncogene
c-*onc* = homologous **c**ellular proto-oncogene

The distinction between **proto-oncogenes** (the c-*onc* genes) and the term **oncogene** is gradually becoming outmoded; the term oncogene is often used for the normal gene and the term activated oncogene is used for the abnormal version (see *Chapter 18* of Strachan and Read (2004)).

*According to the 1987 Guidelines for human gene nomenclature, human genes are written in italicized capitals. Their gene products are written in non-italicized capitals (see also *Chapter 2*).

Table 8.1 Classes of oncogenes

Group	Comments	Example
Growth factors	The viral oncogene v-*sis* is homologous to part of the human growth factor gene *PDGFB*	*SIS*
Growth factor receptors	These cell surface receptors receive signals from the growth factors	*ERBB*
Signal transduction systems	Involved in intracellular signalling	*HRAS*
Nuclear proteins	Bind DNA (e.g. transcription factors)	*MYCN*
Cell cycle-related	Play a normal role in progression through the cell cycle (e.g. cyclins)	*MDM2*

finding that proto-oncogenes may code for growth factors and their receptors (see *Table 8.1* for more details) supports this.

Oncogenes tend to be grouped into classes based on either the normal genes they code for, or where their proteins are found in the cell.

How oncogenes exert their effect

Oncogenes exert their effects in two main ways (*Figure 8.1*):

- increasing the amount of normal protein
- production of an altered protein

A proto-oncogene may be induced to produce an increased amount of product. This may also be in the wrong cell type. The result will be **uncontrolled overexpression.**

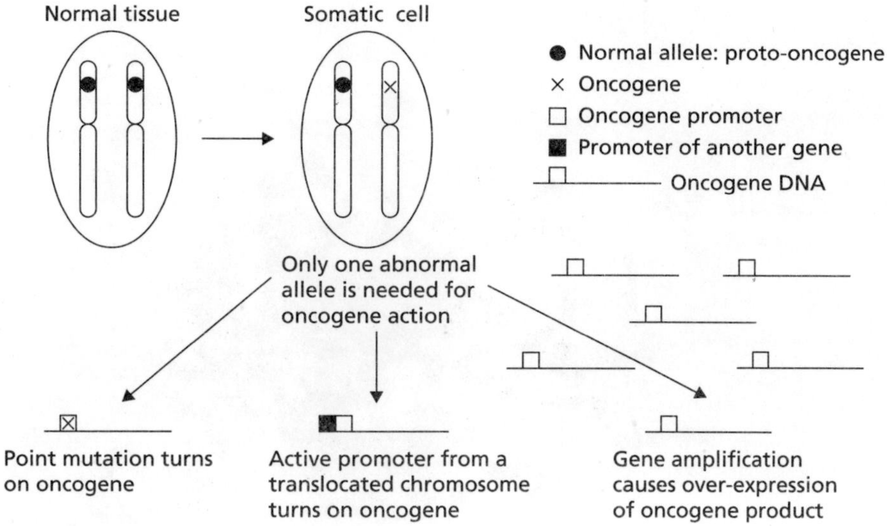

Figure 8.1
Mode of action of oncogenes.

Proto-oncogenes may be amplified such that there are tens or hundreds of copies of the resultant cellular oncogene in the cell nucleus. For example, the *MYC* gene may be amplified in response to exposure to methotrexate, a change which confers resistance to this chemical.

In the case of the childhood thoracic/abdominal tumour neuroblastoma, the amplification of the oncogene *MYCN* (on chromosome 2q) is sometimes visible cytogenetically in chromosome preparations as tiny paired spots of DNA called double minutes (*Figure 8.2*). The amplified *MYCN* can also appear as non-banding pieces of translocated DNA called HSRs (homogeneously staining regions).

Normally inactive proto-oncogenes can be activated by moving them next to an active promoter. This happens in Burkitt lymphoma, where the chromosomes may display three types of translocation. The most common is the t(8;14), but t(2;8) and t(8;22) are also seen. The *MYC* oncogene on chromosome 8 is translocated next to the active promoter of the immunoglobulin heavy chain gene on chromosome 14, and is subsequently activated itself. The loci on chromosomes 2 and 22 are genes coding for the immunoglobulin kappa and lambda light chains, respectively.

Gene activation may also occur through the creation of a hybrid or chimeric gene, caused by one translocated gene fusing with another. An example is seen in chronic myeloid leukaemia (CML), where the *ABL* gene on 9q34 is reciprocally translocated to chromosome 22, and fuses with the breakpoint cluster region (*BCR*) at 22q11.

A proto-oncogene may be turned into a cellular oncogene by mutations in the DNA. Point mutations have been found in the *RAS* oncogene, for example.

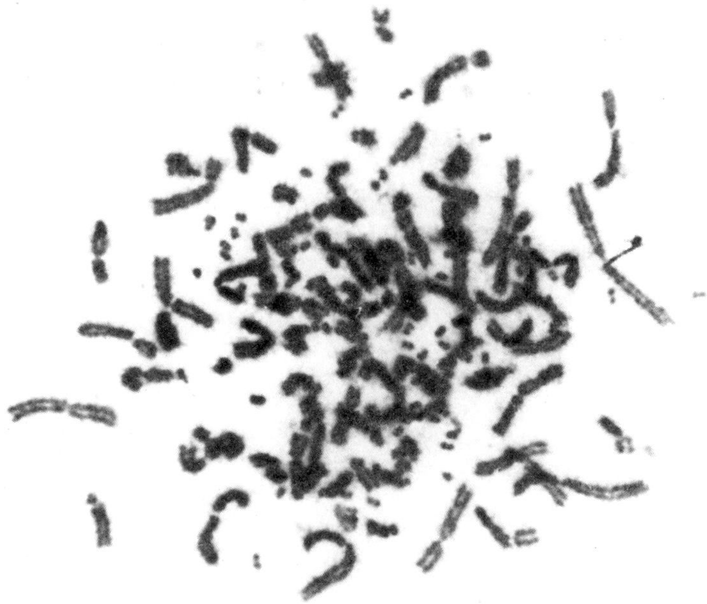

Figure 8.2
Chromosomes from a metaphase cell of a neuroblastoma showing double minutes.

Summary

Oncogenes appear to act in a dominant fashion; they have been compared to the accelerator on a car – once activated, a direct effect results in abnormal cell proliferation. Oncogene mutations are usually acquired, not inherited. One exception is the *RET* proto-oncogene, in which dominantly inherited point mutations may cause multiple endocrine neoplasia type 2a (MEN2a).

8.3 TUMOUR SUPPRESSOR GENES

Tumour suppressor genes act in a recessive manner such that **both** copies (or alleles) of the normal gene have to be lost on homologous chromosomes before the onset of malignancy. If malignancy is the uncontrolled growth or proliferation of cells, then the normal function of these genes is to suppress or control cell growth and progression.

The existence of tumour suppressor (**TS**) genes can be revealed in two ways. The first uses hybrid cell lines, whereby normal cells are mixed with cells from a malignant cell line; the hybrid line resumes normal control of cell growth, and the excess growth of the tumour cell line is suppressed. This implies that normal gene alleles can override absent or abnormal TS function.

The point at which uncontrolled growth occurs can sometimes be related to the loss of whole or parts of certain chromosomes. Again, this implies that TS genes were present on a particular chromosome and have now been lost, resulting in uncontrolled cell proliferation.

Knudson's two-hit hypothesis

In 1971, Alfred Knudson statistically analysed children with the eye tumour retinoblastoma, and provided the first practical example of a TS gene using his theory which he formulated from his epidemiological data.

Retinoblastoma is a childhood tumour of the eye that usually appears before the age of three years. It can be found in one eye (unilateral) or both eyes (bilateral), and comprises a mass of undifferentiated retinal cells. Knudson compared the incidence of unilateral or bilateral tumours in patients with or without a family history, and also with the age of onset. In patients with a family history (around 40%), the tumour often arose in both eyes, and had an earlier age of onset. In patients with no family history (around 60%), the tumour was more likely to be unilateral, and have a later age of onset. Knudson suggested that the disease occurred as a result of two consecutive mutations arising in two different ways in families with a history (i.e. hereditary) and without a history (sporadic).

Deletions or mutations of the retinoblastoma gene (*RB1*) lead to absence of the retinoblastoma protein pRB. This can happen in one of two ways.

- In the familial form, the first mutation is already present in the retinoblasts, having been **inherited** from the parental germline – the mutation is therefore **constitutional**. The second mutation is **somatic** and occurs in a retinal cell already bearing a constitutional mutation, therefore **both alleles are lost** (*Figure 8.3*). The presence of a constitutional mutation (the first hit) predisposes a cell to a second somatic mutation (the second hit), therefore the age of onset of retinoblastoma is earlier and the eye tumours are more likely to be bilateral.
- In the later onset **sporadic** unilateral tumours, **both mutations** (or hits) **arise somatically**; the chance of two allelic mutations eventually occurring in the same cell is low and therefore rare. The probability of a second hit is higher than the probability of the first hit, however, as some of the mechanisms leading to the second hit depend on the existence of the first hit (see section on *Loss of heterozygosity* below, and *Box 8.2*).

The two-hit theory was later shown to be true when microscopically visible deletions on the long arm of chromosome 13 (13q14) were found in

Box 8.2 Malignant changes following sunburn

A good reason for not letting young children get sunburned early in life is because their skin cells may receive their first 'hit' at this time from the sun's ultraviolet light, which damages their DNA. This increases their chance of a second 'hit' later in life if they are often exposed to the sun, and may set in motion the malignant changes found in, for example, melanomas.

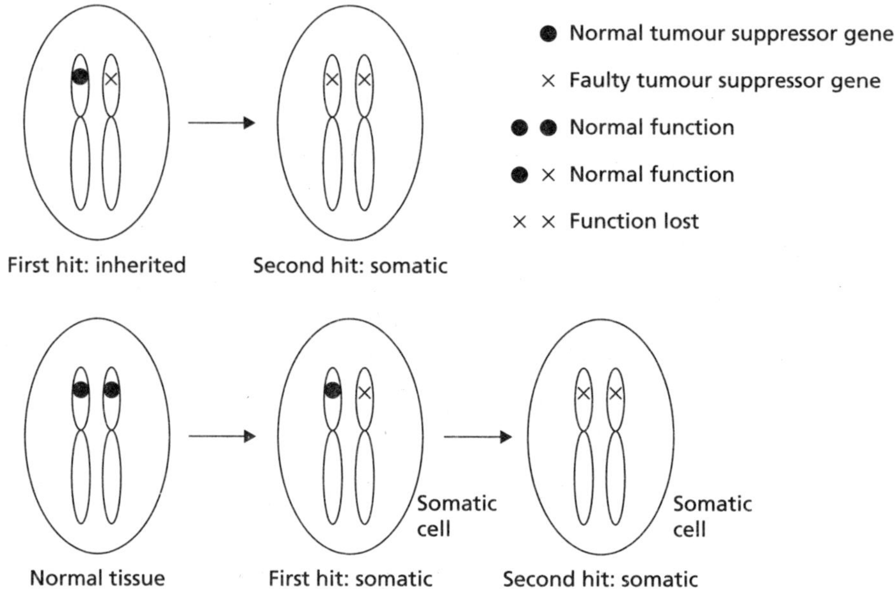

Figure 8.3
Knudson's two-hit hypothesis.

retinoblastoma patients. Non-visible deletions and mutations were eventually confirmed by molecular means, and the retinoblastoma gene is now fully characterized (see *Box 8.3*).

Loss of heterozygosity

By comparing the patient's somatic cells with their tumour cells, there should be a difference at the molecular level, such that there is no working copy of *RB1* remaining in the tumour cells. This may arise by various mechanisms (*Figure 8.4*), but is sometimes detectable using molecular DNA markers. It can be shown, for example, that an individual may be heterozygous for a certain polymorphic molecular marker in their somatic cells, but be homozygous for that marker in the tumour cells due to loss of the whole chromosome or deletion of that marker. This phenomenon is known as **loss of heterozygosity** (LOH), and is a common feature of mutated tumour suppressor genes.

Box 8.3 Function of pRB

The retinoblastoma gene is involved with regulation of the cell cycle. If pRB is unphosphorylated, it can bind another factor E2F, suppressing progression through the cell cycle. If the pRB protein is phosphorylated, it releases the E2F transcription factor which causes the cell to cross the G_1 checkpoint and enter S, hence promoting cell growth and division. *RB1* is therefore said to connect the cell cycle clock with the transcriptional machinery.

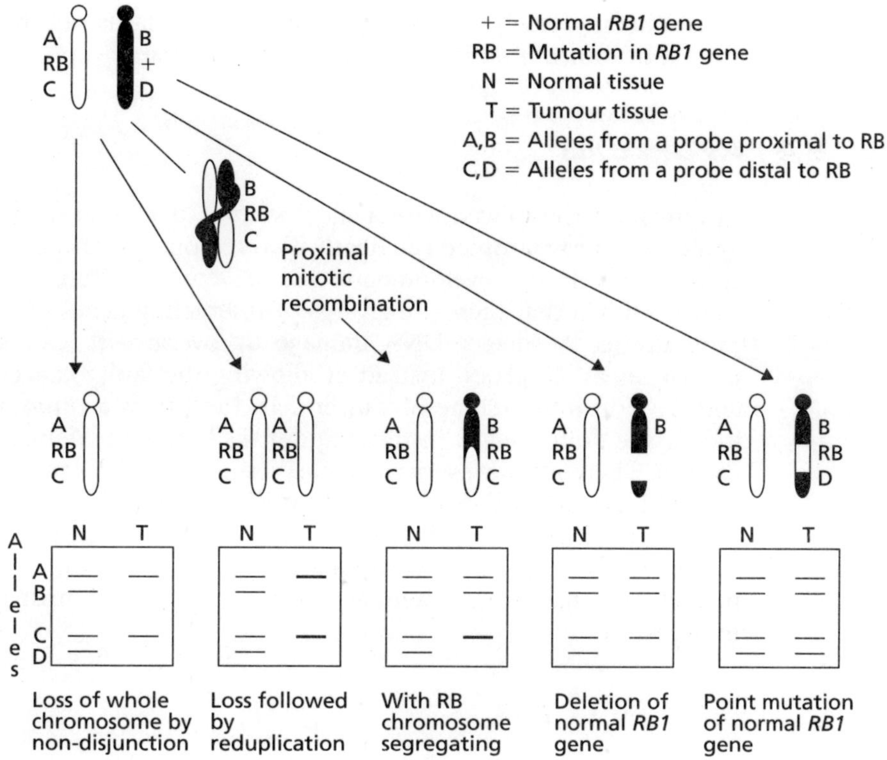

Figure 8.4
Mechanisms leading to the loss of heterozygosity.

A Wilms' tumour is an embryonal kidney tumour of children. Knudson's two-hit theory can also be applied here, but the sporadic tumours comprise 99% of the total compared with 1% of constitutional tumours. There is an association of Wilms' tumours with a deletion of the short arm of chromosome 11 (11p13), and LOH can often be demonstrated. However, there are other Wilms' tumour loci, so the 11p13 locus is usually designated *WT1*.

Neuro-oncology

LOH can be used to predict chemosensitivity in certain brain tumours known as oligodendrogliomas. Markers from chromosomes 1p and 19q are compared in separate DNA samples extracted from blood and tumours. The finding of LOH is strongly associated with chemosensitivity, and is a good prognosis for treatment.

Similarly, DNA methylation analysis of gliomas can be undertaken. A DNA repair protein called MGMT can counteract the effects of some chemotherapeutic agents. In some tumours the MGMT promoter is methylated, therefore the gene is inactivated, and the tumours respond better to chemotherapy. Analysis of MGMT methylation status can therefore be used as an indicator to predict response to chemotherapy.

8.4 APOPTOSIS

Apoptosis (pronounced with a silent second 'p' so that it sounds more like apotosis) is **programmed cell death**. Two reasons why a cell may be forced to commit suicide are developmental and prevention of tumour progression.

It is known that there are groups of interacting genes in a cell which are programmed to detect DNA damage or overgrowth caused by mutated oncogenes or TS genes. Instead of allowing the faulty cancer cell to divide and develop into a clone of cancer cells (and then a tumour), these genes induce the faulty cell to instigate apoptosis before the damage grows out of control. There is always a delicate balance between cell growth, promoted by the cell cycle, and cell death, promoted by apoptosis. In some circumstances, cancer could be seen as a lack of cell death due to failure of apoptosis.

One of these genes is so frequently mutated in human cancer that its normal role has been extensively studied; this gene must be particularly important.

The role of p53

The gene ***TP53*** (**t**umour **p**rotein **53**) codes for the protein product TP53, which is a transcription factor (see *Box 8.4* for an explanation of nomenclature). The gene locus is 17p13.1. TP53 is thought to have three roles in normal cell function; it is involved in:

- cell cycle checkpoints and genetic stability
- apoptosis
- differentiation and development

As a transcription factor, p53 is involved in the cell cycle by regulating genes which cause progression through that cycle. It appears to monitor the cell for DNA proof-reading or repair errors. When the genomic DNA is damaged, which may lead to tumour growth, p53 can activate genes which arrest faulty cells in G_1 and can promote their death or apoptosis. p53 may also participate in the G_2 checkpoint, as there is evidence from mice that lack of p53 disrupts the cell's ability to monitor correct spindle assembly.

p53 probably exerts its effects on many genes, but two examples known to be targeted by p53 are *BCL2* and *BAX* (see *Box 8.5*). Whereas overexpression

Box 8.4 Nomenclature of TP53

TP53 gets its name from the molecular weight of the protein product: 53 000 Daltons. The gene is often written in the commonly accepted form *p53*; the corresponding protein nomenclature is then p53.

Box 8.5 Function of *BCL2*

BCL2 is a member of a family of cell death regulators. Its locus is 18q21, and it is expressed in cells which need a long lifespan (so apoptosis is undesirable) such as memory B cells. In follicular B cell lymphoma, the translocation t(14;18)(q32;q21) creates a *BCL2*/immunoglobulin fusion gene. The immunoglobulin promoter ensures expression of BCL2 and a survival advantage in tumour cells.

of BAX accelerates apoptosis, BCL2 inhibits BAX, and therefore inhibits apoptosis. Expression of p53 causes a decrease in BCL2 and an increase in BAX.

As the tumour cells are killed early on, apoptosis essentially prevents tumour progression (*Figure 8.5*). The third role of p53 is to trigger cell differentiation that consequently restricts the proliferation of genetically damaged cells. Lack of p53 has been known to cause failure of neural tube closure in mice.

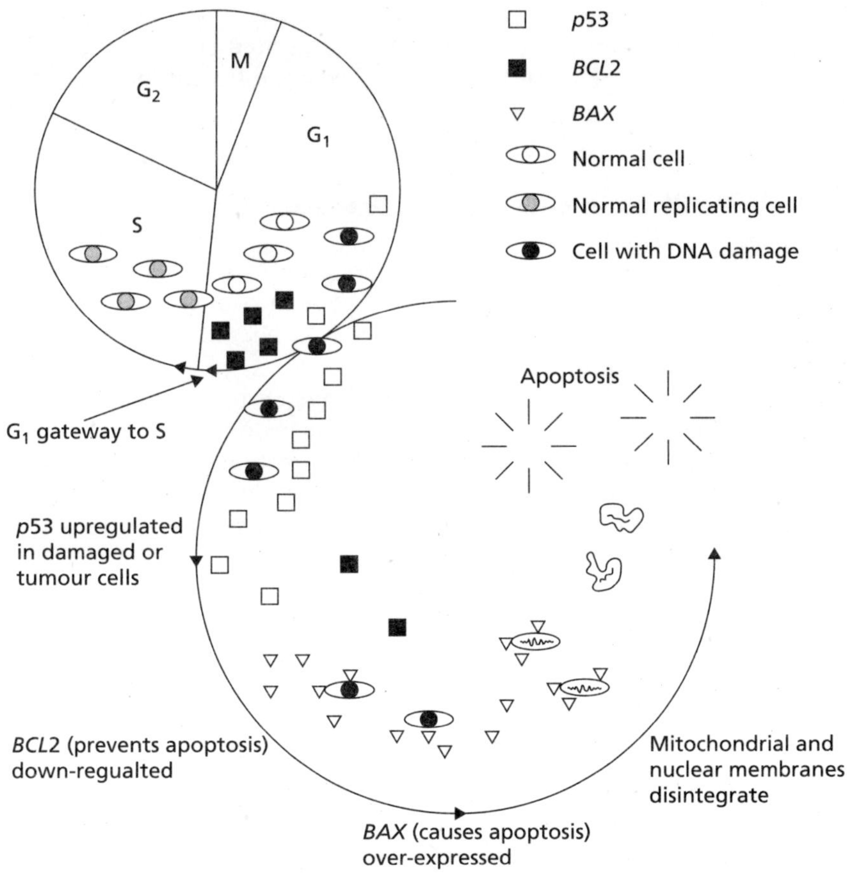

Figure 8.5
p53 and its role in the cell cycle and apoptosis.

The *p53* gene is usually regarded as a TS gene because *p53* disruption causes loss of TS function. Apoptosis does not occur and unregulated tumour progression ensues. Some mutations can lead to the p53 protein acting as a tumour promoter. *p53* is an unusual gene, however, in that it can also act in a dominant manner. The mutant protein product can act in a **dominant negative** fashion in the presence of normal p53 by forming complexes with and inhibiting the wild type (normal) protein.

Because p53 can stop cell proliferation by its action on the cell cycle and stop replication of damaged DNA, it has been called **the guardian of the genome**.

p53 and cancer

Mutations in *p53* are the most frequent secondary changes in many cancers; point mutations and deletions of this gene occur in 70% of all tumours. *p53* mutations are found in 60% of breast tumours, and are frequently found in lung cancer and colorectal cancer.

Li–Fraumeni syndrome is an autosomal dominant disorder whereby *p53* point mutations in one allele are inherited constitutionally (i.e. through the germ line). Half of the carriers develop different kinds of cancers by the time they are 30 years old, compared with just 1% of the general population.

PTCH as a tumour suppressor

The involvement of developmental genes in cancer has been suspected for a long time; the **patched** (***PTCH***) gene introduced in *Chapters 2* and *3* is also a TS gene, producing a cytoplasmic protein which is important in the hedgehog signalling pathway.

Normally the hedgehog morphogen is bound by the PTCH receptor, but if *PTCH* is lost or mutated, other gene products are abnormally expressed due to the presence of the hedgehog morphogen. As PTCH is cytoplasmic, the first effects of a faulty *PTCH* may be abnormal adhesion of cells such that they heap up and proliferate. This then provides a target for other genetic faults.

It has been suggested that *PTCH* may be an example of a **gatekeeper** gene, which directly regulates tumours either by inhibiting growth or promoting death. If a gatekeeper gene is inactivated, a genetic threshold is passed and the tumour process begins – at first in one cell of a particular tissue, closely followed by expansion from that cell. The gatekeeper may be tissue-specific – even if other proto-oncogenes are hit, as long as the gatekeeper of that tissue is intact, progression should not occur.

Because mutation or loss of *PTCH* results in basal cell carcinoma (BCC), patched may be the gatekeeper of BCC – or even of all common skin cancers.

Another example of a gatekeeper is found in the gene for Von Hippel–Lindau (VHL) disease, where mutations may give rise to cancer of the kidneys.

Summary

TS genes appear to act in a recessive fashion such that both alleles have to be dysfunctional or deleted before control of cell suppression is lost – resulting in tumour growth.

If an oncogene is compared to a car accelerator, then TS genes could be seen as a pair of brakes – both of which must be lost before the car (or cell!) goes out of control. TS gene mutations may either be constitutional or somatic.

8.5 MISMATCH REPAIR GENES

A third class of genes associated with cancer, originally seen in yeast and bacteria, have now been found in humans. In the bacterium *E. coli* these are known as mutator genes; in humans as **mismatch repair genes**. These genes are believed to code for enzymes which can 'proof-read' DNA, and are therefore able to detect mismatched base pairs. The mismatching may arise through DNA replication errors, or have an environmental cause (see *Boxes 8.6* and *8.7*).

Inactivation of mismatch repair genes leads to instability of the genome, thus increasing the risk of further mutations. In accordance with the guardian/gatekeeper analogy, these genes have been called **caretakers**. Other examples of caretaker genes include the excision repair genes of xeroderma pigmentosum, the breast cancer genes *BRCA1* and *BRCA2*, and possibly the *ATM* gene of ataxia telangectasia.

Using the car analogy once again, mismatch repair genes have been compared by Strachan and Read (2004) to a vehicle which goes out of control because some of the nuts and bolts are defective (see *Box 8.8*).

Box 8.6 HNPCC and microsatellite instability

In the autosomal dominant disease **h**ereditary **n**on-**p**olyposis **c**olon **c**ancer (HNPCC), new alleles may be seen in the tumour tissue when polymorphic microsatellite markers are used. This is called microsatellite instability. There are six human mismatch repair genes. Two are due to gene mutations in the same chromosome regions as those found in HNPCC patients:

hMSH2 at 2p15–22
hMLH1 at 3p21.3

Box 8.7 Predisposition to colon cancer

In some families, individuals develop colon cancer at a very young age. It may be that this predisposition is constitutionally inherited, and may involve mutations in mismatch repair genes.

Box 8.8 Breast cancer

Breast cancer (see also *Section 8.7* for further details) may arise from mutations at two major gene loci:

- the *BRCA1* locus is 17q21 and is associated with breast and ovarian cancer
- the *BRCA2* locus is 13q12–13 and is associated with female and male breast cancer

However, although these tumour suppressor mutations are inherited in *BRCA1* and *BRCA2*, they are absent in sporadic tumours. This implies that they do not follow Knudson's two-hit hypothesis and is consistent with the theory that breast cancer genes are caretaker genes whose normal gene products (which are expressed at G_1/S) are essential cofactors in DNA repair.

The risk to a British woman of developing breast cancer is between 1 in 8 and 1 in 12. This risk increases if there are close female relatives with breast cancer. Around 5–10% of breast cancer is familial. If a familial mutation has been inherited by an individual, counselling is advisable, as she has an 80–90% risk of the cancer developing (not 100%). If she does not inherit the mutation, her overall risk only falls back to that of the general population – she is not completely free of risk.

The mutations in breast cancer are usually analysed by DNA sequencing (see *Section 7.3*). Strict criteria must be met before an individual is accepted onto a screening programme; this usually requires the female consultand to have two (or sometimes three) affected first degree relatives.

8.6 CYTOGENETICS AND THE INVESTIGATION OF MALIGNANCY

The genetic alterations that disrupt the normal balance between cell proliferation, survival and differentiation, and result in cancer, are often seen as visible changes to the chromosomes. These clonal changes have been described in all major tumour types and more are being discovered using improved cytogenetic and molecular genetic techniques (see *Chapter 6*). A **clone** is defined as a cell population derived from a single progenitor cell.

These alterations are **acquired** during the cancer process and are only present in the tumour cells, while the normal non-cancerous cells retain their **constitutional** somatic karyotype. There is now substantial evidence that these alterations are early or even initiating events in tumourigenesis. For instance, studies of identical twins and retrospective study of neonatal blood spots or Guthrie cards used in newborn screening (see *Chapter 9*) have shown that certain translocations associated with childhood leukaemia can arise *in utero*, years before the appearance of the disease.

It was in 1960 that Nowell and Hungerford described the first specific chromosomal abnormality associated with cancer, the Philadelphia (Ph) chromosome. It is found in almost all patients with chronic myeloid leukaemia (CML) and is the result of a reciprocal translocation between chromosomes 9 and 22 (see *Figure 8.6a*). Since then numerous karyotypic changes of tumour cells have been described, and it has also been noted that they are unevenly distributed throughout the genome. Different chromosomes, regions and bands seem to be preferentially involved in different cancers. However, a steadily increasing number of abnormalities are found to be associated with particular disease or disease subtypes (see *Box 8.9* and *Figure 8.6a–e*). Many different types of chromosome abnormalities may be encountered including numerical aneuploidy, polyploidy, and structural

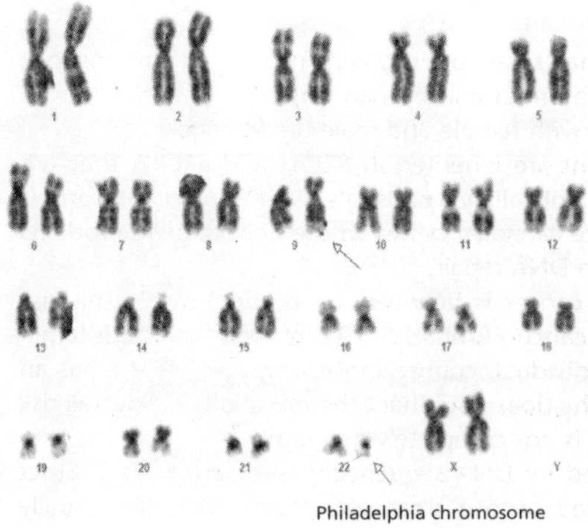

(a)

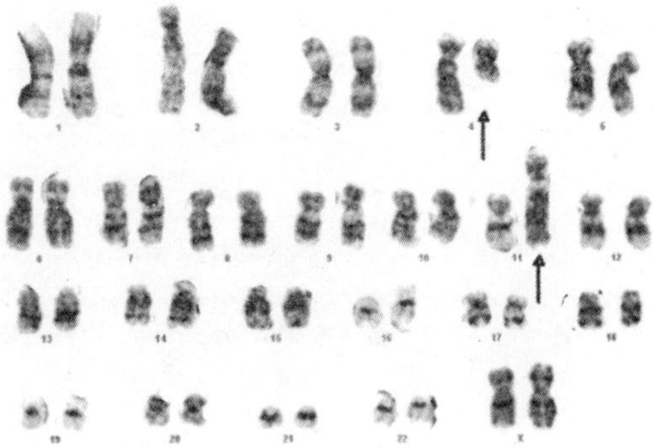

(b)

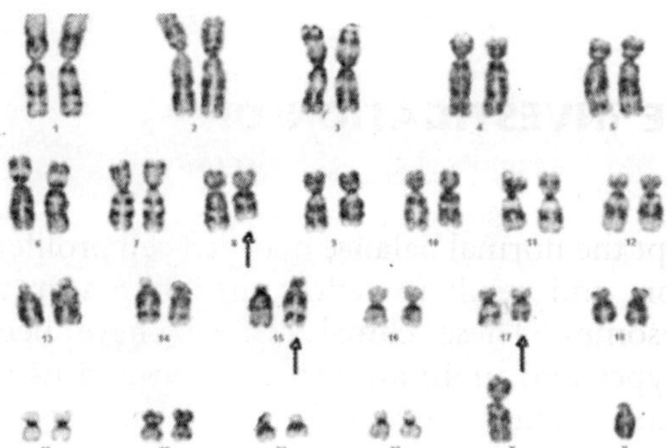

(c)

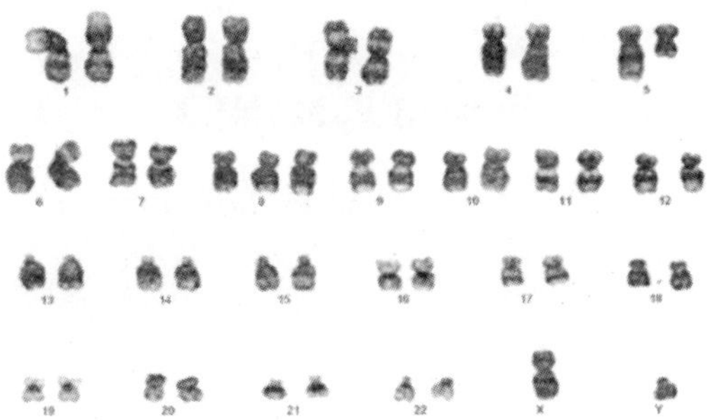

(d)

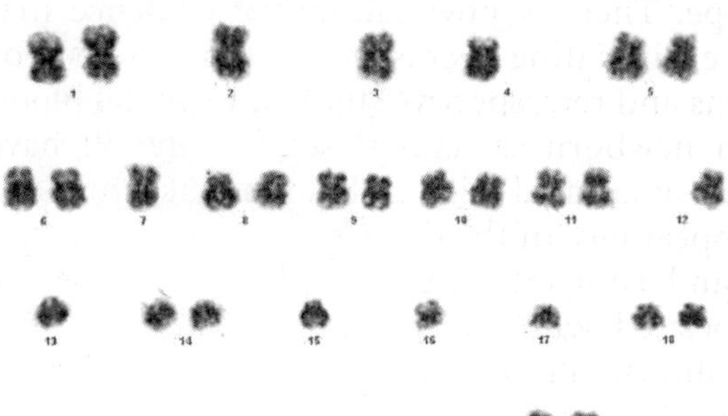

(e)

Figure 8.6
(a) A female karyotype from a patient with chronic myeloid leukaemia and the typical 9;22 translocation resulting in the Philadelphia chromosome.
(b) A female karyotype with a 4;11 translocation in a child with acute lymphoblastic leukaemia.
(c) A male karyotype with a 15;17 translocation in an adult with acute myeloid leukaemia (acute promyelocytic leukaemia). Note also the additional abnormality, a deleted chromosome 8.
(d) A male karyotype with a deleted chromosome 5 in a patient with myleodysplastic syndrome (MDS). Note also trisomy 8.
(e) A hypodiploid karyotype of 37 chromosomes with single copies of chromosomes 2, 3, 4, 7, 12, 13, 15, 16, and 17. This was found in a 12-year-old girl with acute lymphoblastic leukaemia. Hypodiploidy is associated with a poor prognosis and requires more intensive therapy.

Box 8.9 Examples of recurrent chromosomal abnormalities in cancer

For more information on types of chromosome abnormalities see *Chapter 5*.

Type		Disorder	Genes
Genomic losses			
Deletions	del 5q	Acute myeloid leukaemia (AML); myelodysplastic syndromes (MDS)	Unknown
	del(5)(q12–q22)	Colon cancer	*APC*
	del(9)(p13)	Acute lymphoblastic leukaemia (ALL)	*PAX5*
	del 7q	AML/MDS	Unknown
	del(13)(q14.2)	Retinoblastoma	*RB1*
	del(17)(p13.1)	Various cancers	*p53*
Genomic gains			
Isochromosome	i(6)(p10)	Retinoblastoma	*RB1*
	i(17)(q10)	Various cancers	Unknown
Whole chromosomes	+12	Chronic lymphocytic leukaemia	Unknown
	+8	AML, MDS	
Amplification	(17)(q21.1)	Breast and other cancers	*ERBB2*
	(8)(q24.21)	Various cancers	*MYC*
Structural changes			
Inversions	inv(3)(q21;q26)	AML/MDS	
	inv(10)q11.2;q21)	Papillary thyroid cancer	*RET-CCDC6*
	inv(16(p13;q22)	AML	*CBFB-MYH11*
Translocations			
	t(2;5)(p23;q35)	Anaplastic large cell lymphoma	*ALK-NPM1*
	t(8;21)(q22;q22)	AML	*RUNX1-RUNX1T1*
	t(9;22)(q34;q11.2)	CML, ALL, AML	*BCR-ABL1*
	t(11;22)(q24;q11.2)	Ewing sarcoma	*FLI1-EWSR1*
	t(X;18)(p11;q11)	Synovial sarcoma	SYT-SSX
	t(15;17)(q22;q21)	Acute promyelocytic leukaemia (APL)	*PML-RARA*
	t(14;18)(q32.22;q21.3)	Follicular lymphoma	*IGHG1-BCL2*

Adapted from Frohling and Dohner (2008).

alterations such as translocations and deletions. Sometimes a tumour karyotype contains numerous abnormalities, many more than could be tolerated by normal cells.

Cytogenetics and the haematological malignancies

It is the haematological malignancies that have been most widely studied by cytogeneticists. This is due in part to the ease with which it is possible to obtain chromosome preparations in these disorders.

The word **leukaemia** comes from a Greek word which literally means 'white blood'. Leukaemia is often referred to as cancer of the blood. The term refers to a group of closely related malignant conditions affecting the immature blood-forming cells of the bone marrow. In leukaemia, normal control mechanisms break down and the marrow starts to produce large numbers of abnormal white blood cells of an identical type. This disrupts the normal production of blood cells leading to anaemia and a low platelet count. Often the spleen and liver will become enlarged and sometimes the lymph glands as well. There are a number of different types depending on which cells are affected.

Leukaemia accounts for approximately 1% of all cancers and is the most common cancer in children and young adults; it is the main cause of death in children and young adults, and the ultimate cause of death in 1 in 200 of the UK population. Approximately 38% of childhood cancers and 2% of adult cancers are a type of leukaemia.

Leukaemia can be classed as acute, where the disease progression is rapid, or chronic, with slow disease progression sometimes lasting several years. Leukaemia can also be classified as lymphoid, involving T or B lymphoid precursors, or myeloid, involving cells from the myeloid series including granulocyte, monocyte, erythroid or platelet precursors (see section below on *Classification of leukaemia*).

All differentiated blood cells arise from the bone marrow and therefore it is generally bone marrow cells which are cultured in order to prepare the chromosomes for analysis in patients suspected of having leukaemia. In most haematological malignancies the abnormal cancer cells are dividing rapidly *in vivo*. The aim of the analysis is to look for the spontaneous mitotic activity of the abnormal cells carrying the genetic changes, the abnormal clone or clones, amongst the normal cells.

The culture regime can be tailored to the type of leukaemia by changing the length of time in cell culture and the presence or absence of mitogens such as PHA, TPA or pokeweed (see *Chapter 5*). For example, the mitogen TPA is used to investigate mature B cell disorders. By using a range of culture conditions and examining a large number of cells (usually at least 20), the chance of detection of the abnormal clone amongst the normal cells is maximized.

Successful cell culture of solid tumours is more difficult as they often show poor growth in culture. They may also be contaminated with the presence of normal cells which can also grow and may mask the abnormal cells.

When a chromosome study is undertaken on malignant tissue, the starting point is usually conventional G-banded analysis. This can be difficult because the tissue culture often yields poor quality chromosomes, there may be few cells that are suitable for examination, and very complicated chromosome changes may be present. Karyotypic analysis may fail to detect every genetic change due to cryptic rearrangements or failures of cell culture.

Chromosome analysis can also be valuable in some solid tumours, such as synovial sarcoma and the childhood tumours such as Wilms' tumour, neuroblastoma, Ewing sarcoma and rhabdomyosarcoma, where specific changes may be found (see *Box 8.10*). These changes, as for the haematological malignancies, may also indicate prognosis and help with diagnosis.

Box 8.10 Examples of solid tumours where cytogenetic analysis is helpful

Neuroblastoma comprises 6–10% of all childhood cancers and 15% of cancer deaths in children. There are about 650 new cases per year. It is a neuroendocrine tumour arising from any neural crest element of the sympathetic nervous system.

Synovial sarcoma is a rare form of cancer that usually occurs near to the joints of the arm or the leg. It occurs most commonly in the young with a peak incidence before the 30th birthday.

Ewing sarcoma is a malignant round cell tumour, in which cancer cells are found in the bone or in soft tissue. It occurs most frequently in male teenagers but is very rare, with an average of only six new cases per year in the UK.

Wilms tumour is a tumour of the kidney that typically occurs in children, and only rarely in adults. The majority occur in otherwise normal children and a minority (25%) are associated with other developmental abnormalities (see WAGR in *Appendix: Glossary of disorders*).

Significance of chromosome abnormalities in haematological malignancies

Analysis of chromosomes from leukaemia patients is valuable for a number of reasons. The presence of an abnormal clone itself indicates that there is a malignancy present. Furthermore, some rearrangements are consistent with particular types of leukaemias. For example, the translocation between chromosomes 15 and 17 is found in acute promyelocytic leukaemia. The presence of a balanced rearrangement, for example a translocation in every cell, especially one that is not a change known to be associated with cancer, may indicate that the rearrangement is constitutional and not due to the cancer. The finding of this in every cell analysed from a blood sample would confirm this suspicion. This karyotype would be designated with a 'c' e.g. 46,XX,t(2;4)(q23;q22)c.

Certain chromosome changes can indicate the prognosis for the patient, as some abnormalities are associated with favourable, intermediate or poor prognosis. This information can inform the decision on the treatment given (see *Box 8.11*). Other factors that affect prognosis include age and sex of the patient, whether the patient achieves remission, whether the patient undergoes relapse, and the presence of leukaemia cells in the blood.

Box 8.11 Examples of prognostic significance of chromosome rearrangements in acute myeloid leukaemia

Prognosis	Chromosome change	Alive at 5 years (%)
Favourable	t(8:21)	68
	Inv(16) or t(16;16)	60
	t(15;17)	64
	Normal karyotype	41
Intermediate	Del 9q	60
	+22	55
	+8	49
	+21	47
	11q23 rearrangements	44
	Del 7q	25
	Normal karyotype	41
Unfavourable	Complex karyotype	20
	Abnormalities of chromosome 3	15
	–7	10
	5q–	10
	–5	4
	Normal karyotype	41

Data adapted from Medical Research Council AML10 trial.

Cytogenetic analysis continues at regular intervals once a diagnosis of leukaemia is made to follow the outcome of treatment. Chromosome analysis can identify patients in remission after treatment; this is indicated by the loss of the original chromosome abnormalities found at diagnosis. This does not necessarily mean that the patient is cured, as there may be very small number of disease cells still present but not detectable, i.e. minimal residual disease (see below). Patients who have relapsed can be identified by the return of the original acquired abnormalities noted at diagnosis and/or the presence of new abnormalities.

Progression of the disease can be identified by additional abnormalities observed due to evolution of the clone which may then show more genetic changes (see *Box 8.12*).

Classification of leukaemia

Leukaemia can affect any of the cells in the blood or bone marrow. The World Health Organization (WHO) has devised a classification of tumours which is based on their histopathological and genetic features, as follows:

Box 8.12 Cytogenetic evolution in classic Ph-positive CML

Although occasional CML patients have other changes early in the disease, in typical cases the 9;22 translocation remains the sole abnormality throughout most of the chronic phase. When disease progression occurs, however, 75–80% develop additional chromosome aberrations. These secondary changes may sometimes precede the haematological and clinical manifestations of more malignant disease by several months, and thus may serve as valuable prognostic indicators. Chromosomes 8, 17 and 22 are by far the ones most often involved in karyotypic evolution, usually due to trisomy 8, isochromosome 17q and extra Ph chromosomes.

- chronic myeloproliferative diseases
- myelodysplastic/myeloproliferative diseases (MPD)
- myelodysplastic syndromes (MDS)
- acute myeloid leukaemias (AML)
- B cell neoplasms
- T cell neoplasms

Acute myeloid leukaemia. Acute myeloid leukaemia (AML) is often classified into sub-types depending on the morphology of the leukaemic cells; specific chromosome abnormalities are associated with particular sub-types. Taken in conjunction with the abnormal cell morphology, the chromosome abnormality is a dependable indicator of the correct diagnosis.

In terms of prognosis for AML, chromosome changes can be divided into three major groups: favourable, intermediate and unfavourable. Examples of three chromosome changes that tend to be associated with a relatively favourable prognosis include an 8;21 translocation, a 15;17 translocation and a pericentric inversion of chromosome 16. Rapid confirmation of the 15;17 translocation is important, as patients with this rearrangement and AML are susceptible to coagulation disorders and require a special regime of treatment.

Other chromosome abnormalities in AML include rearrangements of the *MLL* locus on chromosome 11q23, deletions of the long arm of chromosome 5, and deletions or loss of chromosome 7. With these abnormalities, many of the patients have had previous exposure to cytotoxic agents, for example, treatment for some other malignant condition, and often the prognosis is relatively poor.

In AML, 90% of children and 85% of adults will achieve remission and 60% of children and 45% of adults will be in remission at 5 years.

Acute lymphoblastic leukaemia. Acute lymphoblastic leukaemia (ALL) is more frequent in children, and also has its own characteristic chromosome changes, providing valuable information with respect to diagnosis and prognosis. Amongst these, **hyperdiploidy** with more than 50 chromosomes is frequent: the chromosomes gained are not always exactly the same, but most often include 4, 6, 10, 14, 18 and 21. The Philadelphia translocation is sometimes seen in ALL, conferring a poor prognosis. A 4;11 translocation is found primarily in infants, and involves the *MLL* locus, mentioned above with regard to AML.

In ALL, 98% of children will achieve remission and 75% will be in remission at 5 years. However, this figure disguises success rates that vary from 10 to 90% with different biological subtypes of the disease. In adults with ALL, 85% achieve remission and at 5 years this figure falls to 45%.

Applications of molecular cytogenetics techniques in malignancy

The presence of a normal karyotype, or failure to detect any abnormalities, does not mean that no genetic changes have occurred. A growing number can only be recognized using the more sensitive molecular cytogenetics techniques such as FISH and the PCR-based techniques (see *Chapters 6* and *7*). Also, variants of known rearrangements may occur. The Ph marker chromosome originates through other rearrangements in 5–10% of CML cases. The majority involve three chromosomes, although four- and five-way variant translocations have been described. All chromosomes with the exception of the Y have been described in variant CML translocations, although the distribution pattern is non-random, with 3p21, 11q13, 12p13, 17q25 and 22q13 involved most commonly.

In the investigation of genetic changes, it is important to note that overall there is no one single optimal strategy, and the availability of a range of methods ensures best results (see *Box 8.13*). A number of examples will now be considered in more detail.

The application of FISH for the detection of the *BCR–ABL* rearrangement in CML has been described in *Chapter 6*. FISH probes for the *BCR* and *ABL* loci can confirm the Philadelphia translocation, not only in metaphase chromosomes but also in interphase nuclei (*Figure 6.4*). This means that it is not necessary to culture the cells first or for the cells of interest to be dividing.

A similar strategy is available for confirmation of the 12;21 translocation found in ALL. As this rearrangement involves an exchange of small pale-staining segments of similar length from the tips of the short arm of 12 and the long arm of 21, it is impossible to see using ordinary banding, and is only visible by FISH.

Another example is the FISH probe spanning the *MLL* locus at 11q23 which is split when there is a translocation, resulting in one large signal on the normal chromosome 11, and two smaller signals, one on each of the translocation products.

For other tumours, interphase FISH can be very helpful using tissue

Box 8.13 Examples of genetic changes

G band visible	G band cryptic FISH detectable	Completely cryptic FISH invisible, PCR detectable
t(9;22)	t(12;21)	*FLT3*
t(8;21)	Ph–ve *BCR–ABL*	*JAK2*
inv(16)	Variants of known rearrangements	

sections from the tumour biopsy or by making imprints by touching the cut surface of the tumour onto a glass slide and hence transferring cells for analysis.

It is also possible for many of these recurrent gene fusions to be identified using PCR-based techniques and to combine these together to test for multiple changes in a single experiment.

Bone marrow transplantation

Bone marrow transplantation (BMT) is the treatment of choice for a wide range of haematological and non-haematological disorders, and cancers.

There are two categories of transplant:

- autograft – where the patient's own bone marrow or stem cells are used
- allograft – where the bone marrow or stem cells from a donor are used

There are also different types of donor:

- sibling – brother or sister
- matched unrelated donor (MUD)
- alternative family donor – a parent, cousin or child
- syngeneic – an identical twin

The patient's own bone marrow is destroyed and replaced with the donor bone marrow. It is important when selecting a donor that the one with the closest HLA tissue type is used to minimize immune system reactions such as graft-versus-host disease or rejection of the graft.

Using a donor and recipient of opposite sex (sex mismatched) results in a haematopoietic chimera. It is relatively easy in these cases to identify the presence of donor and recipient cells using interphase FISH or QF–PCR and hence determine the true level of persistence of the cancer.

The JAK2 V617F mutation

Conventional cytogenetics in myeloproliferative disease (MPD) only detects abnormalities in a small number of patients. A normal karyotype, however, does not exclude the diagnosis of MPD.

The acquired dominant mutation at the *JAK2* locus at 9p24 is present in a high proportion (50%) of patients with MPD and can result in too many blood cells being produced. The *JAK2* gene codes for an enzyme that is present in everyone and is part of a messaging system inside stem cells in the bone marrow. It is a protein kinase of the non-receptor type that associates with the intracellular domains of cytokine receptors. *JAK2* is the predominant *JAK* kinase activated in response to several growth factors and cytokines such as *IL-3*, *GM–CSF* and erythropoietin. *JAK2* belongs to the janus kinase subfamily and so far, four mammalian *JAK*s have been identified.

V617F is a mutation that causes the substitution of phenylalanine for valine at position 617 of the *JAK2* gene. This mutation leads to deregulation of the kinase activity, and thus to constitutive tyrosine phosphorylation activity. The finding of the *JAK2* mutation is useful as it helps to distinguish between true MPD and other conditions affecting the blood cells such as infections. When the V617F mutation is found in MPD patients, it tends to occur in the heterozygous form, and is homozygous only in a minor subset.

Mitotic recombination probably causes loss of heterozygosity (LOH) and the transition from heterozygosity to homozygosity. The V617F mutation seems to occur exclusively in haematopoietic malignancies of the myeloid type.

This mutation is detected with an ARMS PCR method (see *Section 7.3*), using primers designed to discriminate between sequences that differ by a single nucleotide, and will detect both the normal and mutant sequence (*Figure 8.7*). The mutation can be detected at a level as low as 2.5% cancer cells in a blood sample in an affected patient.

A further group of mutations have been identified in the *JAK2* gene in exon 12, in some patients who are negative for the *JAK2* V617F mutation. A number of mutations have been described which can be detected as they result in a change in the number of base pairs in exon 12.

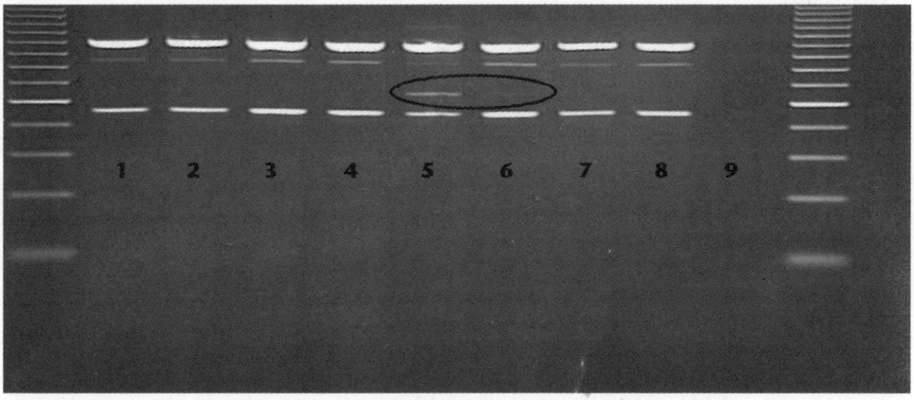

Figure 8.7
ARMS PCR result for detection of the JAK2 V617 mutation.
Lanes 1–4 and 7–8 are normal.
Lane 5 is positive for the JAK2 mutation.
Lane 6 is the positive control with the lowest level of detectable mutant allele for this test at 2.5%.
Lane 9 is the negative control and should be clear with no bands.

FLT3

FLT3 encodes a tyrosine kinase receptor which is expressed in haematopoietic stem cells and is involved in stem cell proliferation. This gene is located at 13q12. The mutations identified within it are mostly internal tandem duplications (ITDs) which are cytogenetically cryptic, but can be seen with FISH in some cases and can be found in otherwise normal or abnormal karyotypes. The *FLT3* ITD can be detected using semi-quantitative PCR of genomic DNA and then analysis of the fragment sizes. The wild-type allele is 330 base pairs. A mutant ITD sequence should produce a fragment greater in size than the wild-type (*Figure 8.8*).

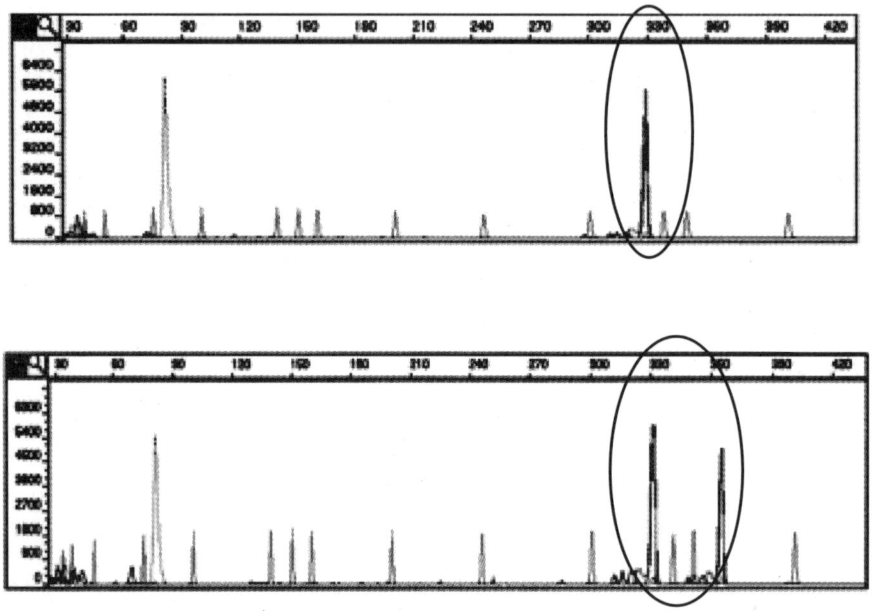

Figure 8.8
Detection of the FLT mutation.
The normal wild-type result has both alleles of the same size.
The mutation is shown as one allele of a larger size due to internal tandem duplication.

In 7% of cases, point mutations have also been described. *FLT3* ITDs have been found in 25% of adult and 15% of childhood AML cases. In all types of AML, the presence of an *FLT3* mutation is a powerful adverse risk factor with an increased risk of disease relapse, and decreased survival.

NPM1

NPM1 codes for a nucleophosmin protein located at 5q35. Mutations are found in 25–35% of AML cases and are usually associated with an apparently normal karyotype. Mutations of *NMP1* are most commonly restricted to exon 12. Approximately 40 *NPM1* mutations have been identified in exon 12 so far. The most common mutation (Type A) is a duplication of the TCTG at position c956–959 and accounts for 75–80% of cases. Type B (CATG duplication) accounts for 10% of all cases, and Type D (CCTG duplication) occurs in about 5% of cases. Other mutations are very rare and these are usually 4 base pair insertions in exon 12 resulting in a frameshift and an elongated protein retained in the cytoplasm.

The presence of the *NPM1* mutation is strongly associated with a normal karyotype, detectable in 60% of patients with AML. Its presence is prognostically significant, with patients having a higher remission rate, longer event-free survival and a stronger trend to overall survival compared to non-mutants.

NPM1 displays a significant association with *FLT3*, in that patients with mutated *NPM1* but normal *FLT3* have a lower relapse rate, and a better treatment response. From the recent MRC AML 12 clinical trial, three prognostic groups have now been stratified:

- **Good** – patients who are negative for *FLT3* ITDs, but positive for *NPM1* mutations
- **Intermediate** – patients who are negative for *FLT3* ITDs and negative for *NPM1* mutations, or positive for *FLT3* ITDs and positive for *NPM1* mutations
- **Poor** – positive for *FLT3* ITDs and negative for *NMP1* mutations

Clinical trials

There are a number of organizations involved in clinical trials including the Medical Research Council, Cancer Research UK and the British Heart Foundation. The aim of clinical research is to discover the causes of human diseases and how they can be treated as well as how disease can be prevented in the first place. A key branch of clinical research is clinical trials to test new treatments for safety and effectiveness. Clinical trials aim to show the benefits and risks of new drugs or treatments, usually by comparing them with the standard treatments in use. No matter how promising a new treatment may seem in the laboratory, it must be carefully tested through clinical trials so that its effects on patients can be more fully understood.

Vital treatments now commonly in use in the NHS were all developed and tested through clinical trials, including chemotherapy. Most clinical trials are randomized controlled trials, which means that they are designed to compare two or more treatments as fairly as possible by reducing the likelihood of bias.

The MRC has been funding clinical trials for more than 70 years and the Clinical Trials Unit runs many studies, most of which are large prospective randomized controlled clinical trials comparing two or more treatments.

A number of trials involving leukaemia patient groups have been completed and new ones continue. They afford the opportunity to determine independent prognostic significance of pretreatment cytogenetics in the context of large patient groups receiving comparable therapy.

Reverse transcriptase PCR

Reverse transcriptase PCR uses RNA as the template rather than DNA. It is a sensitive method as RNA is only produced when a gene is actively expressing a product. The RNA is first converted to cDNA by the enzyme reverse transcriptase (see *Box 8.14* for a note about nomenclature). The cDNA can then be amplified by the usual PCR reaction. It can detect 1 cell in 10 000 cells compared to the 1 cell in 20 that can be detected by conventional cytogenetics, and the 1 cell in 200 by FISH.

The presence or absence of specific sequences may give information about a particular pathogenic state, for example, minimal residual disease (see below). In leukaemia, the best targets for detection of disease are those primary genetic lesions that are specific for cells derived from the malignant

Box 8.14 RT–PCR terminology

Historically, the transformation of mRNA to cDNA using the enzyme reverse transcriptase and standard PCR techniques was called RT–PCR. However, following the emergence and use of the real-time PCR method in molecular genetics laboratories (see *Section 7.3* and *Figure 7.14*), this has also become abbreviated to RT–PCR. Care must therefore be taken when using these abbreviations – it is preferable to use RT–PCR to mean 'reverse transcriptase' PCR and qPCR for real-time PCR.

clone, which usually means chromosomal translocations or their products. In most chromosomal translocations the genomic DNA breakpoints are too widely dispersed to enable routine detection by standard PCR. However, there are a limited number of ways in which two genes can be spliced together to yield a functional oncogene and so most fusion genes can be detected readily by reverse transcriptase PCR using a limited number of primer sets.

Simple detection or non-detection of a particular fusion transcript after treatment is of very limited value. Instead, **quantitative** assays are required that enable the kinetics of residual disease to be monitored over time, or the level of disease to be determined at key time points.

The number of gene transcripts in a specimen is typically normalized to the number of transcripts of a housekeeping gene which serves to control for the integrity and quantity of patient cDNA. The control gives a clear indication of sensitivity with which residual disease can be excluded for that particular specimen.

An International Scale (IS) for quantitative measurement of *BCR–ABL* mRNA has been established that is anchored to two key points defined in the IRIS trial: a common baseline (100% $BCR\text{–}ABL^{IS}$) and major molecular response (0.1% $BCR\text{–}ABL^{IS}$). A more robust definition of the IS is under consideration along with the development of internationally accredited reference reagents.

BCR–ABL monitoring in patients with CML

Reverse-transcriptase real-time quantitative PCR is routinely used for molecular monitoring by quantifying levels of *BCR–ABL* mRNA transcripts in peripheral blood and bone marrow samples from CML patients. The technique can determine accurately the response to treatment and is particularly valuable for patients who have achieved complete chromosomal remission (i.e. no evidence of the Ph chromosome by cytogenetic analysis using FISH; see *Figure 8.9*).

Minimal residual disease analysis

There has been great progress in improved survival of patients with leukaemia in recent years. One example of a success story is that of childhood ALL, which is the commonest cancer of childhood and has an incidence of approximately 450 new cases each year. Over 80% of children with ALL can be now be cured with first-line treatment given over a period of 2–3 years.

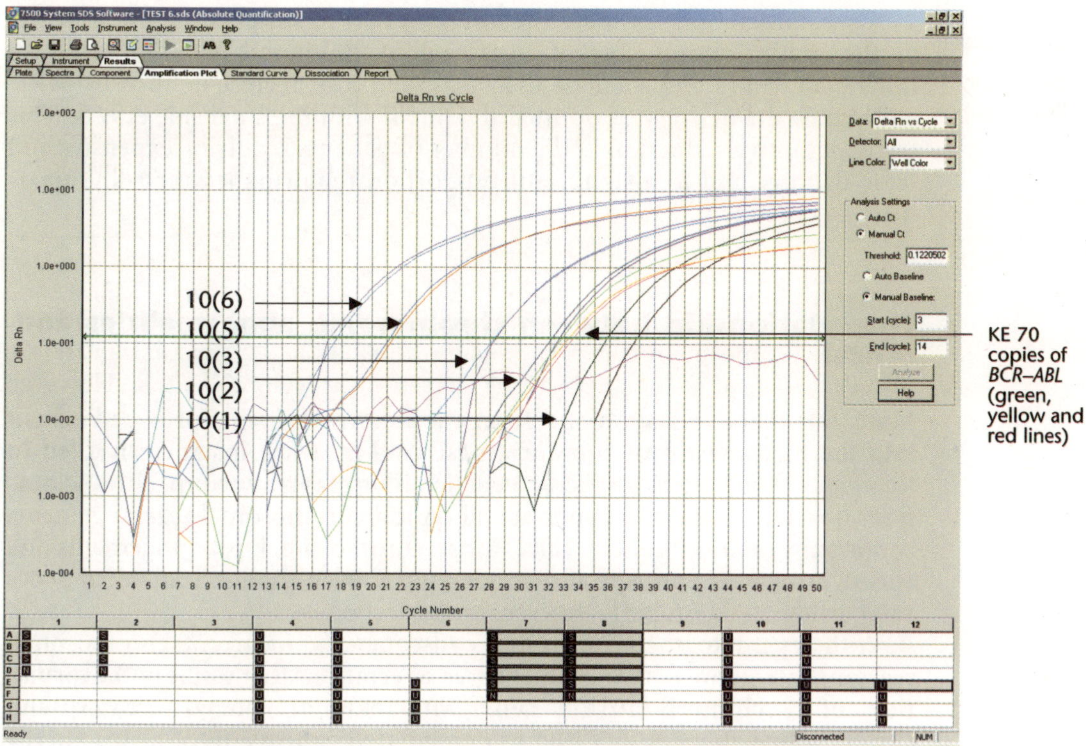

Figure 8.9

Monitoring response to drug treatment in a patient with chronic myeloid leukaemia. Quantitative *BCR–ABL* analysis using real-time PCR. Arrows indicate $\log_{10}$ dilutions of a standard control curve. The patient KE has 70 copies of the *BCR–ABL* transcript; this is a reduction from the result found at diagnosis, showing that she is responding to her imatinib treatment.

Minimal residual disease (MRD) analysis is a highly sensitive test for quantifying residual leukaemia in the bone marrow of patients with ALL who are classed as being in haematological remission by conventional microscopic analysis of the bone marrow. This is defined as having the presence of less than 5% blasts. A patient in remission may still harbour malignant cells, hence minimal residual disease.

MRD is measured using real-time **quantitative** PCR (RQ–PCR) with patient-specific gene rearrangements such as the immunoglobulin and T-cell receptor gene rearrangements. The methodology identifies unique immunoglobulin and/or T-cell receptor gene rearrangements that are clone specific. About 98% of patients will have at least one clonal rearrangement. Two patient-specific RQ–PCR assays are designed for each patient and the quantity of the transcript measured by comparison against logarithmic dilution series curves. It has been found that the level of MRD after one month of treatment is the best predictor of an individual patient's chance of cure.

A substantial body of literature already exists to support the use of MRD in the routine monitoring of treatment of childhood ALL. Tailoring an individual child's treatment to their precise risk of relapse defined by MRD will enable the delivery of optimal treatment with minimal side effects. This translates into a reduction in the number of patients requiring expensive and toxic therapy, and to a significant proportion requiring less first-line treatment to cure them.

The relationship between chromosome abnormalities and cancer

Molecular characterization of cytogenetic abnormalities has provided insight into the mechanisms of tumourigenesis and in a few instances has led to treatment that targets a specific genetic abnormality. Reciprocal translocations, inversions and insertions are all typical chromosomal rearrangements commonly seen in cancer cells. Most chromosomal rearrangements are closely associated with specific tumour types, even though individual genes can participate in multiple different translocations.

With regard to their functional consequences, recurrent chromosomal rearrangements are of two general types: aberrations that result in the formation of a **chimeric fusion gene** with new or altered activity, and chromosomal changes that lead to **deregulated expression** of a structurally normal gene. Genomic rearrangements that juxtapose two genes have been found to play a role in haematological and other cancers. It is possible that similar rearrangements exist in many cancers but have escaped notice because of technical problems in growing tumour cells for chromosomal analysis, or because they are cytogenetically invisible or masked by multiple complex and non-specific karyotypic changes.

Chimeric fusion genes

The majority of chromosomal rearrangements result in the formation of a chimeric gene through the fusion of two genes. The main two groups of genes that participate in such fusions are those encoding **tyrosine kinases** and **transcription factors**.

The classic example of a cytogenetic abnormality leading to the formation of a chimeric fusion gene is the Ph chromosome, which is the result of a reciprocal translocation t(9;22)(q34.1;q11.23) in which sequences of the *BCR* gene on band 22q11.23 are joined to portions of the gene encoding the cytoplasmic *ABL1* tyrosine kinase on band 9q34.1, promoting aberrant tyrosine kinase activity (see *Box 8.15*).

Deregulation of expression of normal genes

Chromosomal rearrangements that juxtapose tissue-specific regulatory elements, such as gene promoters or enhancer sequences, to the coding sequence of a proto-oncogene, deregulate expression of the proto-oncogene.

Box 8.15 Fusion proteins and aberrant expression

Chromosomal rearrangements that disrupt transcription factor genes can result in fusion proteins with enhanced or aberrant transcriptional activity or fusion proteins that mediate transcriptional activity.

A fusion protein with enhanced or aberrant transcriptional activity is present in virtually all cases of Ewing sarcoma, in which unique translocations t(11;22)(q24.1–q24.3;q12.2) and t(21;22)(q22.3;q12.2) fuse the *EWSR1* gene on band 22q12.2 to a gene encoding a member of the ETS family of transcription factors, most frequently *FLI1* on band 11q24.1–q24.3 (in approximately 85% of patients) and *ERG* on band 21q22.3 (in approximately 10% of patients). This results in the induction of various genes, whose aberrant expression appears to be required for *EWSR1–ETS*-mediated tumour growth.

Chromosomal rearrangements that entail aberrant transcriptional repression occur in a substantial proportion of patients with AML. For example, the chimeric proteins resulting from fusion genes such as *PML–RARA, RUNXi–RUNX1T1* and *CBFB–MYH11* all contain a transcription factor that retains its DNA binding motif and an unrelated protein that interacts with inhibitors of gene transcription. Aberrant transcriptional repression contributes to the accumulation of immature myeloid cells in AML. In acute promyelocytic leukaemia, all-trans retinoic acid and arsenic trioxide reverse the transcriptional repression caused by the *PML–RARA* fusion protein, by forcing the release of transcription inhibitors from the fusion protein or stimulating degradation of *PML–RARA,* or both. These two drugs are remarkably effective in APL.

In Burkitt lymphoma, the enhancement of immunoglobulin genes (*IGHG1*, band 14q32.33; *IGKC*, 2p12; and *IGLC1*, 22q211.2) drives the constitutive expression of the gene encoding the *MYC* transcription factor on band 8q24.21. Chromosomal changes that cause over-expression of structurally normal genes occur in other cancers of B-cell or T-cell origin and have also been discovered in prostate cancer.

Chromosomal imbalances

Chromosomal imbalances, leading to gains or losses of genetic material, can range from alterations spanning entire chromosomes to intragenic duplications or deletions. Most chromosomal imbalances have functional consequences that are unknown, due in part to the large numbers of genes involved in some cases, and to the genetic complexity frequently observed.

Most genomic gains probably contribute to tumourigenesis by enhancing the activity of specific genes in affected chromosomal regions. Some of these genes encode proteins that can be specifically targeted by new anti-cancer agents. Genomic gains commonly involve whole chromosomes arising from non-disjunction or from unbalanced translocations, which cause complete or partial chromosomal trisomies (see *Box 8.9*), or from amplification events affecting DNA segments of different sizes. Numerous examples of large-scale genomic gains are associated with specific cancer types. Gains affecting small genomic regions or even single genes have also been described but less frequently than large gains.

It is now possible to identify focal gains by scanning cancer genomes for variations in DNA copy numbers with high resolution methods such as array CGH (see *Chapter 6*).

HER2 (ERBB2) in breast cancer

Amplification of a gene on 17q21.1 that encodes the *ERBB2* receptor tyrosine kinase occurs in approximately 30% of women with breast cancer. This can be detected using FISH as described in *Chapter 6*. The resulting over-expression of *ERBB2* represents a target for the monoclonal antibody trastuzumab. Patients with over-expression respond better to treatment with Herceptin than those with normal levels of *HER2*.

Genomic losses

The spectrum of genomic losses ranges from cytogenetically visible alterations such as complete or partial chromosomal monosomies to single gene or intragenic deletions. Most recurrent genomic losses probably contribute to malignant transformation by reducing the function of specific genes in the affected regions.

Larger genomic deletions affecting multiple genes are frequent in tumours, making it difficult to identify which lost genes contribute to the development of the cancer. Examples of such important tumour suppressor genes include *RB1* at 13q14.2 in retinoblastoma, *APC* at 5q21–15q22 and *ATM* (11q22–11q23).

In many cases the critical genes are unknown. However, some gene deletions have been found to be of great value for determining prognosis and hence guiding treatment decisions such as deletions of 5q in AML (indicates favourable prognosis), and deletions of 11q, 13q and 17p in CLL.

Cancer-associated chromosomal losses may act through inactivation of genes that do not encode proteins. For example, several genomic regions that are recurrently deleted in a variety of tumours contain microRNA genes. These genes encode small RNAs involved in post-transcriptional regulation of gene expression and there is growing evidence that the loss of specific microRNAs with tumour suppressive activity may contribute to tumourigenesis.

Role of genetic changes in the development of cancer therapies

Conventional chromosomal analysis and molecular cytogenetic techniques such as FISH have been useful in guiding the development of anti-cancer agents. The aberrant tyrosine kinase signalling in CML has led to the use of a selective tyrosine kinase inhibitor, imatinib mesylate, to treat the disease. This is currently the front-line therapy for treatment of CML and the overall survival rate has increased with more patients dying from causes unrelated to CML than from their disease. However, imatinib-resistant kinase domain mutations have been identified as a major cause of relapse during imatinib therapy and this has now led to the development of second-generation *BCR–ABL1* inhibitors, such as dastinib and niltinib. Both *BCR–ABL*-dependent and -independent resistance mechanisms have been described. The most common mechanism is point mutations of the *BCR–ABL* kinase domain, but *BCR–ABL* amplification at the genomic or transcript level also occurs in around 10% of patients. Kinase domain mutations are found in approximately 50% of imatinib-resistant patients. Such point mutations render the kinase insensitive to imatinib, while preserving the transforming capabilities of the enzyme. Over 100 different amino acids have been shown to be mutated, with 16 of these accounting for 87% of reported mutations.

Expression arrays

Expression arrays are based on total RNA in cells and are being used to identify diagnostic/prognostic groups and to investigate the fundamental biology of cancer. During the last 10 years the number of papers describing data obtained by microarray technology has increased exponentially. Most of the data generated have been cancer-related. Some of the issues related to the construction of DNA arrays, quantities and heterogeneity of probes and targets, the consequences of the physical characteristics of the probes, data extraction and data analysis, as well as the applications of array technology are discussed in *Chapter 6*.

Unquestionably, array technology will have a great impact on the management of cancer, and its applications will range from the discovery of new drug targets, through new molecular tools for diagnosis and prognosis, to a tailored treatment that will take into account the molecular determinants of a given tumour.

8.7 CANCER AND THE FAMILY

Cancer is a common disease. Everyone has a certain risk of developing cancer in their lifetime, even if there is no family history. It is estimated that in the UK about 1 in 3 people will get cancer during their lives. Most people who get cancer are over 65 and it is relatively rare for young people (under 50) to get cancer. Only a minority of cancers (5–10%, or fewer than 10 out of 100 cases) are clearly linked to an inherited gene mutation.

Some situations make it more likely that a certain cancer is constitutional rather than sporadic. A cancer is more likely to be inherited if:

- there are two or more close blood relatives on the same side of the family affected by the same type of cancer
- a close relative had more than one primary tumour; this means that a person has had cancer twice, but the second cancer is not due to the first cancer spreading to another part of the body, but is in fact a new cancer (for example, breast and then womb cancer)
- members of a family get cancer at a younger than normal age (under 60 years)
- certain cancers happen together in a family

There are two main patterns where cancers occur together. First, there is breast and ovarian cancer, which sometimes run together in a family. Secondly, there is bowel and womb (endometrial) cancer, which sometimes occur together with other cancers such as stomach, kidney, ovarian or pancreatic cancer.

In some cases, people who inherit a familial cancer gene will not definitely get cancer (see *Section 4.6* on reduced penetrance). Inheriting a cancer gene usually means that a person has a significantly increased risk of developing cancer compared to other people in the population.

Inherited breast cancer

Breast cancer occurs with a frequency of 1 in 8–12 of the general female population. Breast cancer is caused by a combination of many different factors, with the three main risk factors being gender, increasing age and significant family history.

Breast cancer is the most common cancer in women in the UK with over 44 000 new cases each year. Around 300 men each year are also found to have breast cancer. Most breast cancers (around 80%) occur in women over the age of 50. Most men who get breast cancer are over 60. Only about 5–10% of breast cancers are due to inheriting abnormal genes. Most breast cancers are found in women who have not inherited a familial mutation.

Although familial mutations constitute a relatively small proportion of cases they are very important, as identification of these high risk patients allows early screening and intervention that can save lives. Although having a faulty breast cancer gene does not mean an individual will definitely get breast cancer, penetrance may be as high as 80%, so the risk of developing breast cancer is still higher than someone who does not carry a familial mutation. A negative result for a familial mutation reduces the risk back to that of the population; it does not become zero.

The two main genes that are known to increase the risk of breast cancer in families are *BRCA1* and *BRCA2*. Someone who has a faulty *BRCA1* or *BRCA2* gene has one normal and one faulty copy of the gene. As these genes act in an autosomal dominant manner, each child has a one in two chance of inheriting the fault if one of their parents has a gene mutation.

Both men and women can have faulty breast cancer genes, so it is possible to inherit a breast cancer gene from your mother or your father. Men with a faulty *BRCA2* gene have a much higher risk of developing breast cancer than men on average; however, breast cancer in men is very rare.

There are two other genes in which mutations can lead to breast cancer in families. They are much rarer than *BRCA* and probably account for only 1–2% of all familial breast cancers. These are the *p53* gene and the *ATM* (ataxia telangiectasia) gene.

BRCA1

BRCA1 (**BR**east **CA**ncer **1**) is located at 17q21. It is a large gene of 24 exons (of which 22 are coding) and codes for 1863 amnio acids. It is a nuclear phosphoprotein with two recognizable domains: the N-terminal RING finger domain which facilitates protein–protein and protein–DNA interactions, and a C-terminal BRCT domain, as found in DNA repair proteins.

The *BRCA1* gene, when mutated, significantly increases a person's risk of developing breast and ovarian cancer. Mutations in *BRCA1* can also increase a person's risk of developing prostate, and possibly colon cancer. In addition to having a higher risk of developing breast cancer, women with mutations in *BRCA1* tend to develop the disease at a much earlier age than women in the general population.

BRCA2

BRCA2 (**BR**east **CA**ncer **2**) is located at 13q12.3. It has 27 exons (of which 26 are coding) and codes for 3418 amino acids. It is a nuclear protein with no recognizable motifs, but it shares a number of functional similarities with *BRCA1*. The gene, when mutated, significantly increases a person's risk of developing breast and ovarian cancer. Mutations in *BRCA2* can also increase a person's risk of developing prostate and pancreatic cancers. In addition, researchers suspect that mutations in *BRCA2* carry with them an increased risk for cancer of the lung, larynx (voice box), and skin (melanoma).

Both *BRCA1* and *BRCA2* have 'caretaker' functions and maintain genome integrity. *BRCA1* is involved in DNA repair and cell regulation, and *BRCA2* is involved in DNA double strand break repair in conjunction with *BRCA1*.

Testing for BRCA mutations

Guidelines issued by the National Institute for Health and Clinical Excellence (NICE) stipulate that all women with a 20% or greater risk of carrying *BRCA* gene mutations from their family history must have access to testing. The aims of genetic testing are to confirm the clinical diagnosis and to offer presymptomatic testing to 'at risk' relatives, and thereby help to prevent early death. Genetic testing and counselling are discussed further in *Chapter 10.*

However, there are a number of issues associated with the test. The genes are large and, in order to identify mutations, all of the 48 coding exons (16 190 base pairs) need to be sequenced which is very time-consuming. A proven mutation that is known to be linked to cancer risk may be found following DNA analysis, and this will increase a person's risk of developing breast or ovarian cancer. If such a mutation is found in the affected family member, other family members can then be tested for the same mutation. This is called **predictive testing.** Testing other family members will take less time as the familial mutation is known, and the whole gene does not have to be analysed. This process of testing is called **site specific analysis** or **single site analysis.**

The result of a predictive test in a relative of the affected family member may be either positive or negative. **If the result of the predictive test is positive,** this means that the family member has inherited the gene mutation and therefore has an increased risk of developing breast or ovarian cancer. They can also pass on the mutation to their children. **If the result of a predictive test is negative,** that person has not inherited the gene mutation known to be associated with the cancer **in their family**. Their risk of developing breast or ovarian cancer reduces to that of the general population.

Many mutations are private, i.e. unique to each family. When a mutation is found, the interpretation needs to be carefully considered to determine whether or not it is likely to be pathogenic. Possible mutations where the link with cancer is not so clear-cut may be found. Sometimes, DNA analysis reveals a polymorphism or unknown variant (see *Chapter 7*). Testing other family members who also have breast or ovarian cancer may be helpful in this situation.

Broadly speaking, there are two types of mutation that, if found, could give an uncertain result. These are polymorphisms and unclassified variants.

When no mutation is found

In many cases of DNA analysis, no mutation is found in any of the genes known to be involved in hereditary breast/ovarian cancer, even if there is a strong family history. Overall, mutations are only found in about 10 to 20 of every 100 tests. However, this does not mean that the family in question is not at an increased risk. In fact, if there is a very strong family history of breast or ovarian cancer, clinicians assume that the members of this family are at increased risk of that cancer type, even if the laboratory cannot find a specific gene mutation.

There are several possible reasons for a test not showing a mutation:

- the test might have missed the mutation, although this is unlikely with modern sequencing technologies
- it is possible that there is a gene change/mutation present in a family with a strong history of breast/ovarian cancer, but in a breast cancer gene that has not yet been identified
- the cancer in the family, or person, is not constitutional, and the clustering of cases of cancer has occurred by chance

Advantages of DNA analysis

DNA analysis can provide important information for families who have a very strong history of breast or ovarian cancer. It enables patients and relatives to make important life decisions, according to the knowledge that they have gained. Unaffected members will be reassured as they and their offspring will only have the general population risk.

Disadvantages of DNA analysis

Unfortunately, genetic testing does not remove uncertainty for everyone. For example, finding a mutation of unknown significance, or even no mutation, in the affected family member can leave people no wiser regarding their possible risk. Going through the process of genetic testing and dealing with the uncertainty of it all can be a stressful experience.

8.8 THE MULTISTEP NATURE OF CANCER

The development of cancer can be seen as a series of steps, each of which gives a growth advantage to a cell. One particular colon cancer has been used as a model as it has a well-defined pattern of progression. Familial adenomatous polyposis coli (**FAP** or FAPC) is an autosomal dominant colon cancer, which first appears in an affected person in late childhood to teenage years. Hundreds or thousands of growths called adenomatous **polyps** develop in the large intestine. In this disorder it is almost certain that one or more polyps will become malignant, passing through the early, intermediate and late stages of **adenoma**, finally culminating in malignant **carcinoma**.

Fearon and Vogelstein (1990) described each progressive stage with respect to the interaction of the various oncogenes and tumour suppressor genes believed to be implicated in each step (*Figure 8.10*). The gene which is mutated or lost in FAP is called ***APC*** (**a**denomatous **p**olyposis **c**oli), and is located at 5q21. This may be lost constitutionally and, although only involving a single copy, it is sufficient to initiate polyp formation by giving the colon epithelial cells a proliferative advantage. The *APC* gene acts like a tumour suppressor gene in that the final stage carcinomas often show LOH of *APC*, however, this is rarely seen in early stages. Thus the distinction between dominant oncogenes and recessive tumour suppressor genes (as in *p53*) is not as distinct as previously thought.

Although not every colon cancer may develop in this manner, the steps leading to colon cancer may begin with the loss or mutation of *APC*. Each stage in adenoma progression is then dependent on another gene fault, such as activation of the *KRAS* oncogene, loss of a tumour suppressor gene on 18q (*SMAD2/4*), and loss or mutation of *p53*, which allows progression from G_1 to S phase instead of apoptosis. Any mutations in the mismatch repair genes act in the background, predisposing the adenomas to progress by increasing the general mutation rate (see *Box 8.16*).

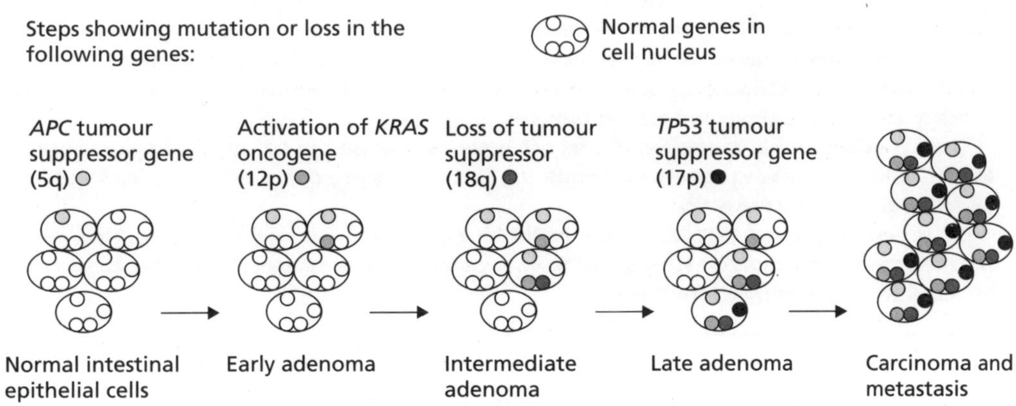

Figure 8.10
Steps leading to colorectal carcinoma.

8.9 SUMMARY

Although it would be ideal to have a universal theory for the development of cancer, it is difficult to find a general explanation for neoplasias which range

Box 8.16 Genes in colorectal cancer

Looking at the gene mutations found at each stage of adenoma/carcinoma can support the multistep theory. It is common to see *KRAS* mutations in intermediate to late adenomas and carcinomas (50%), but rare in the early ones (10%). It is common to see LOH of 18q in late adenomas and carcinomas (50%), but rare in the early stages. Many carcinomas have *p53* mutations which are not seen in adenomas.

from the purely genetic inherited cancers such as FAP to the acquired cancers such as leukaemia (see also *Box 8.17*).

However, the human body does seem to have multiple safeguards against DNA damage leading to carcinogenic changes. It may be that the caretaker genes, such as the mismatch repair genes, require two hits to make the cell genome unstable, followed by another two hits of a gatekeeper gene to begin the process of tumour formation in a particular cell type. Even then, we have the 'guardian angel' gene *p53*, which would also need to be knocked out in order for tumour cell proliferation to proceed.

It is now believed that models such as that for colon cancer based on a linear progression may be an oversimplification of the many different routes and changes seen in various cancers. The neoplastic process may change pathways or even regress before proceeding to its final conclusion, requiring a more flexible theory of genetic changes that lead to cancer.

Box 8.17 Telomerase

Conventional DNA polymerases cannot replicate telomeric sequences, which cap chromosome ends. Each time somatic fibroblast cells divide, telomeres shorten and the cells gradually enter senescence.

An enzyme called telomerase is present in germ cells and is able to add telomerase repeats onto chromosome ends. Although its activity is not detected in most normal somatic cells, it is detectable in many immortal cell lines and human tumours.

There is evidence that in some epithelial cell types telomerase can be activated by oncogenes such as *MYC*, and by increasing telomerase length the lifespan of these cells is extended, leading to immortalization and tumour formation.

It has been suggested that telomerase might be employed as an 'anti-ageing' enzyme, or that an 'anti-telomere' drug could be synthesized to fight cancer, despite evidence that other factors may also be required for cell immortality *in vivo*.

SUGGESTED FURTHER READING

Cook, D.J. (2006) *Cellular Pathology*, 2nd Edition. Bloxham: Scion Publishing.

Davies, K. (1996) Cancer and development patched together. *Nature Genetics*, **13**: 258.

Druker, B.J. (2008) Translation of the Philadelphia chromosome into therapy for CML. *Blood*, **112**: 4808–4817.

Frohling, S. and Dohner, M.D. (2008) Chromosomal abnormalities in cancer. *N. Engl. J. Med.*, **359**: 722–734.

Gottlieb, T.M. and Oren, M. (1996) p53 in growth control and neoplasia. *Biochim. Biophys. Acta*, **1287**: 77–102.
Hesketh, R. (1997) *The Oncogene and Tumour Suppressor Gene Facts Book*, 2nd Edition. London: Academic Press.
Kaelin, W.G. and Maher, E. (1998) The VHL tumour-suppressor gene paradigm. *Trends Genet.*, **14**: 423–426.
Lane, D. (1998) Awakening angels. *Nature*, **394**: 616–617.
Macdonald, F., Ford, C.H.J. and Casson, A. (2004) *Molecular Biology of Cancer*. Oxford: Bios Scientific Publishers.
Nowell, P.C. and Hungerford, D.A. (1960) A minute chromosome in human chronic granulocytic leukaemia. *Science*, **132**: 1497.
Sidransky, D. (1996) Is human patched the gatekeeper of common skin cancers? *Nature Genetics*, **14**: 7–8.
Skirton, H. and Patch, C. (2009) *Genetics for the Health Sciences*. Bloxham: Scion Publishing.
Strachan, T. and Read, A.P. (2004) *Human Molecular Genetics*, 3rd Edition. Oxford: Garland Science.
Swerdlow, S.H., Campo, E., Harris, N.L., *et al.* (2008) *World Health Organization Classification of Tumours of Haematopoietic and Lymphoid Tissues*, 4th Edition. Geneva: WHO Press.
Turnpenny, P. and Ellard, S. (2005) *Emery's Elements of Medical Genetics*, 12th Edition. Oxford: Elsevier Churchill Livingstone.

Atlas of Genetics and Cytogenetics in Oncology and Haematology – http://AtlasGeneticsOncology.org
Mitelman Database of Chromosome Aberrations in Cancer (2008) Mitelman, F., Johansson, B. and Mertens, F. (Eds) – http://cgap.nci.nih.gov/Chromosomes/Mitelman

SELF-ASSESSMENT QUESTIONS

1. What is the name given to tumours which are acquired during one's lifetime rather than inherited? Name three environmental factors which might contribute to the development of malignancies.
2. Why are growth factor genes likely to show oncogenic properties when mutations occur?
3. Why is the t(8;14) translocation in Burkitt lymphoma significant? Hint: think about what happens to one of the translocated genes when it changes position.
4. There is a syndrome called the WAGR complex (**W**ilms tumour, **a**niridia, **g**onadoblastoma and mental **r**etardation). When a baby is born with aniridia (lack of the iris, the coloured part of the eye), why does the Clinical Geneticist ask for chromosome 11 to be checked? The baby's kidneys are going to be scanned regularly for the first two years – why is the baby being checked at such an early age? What sort of cancer gene may be involved?
5. A protein called MDM2 regulates p53 by degrading it if there is too much. What would happen to the amount of p53 if the action of MDM2 was blocked by DNA damage? What effect would this have on the affected cells?

6. Why do you think that the *RB1* and *APC* genes are always said to act in a dominant manner in familial cases, when they are supposed to be 'recessive' tumour suppressor genes?
7. Why have mismatch repair genes been described as caretakers? Briefly explain what might happen if they are mutated.

CHAPTER 9

Prenatal diagnosis and screening

Learning objectives

After studying this chapter you should confidently be able to:

- **Outline the ways in which abnormality and genetic disease can be detected early during pregnancy and at birth**
  Genetic disease can be detected by identifying couples who are carriers of a genetic disorder by screening selected populations, by identifying pregnancies at risk through prenatal screening and prenatal diagnosis, and by early treatment at birth after newborn screening.
- **Explain what is meant by prenatal diagnosis**
  Prenatal diagnosis is the detection or exclusion of abnormality in the fetus during pregnancy.
- **Discuss the criteria for effective prenatal diagnosis**
  The criteria that must be met before prenatal diagnosis is considered are that the disorder should be severe, a reliable test should be available, no treatment is available, or there are benefits in early diagnosis allowing treatment.
- **Discuss how pregnancies at risk of fetal abnormality are identified**
  Pregnancies at risk of fetal abnormality can be identified by considering factors including:
  - maternal age
  - maternal serum biochemistry
  - ultrasound scanning
  - previous child with Mendelian disorder
  - family history of Mendelian disorder
  - previous child with a chromosome abnormality
  - previous child with a DNA mutation
  - parent carries chromosome abnormality
  - carriers of autosomal recessive disorders identified by population screening
- **Detail the diagnostic methods of prenatal diagnosis**
  The methods of prenatal diagnosis are amniocentesis, chorionic villus sampling and fetal blood sampling. These specialist techniques all carry a 1–2% risk of miscarriage.
- **Outline the problems and limitations associated with each method**
  Problems with the methods of prenatal diagnosis include the finding of an unexpected chromosome abnormality, contamination with maternal cells, mosaicism, and uniparental disomy.

- **Explain what is meant by screening**
  Screening is a public health service in which members of a defined population, who do not necessarily perceive they are at risk of, or are already affected by a disease or its complications, are asked a question or offered a test, to identify those individuals who are more likely to be helped than harmed by further tests or treatments to reduce the risk of a disease or its complications.
- **Be aware of the criteria for appraising a screening programme**
  All decisions to implement a particular programme must be evidence-based. For a screening programme to be initiated it must be for a health condition that is important, have a test that is simple, safe, precise and acceptable to the population, and there must be effective treatment or intervention.
- **Be aware of alternatives to prenatal diagnosis**
  Alternatives to prenatal diagnosis include artificial insemination by donor, egg donation and pre-implantation genetic diagnosis. The growth in non-invasive prenatal testing is likely to make prenatal diagnosis more widely acceptable as it can be done early in pregnancy and carries no risk to the fetus.

9.1 INTRODUCTION

The desire of all parents is that any children they have will be healthy and normal; fortunately most are. However, some children are born with varying degrees of problems. About 14% of infants are born with a single minor **malformation** such as an ear tag, 3% have a single major malformation such as isolated spina bifida, and 0.7% have multiple malformations, many of which will result in reduced life expectancy or severe disability; some children will have mental handicap with learning difficulties, behavioural problems or autism.

A malformation is a primary error of normal development or morphogenesis of an organ or tissue. All malformations are congenital, i.e. present at birth, although they may not all be diagnosed until later in life if they are very small or if internal organs are involved.

The cause of many malformations is unknown, but genetics plays a part in over one-third of those where the cause (or aetiology) is known (see *Box 9.1*).

Box 9.1 Aetiology of major congenital malformations

Idiopathic (no known cause)	60%
Multifactorial	20%
Monogenic	7.5%
Chromosomal	6%
Maternal illness	3%
Congenital infection	2%
Drugs, X-rays, alcohol or teratogens	1.5%

This may be due to a **chromosome** abnormality (see *Chapter 5*) or to a **Mendelian disorder**. Mendelian disorders are those that are the result of a single mutant gene and generally follow simple patterns of inheritance such as autosomal dominant, autosomal recessive or X-linked recessive (see *Chapter 4*). Over 5000 Mendelian disorders are known in humans.

In many cases, the birth of a child with health problems is totally unexpected and can be a very difficult experience for the couple and their families. This chapter looks at the ways in which the birth of a child with a genetic disorder can be avoided or detected early, allowing couples to make informed choices about reproduction and to facilitate early treatment where available. Early detection can be achieved by identifying couples or pregnancies at risk through **population screening,** and by prenatal diagnosis, and at birth through **newborn blood spot screening** of the child.

Prenatal diagnosis is not about eugenics and creating perfect babies, but it is about giving parents information and allowing them the choice as to whether or not they wish to continue with a pregnancy in which severe abnormality has been detected. Most parents would terminate an affected pregnancy, but some choose to use the information to prepare for the birth of an affected child. When testing shows the fetus to be normal, it allows the parents to continue the pregnancy reassured.

Population screening entails the testing of a whole population in order to detect those at risk of a disease for themselves or their offspring. Prenatal diagnosis is the detection or exclusion of abnormality in the fetus during pregnancy; it is important in detecting and preventing genetic disease.

9.2 PRENATAL DIAGNOSIS

A normal pregnancy lasts for 40 weeks calculated from the first day of the last menstrual period (LMP). Pregnancy is divided into three parts of approximately three months each, called the first, second and third trimesters.

Although prenatal testing is now commonplace, it is not without risk to the fetus due to the procedures used to obtain material for analysis. Therefore, before undertaking prenatal diagnosis, several criteria should be met; amongst these are that there is a high risk of abnormality, the disorder being tested for is severe, there is either no treatment or there is an advantage to be gained from early treatment, and any test must be very reliable (see *Box 9.2*).

Identification of pregnancies at risk

There are various ways of identifying pregnancies at risk of fetal abnormality and these include:

- maternal age
- maternal serum biochemistry
- ultrasound scanning
- previous child with a Mendelian disorder

- family history of a Mendelian disorder
- previous child with a chromosome abnormality
- previous child with a DNA mutation
- parent carries chromosome abnormality
- carriers of autosomal recessive disorders identified by population screening

Box 9.2 Criteria for prenatal diagnosis

- The pregnancy has been identified as being at **high risk of fetal abnormality**.
- **The disorder must be severe**. Many genetic disorders lead to death *in utero,* infancy or childhood. However, many can be compatible with survival for years even with handicaps such as Down syndrome, spina bifida and cystic fibrosis.
- **There is no available treatment**. Many genetic disorders cannot be treated even if the basis for the disease is known. Research into the use of gene therapy may help to change this.
- **The disorder can be treated effectively,** as in phenylketonuria (PKU), or if early treatment such as surgery is beneficial. Prenatal diagnosis will allow the child to be delivered in a centre with specialist services if appropriate and early treatment commenced.
- **The prenatal test must be reliable enough for a decision to be made**. Over 200 Mendelian disorders can now be detected prenatally by DNA or biochemical analysis and the number is increasing.

Previous child with a Mendelian disorder

For most couples the finding that they are carriers of an inherited disorder is not, unfortunately, made until the birth of an affected child. The risk in subsequent pregnancies for an autosomal recessive disorder is then 1 in 4. Carriers of autosomal dominant disorders have a risk of 1 in 2. Carriers of X-linked recessive disorders have a 1 in 2 chance that a male fetus will be affected. Calculation of risk is described in *Chapter 4*.

At present, tests are only available for a limited number of disorders and therefore prenatal diagnosis may not always be feasible. Fortunately, this number is increasing rapidly due to advances in technology, particularly in the field of molecular genetics (see *Chapter 7*). In order for DNA analysis to be possible it is necessary for the mutation to have been identified within a family or for closely linked informative markers to have been identified (*Chapter 7*). For biochemical analysis of either a specific enzyme or other chemical, there must be a clear distinction between carriers and affected people otherwise it will not be possible to give accurate risks for the fetus (see *Chapters 4* and *7*). Some examples of prenatal diagnosis for genetic disorders available at present are listed in *Box 9.3*.

Family history of a Mendelian disorder

Genetic counselling for families where a Mendelian disorder is known is necessary in order to identify the risk in a pregnancy. This will be based on the mode of inheritance, population frequency of the disorder and tests to

Box 9.3 Examples of genetic disorders that can be diagnosed during pregnancy

Disease	Inheritance	Method of analysis
Cystic fibrosis	AR	DNA
Hurler syndrome	AR	Biochemical
Spinomuscular atrophy type 1	AR	DNA
Myotonic dystrophy	AD	DNA
Huntington disease	AD	DNA
Achondroplasia	AD	DNA
Haemophilia A	XL	DNA/biochemical
Fragile X syndrome	XL	DNA
Duchenne muscular dystrophy	XL	DNA
Tay–Sachs disease	AR	Biochemical
Ornithine carbamoyltransferase	AR	Biochemical
Sanfilippo syndrome	AR	Biochemical
Lesch–Nyhan syndrome	XL	Biochemical

AR – autosomal recessive; AD – autosomal dominant; XL – X linked.

determine the carrier status of the couple. Calculation of risks is discussed more fully in *Chapter 4*, and genetic counselling is discussed in *Chapter 10*.

Carriers of inherited disorders identified by population screening programmes

Population screening is now in place for a small number of disorders to identify couples at risk of having an affected child. This is discussed in more detail later in this chapter.

Previous child with a chromosome disorder

Parents who have had a child with trisomy 21 or any other trisomy are at increased risk of having a trisomy in subsequent pregnancies. This is approximately 1% above the maternal age-related risk and many of these couples request prenatal diagnosis. This increased risk may be the result of one parent being a low-level mosaic in some cells, sometimes even only in the germ cells (gonadal mosaicism) for the chromosome involved. Alternatively there may be a genetic predisposition to non-disjunction in one parent. Risks from any other previous *de novo* chromosome abnormality are extremely low, but parents may request prenatal diagnosis for reassurance.

Inherited chromosome rearrangements

Carriers of a chromosome rearrangement are at increased risk of producing **chromosomally unbalanced** offspring as described in *Chapter 5*. The risk in any individual family depends on several factors, including the actual chromosomes involved and the size of the potential imbalance. Many imbalances will result in non-viable conceptions. The occurrence of a chromosomal imbalance as an abnormal live-born infant indicates that this is a possible outcome in any pregnancy, although some unbalanced offspring may also miscarry. Some couples may choose to defer prenatal diagnosis until the second trimester, because unbalanced offspring may be lost naturally early in pregnancy, rather than being faced with the decision as to whether to terminate the pregnancy.

Screening for Down syndrome

Trisomy 21, or Down syndrome, is the most common chromosome abnormality at birth in humans. It affects both boys and girls equally. There is no such thing as a typical person with Down syndrome; like all people they vary a lot in appearance, personality and ability. A number of health problems are linked to Down syndrome including heart, hearing and vision problems, learning difficulties, and a higher than average incidence of leukaemia. Down syndrome occurs with an incidence of 1 in 650 at birth. Children with Down syndrome can be born to parents of any age (see *Box 9.4*); however, the risk rises steeply once the mother's age is over 35 years (*Figure 9.1*). The incidence is lower at birth than at amniocentesis as three-quarters of Down syndrome fetuses are lost during pregnancy.

Box 9.4 Approximate risk for trisomy 21 at birth by maternal age

All ages	1 in 650
Age 20	1 in 1540
Age 25	1 in 1350
Age 30	1 in 895
Age 35	1 in 350
Age 40	1 in 100
Age 45	1 in 20

The maternal age effect

In 1934, Penrose first recognized that children with Down syndrome are more often born to older parents. This was later attributed to the age of the mother, with paternal age noted to have little effect. This is thought to be due to the effect on the oocytes, which arrest in meiosis I from before birth until they are released at ovulation. The suspension at this stage makes them susceptible to damage from the environment such as drugs and radiation.

Trisomy arises by meiotic non-disjunction, most commonly in the first meiotic division in the mother (see *Chapter 5*). Non-disjunction in female meiosis II and in male meiosis is a much less common mechanism.

The development of the techniques of amniocentesis and tissue culture to provide chromosome preparations for fetal karyotyping enabled prenatal diagnosis to be offered to detect pregnancies with Down syndrome from the 1960s.

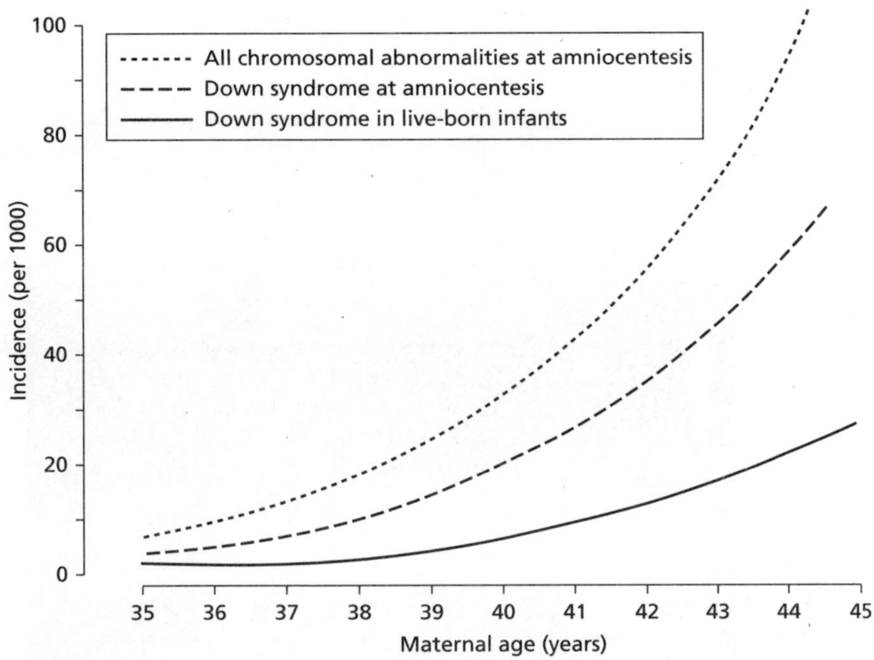

Figure 9.1
Incidence of chromosomal abnormalities and Down syndrome by maternal age. Note the rapid rise in incidence in older mothers.

The risks in the first and second trimester are higher than at birth because many affected pregnancies will be lost as miscarriages and stillbirths. Approximately 75% of all Down syndrome conceptions are lost prior to term. The increased risk associated with maternal age is also observed for other autosomal trisomies (of which only trisomies 13 and 18 are seen at birth), for marker chromosomes and for additional copies of the X chromosome, but these are much less frequent findings at birth. Types of chromosome abnormalities are described in more detail in *Chapter 5*.

Prenatal diagnosis based on maternal age has only had limited success in detecting pregnancies with Down syndrome because, although the risk of Down syndrome is higher in older mothers, most babies are born to mothers less than 35 years of age. The detection rate obtained using maternal age alone is only in the region of 30%. The detection rate has been much improved with the introduction and subsequent improvements in population screening methods to identify the pregnancies at risk.

Abnormal finding by ultrasonography

Ultrasonography involves using high frequency sound waves inaudible to the human ear to produce a picture of the baby. The waves are directed through a transducer into the body and are reflected back as echoes. These are fed back into a computer which uses the information to build up a series of black and white pictures which are linked together to produce a moving image (*Figure 9.2*).

General uses of ultrasonography include checking fetal viability, diagnosis of multiple pregnancies, estimating gestation from fetal size, and monitoring fetal growth during pregnancy. Ultrasonography is also used to screen for abnormalities in the fetus as discussed later.

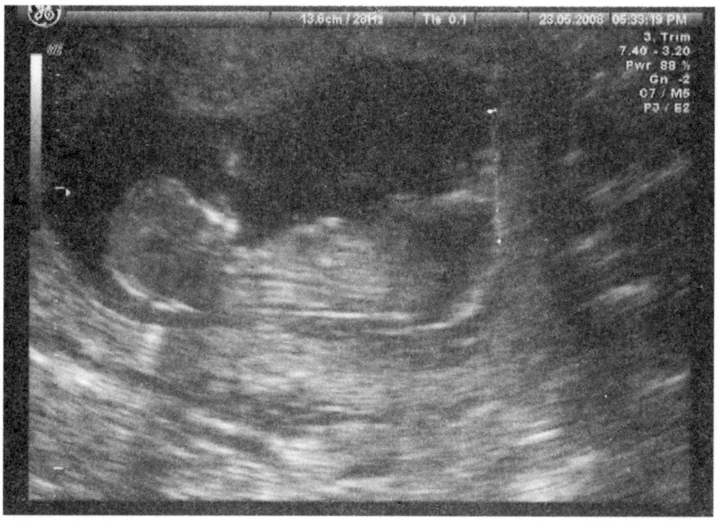

Figure 9.2
An ultrasound scan of a normal fetus at 11 weeks' gestation. The fetus is approximately 7 cm in size and can be clearly identified. At this stage in pregnancy the nuchal translucency (fluid at the back of the neck) can be measured as an early indication of the fetus being at risk of Down syndrome.

9.3 POPULATION SCREENING

Screening is a public health service in which members of a defined population, who do not necessarily perceive they are at risk of or are already affected by a disease or its complications, are asked a question or offered a test, to identify those individuals who are more likely to be helped than harmed by further tests or treatments to reduce the risk of a disease or its complications. When combined with genetic counselling (see *Chapter 10*), screening allows couples to be forewarned of any risk that their children will be affected.

In the UK screening is available for a wide range of conditions including cancers, vascular diseases, antenatal and newborn diseases, sexually transmitted infections, and diseases in childhood (see *Box 9.5*).

The UK National Screening Committee is an independent body that advises the health departments of the four UK countries about all aspects of screening policy and supports the implementation of the programmes. They work to internationally accepted criteria for appraising the viability,

Box 9.5 Examples of screening programmes

- Bowel cancer
- Breast cancer
- Cervical cancer
- Chlamydia
- Cystic fibrosis
- Down syndrome
- Fetal anomaly
- Newborn blood spot
- Newborn hearing
- Sickle cell and thalassaemia

effectiveness and appropriateness of a screening programme (see *Box 9.6*). The balance in deciding on a screening programme is to do more good than harm (see *Table 9.1*). Some of these issues are discussed in more detail in *Chapter 10*.

Box 9.6 Principles of a screening programme

- The condition is an important health problem.
- Its natural history is well understood.
- A suitable test exists that is simple and safe.
- A test exists that is acceptable to the patient.
- The disorder occurs at an appreciable frequency.
- There is an effective treatment.
- There is an advantage to early diagnosis in allowing treatment or prenatal diagnosis.
- Low false positive rate (**specificity**).
- Low false negative rate (**sensitivity**).
- Benefits outweigh the costs – the chance of harm is less than the chance of benefit.

Table 9.1 The balance of harm versus benefit

Benefits	Harms
Early detection allowing earlier effective treatment	False positive and false negative test results
Identification of risk allowing preventative measures	Invasive potentially dangerous tests
Greater awareness among individuals of their own health	Over-detection of symptomless diseases may lead to unnecessary treatment
Control of disease at population level	Genetics screening may impact on relatives, raising questions regarding disclosure of information
Identification of carrier status allows informed family planning	Psychological distress due to unfavourable test results
Cost effective means of disease control	

Not all conditions are suitable for screening. Where no good quality research evidence is available, or research has found that screening for a particular condition causes more harm than good, the National Screening Committee will recommend that *routine* screening should *not* take place, for example, screening for autism, Alzheimer disease, and bladder cancer. Ideally all of the following criteria relating to the condition, the test, the treatment and the programme should be met before screening for a condition is initiated.

The condition

- The condition should be an important health problem.
- The epidemiology and natural history of the condition should be understood and there should be a detectable risk factor, disease marker, latent period or early symptomatic stage.
- All cost-effective primary interventions should have been implemented as far as practicable.
- If the carriers of a mutation are identified as a result of screening, the natural history of people with this status should be understood, including the psychological implications.

The test

- There should be a simple, safe, precise and validated screening test.
- The distribution of test values in the target population should be known and a suitable cut-off level defined and agreed.
- The test should be acceptable to the population.
- There should be an agreed policy on the further diagnostic investigation of individuals with a positive test result and on the choices available to those individuals.
- If the test is for mutations, then criteria used to select the subset of mutations to be covered by the screening, if all possible mutations are not being tested, should be clearly set out.

The treatment

- There should be an effective treatment or intervention for patients identified through early detection, with evidence of early treatment leading to better outcomes than late treatment.

The screening programme

- There should be evidence from high quality randomized controlled trials that the screening programme is effective in reducing mortality and morbidity (ill health).
- Where screening is aimed at providing information to allow the person being screened to make an informed choice, the test must accurately measure risk.
- There should be evidence that the complete screening programme (test, diagnostic procedures, treatment/intervention) is clinically, socially and ethically acceptable to health professions and the public.

- The benefit from the screening programme should outweigh the physical and psychological harm caused by the test, diagnostic procedures and treatment.
- The total cost of the screening programme should be economically balanced in relation to expenditure on the medical area as a whole, i.e. it should offer value for money.

The limitations of screening

Screening has important ethical differences from clinical practice as it targets apparently healthy people, offering to help individuals to make better informed choices about their health. However, there are risks involved and it is important that people have realistic expectations of what a screening programme can deliver. Whilst screening has the potential to save lives, or improve quality of life through early diagnosis of serious conditions, it is not a foolproof process. In any screening programme there is an irreducible minimum number of false-positive results (wrongly reported as having the condition) and false-negative results (wrongly reported as *not* having the condition) and screening does not have 100% **sensitivity** and **specificity** (see *Box 9.7*).

Box 9.7 Definitions of terms used in screening

Cascade screening – systematic identification and testing of members in a proband.

Cut-off – term used for a critical score on an assessment marking the boundary between those scores considered as 'pass' and those considered as 'fail'.

Cut-off level – the value of a screening variable which distinguishes screen positive from screen negative results.

Detection rate – proportion of affected individuals with positive results.

False negatives – individuals with a negative test result but who actually have the condition.

False positives – individuals with a positive test result but who do not have the condition.

Negative predictive value – the proportion of individuals who test negative who do not have the target condition.

Positive predictive value – the proportion of individuals with a positive test result who have a target condition.

Screen negative results – a screening result that is less than the specified cut-off level.

Screen positive results – a screening result that is greater than or equal to a specified cut-off level.

Sensitivity (true positive rate) – the proportion of individuals with the target condition in a population who are correctly identified by a screening test.

Specificity (true negative result) – the proportion of individuals free of the target condition in a population who are correctly identified by a screening test.

True negatives – unaffected individuals with screen negative results.

True positives – affected individuals with screen positive result.

Prenatal screening in pregnancy

Down syndrome pregnancies have been found to be associated with changes in certain chemicals detected in the maternal serum. In addition, some findings on ultrasound scan have been observed to occur more frequently in Down syndrome fetuses. These factors are now routinely used to identify pregnancies at risk and today screening tests are available which can provide women of all ages with information about the chance of their baby having Down syndrome.

All pregnant women in the UK now have the choice of screening for Down syndrome. This allows women to be given clear information about the choices available along the screening and diagnostic pathway. Some couples will choose not to have screening as is their right and this decision should be accepted by the health professionals involved in their care.

Screening tests themselves do not give a definitive answer but only indicate a risk. Screening is a two-stage process; initial identification of the population at risk and then the diagnostic test. If a high risk is identified, a **diagnostic test** will be offered, but again the woman may decline the test. In addition, not every couple who find that they have a pregnancy with Down syndrome will choose termination. Approximately 1 in 10 couples will continue with the pregnancy and use the knowledge and time to prepare themselves and their families for the birth of their child.

There are now a range of screening tests available, with improved tests being introduced to increase the number of abnormal cases that are detected and reduce the number of times a normal pregnancy is identified as being at high risk – screen positive (see *Box 9.7*). The first screening tests available had an expected detection rate of at least 60%, with a false positive rate of 5% or less. This is much better than the 30% detection rate from maternal age alone. The UK National Screening Committee has now set standards to improve this with a target detection rate of greater than 75%, and with a false positive rate of less than 3%; this still means that about one-quarter of babies with Down syndrome will not be detected by the screening test. The aim is to improve this up to a 90% detection rate or even better in the future.

Maternal serum biochemistry

During pregnancy the fetal and maternal circulations do not mix, but the exchange of some chemicals can occur. Analysis of the maternal serum biochemistry can give an indication of fetal wellbeing. One of these chemicals is alpha-fetoprotein or AFP (see *Box 9.8*). Maternal serum AFP (MSAFP) concentration can be used to screen for neural tube defects (see *Box 9.9*) such as spina bifida and for Down syndrome.

Screening for neural tube defects. As long ago as 1972, it was reported that a mother carrying a baby with a neural tube defect (NTD) such as spina bifida had an increased level of AFP in her serum. This is usually 2.5 SD (standard deviations) above the mean value expected for the gestation. The increase is due to leakage of fetal serum into the amniotic fluid from exposed fetal capillaries. This observation led to the introduction of a highly successful

Box 9.8 Alpha-fetoprotein

AFP is the major fetal plasma protein, synthesized first in the fetal yolk sac and later in the liver. It is structurally but not antigenically similar to albumin. Some AFP is excreted via the fetal urine into the amniotic fluid from where some diffuses into the maternal circulation. The levels vary during gestation with a peak in fetal plasma at 12–14 weeks and in maternal serum at 30 weeks. The levels in maternal serum are 1000 × lower than in the amniotic fluid.

Box 9.9 Neural tube defects

A neural tube defect (NTD) is the defective closure of the neural tube during early development. The neural tube appears at 20 days and is mostly closed by 23 days. Failure of closure at the cephalic (head) end produces anencephaly, and lower down it produces spina bifida.

The incidence at birth shows marked geographic variation, from 1 in 1000 births in the USA to 8.6 in 1000 births in Ireland.

screening programme for the detection of NTDs. Over 80% of open spina bifida and almost 100% of anencephalic fetuses can be detected.

There is some overlap between the values of AFP found in normal pregnancies and those with a fetus with a NTD (*Figure 9.3*). The presence of a NTD is not the only cause of a raised value for AFP in maternal serum. If a high value is obtained, a repeat sample may be tested and an ultrasound scan of the fetus undertaken to identify the reason for the result (see *Box 9.10*).

The cause of NTDs is not known, but it has been observed that a diet high

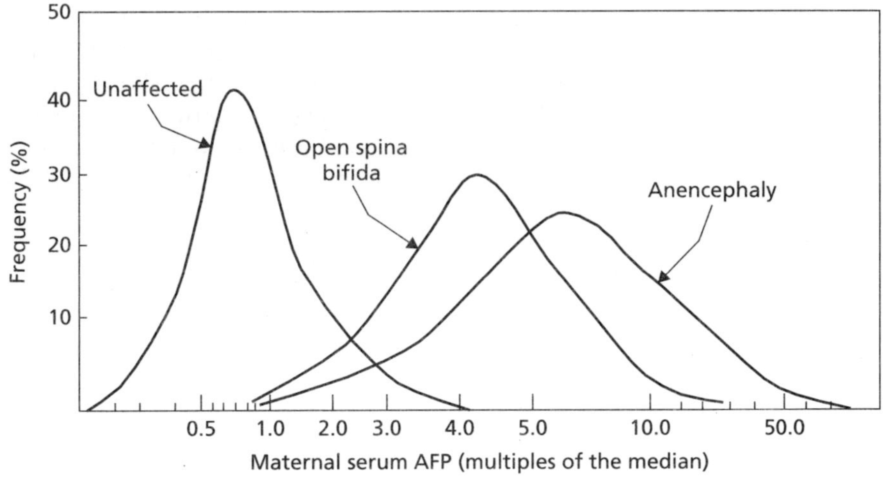

Figure 9.3
Maternal serum AFP expressed as multiples of the median (MOMs) in normal pregnancies and pregnancies affected with open spina bifida and anencephaly (NTDs). Note NTDs tend to have higher levels of AFP than normal pregnancies but there is some overlap.

Box 9.10 Reasons for raised levels of maternal serum AFP

- Neural tube defects
- Incorrect gestational age
- Intrauterine fetal death
- Multiple pregnancy
- Threatened miscarriage
- Abdominal wall defect
- Congenital nephrotic syndrome

in folic acid taken from preconception reduced the risk of a child developing a NTD and led to a significant reduction in their incidence. This is a major breakthrough in that it reduces the occurrence of an abnormality rather than having to consider termination once the abnormality has been detected during pregnancy.

Screening for Down syndrome. The observation that **low** levels of MSAFP were associated with **increased** risk of Down syndrome was first made in the early 1980s. As with the values associated with NTDs, there is an overlap between the values found in normal pregnancies and those where the fetus has Down syndrome (*Figure 9.4*). The combination of this assay with the measurement of human chorionic gonadotrophin (HCG), and sometimes oestriol, has led to an alternative screening strategy for Down syndrome.

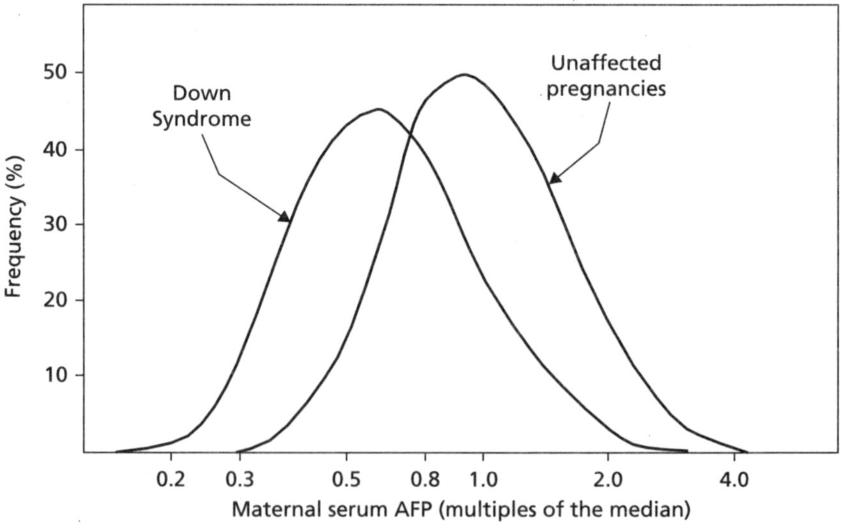

Figure 9.4
Maternal serum AFP expressed as multiples of the median (MOMs) in normal pregnancies and pregnancies affected with Down syndrome. Note that Down syndrome pregnancies tend to have lower levels than normal pregnancies but there is some overlap.

HCG is secreted by the developing trophoblast and later by the chorion and placenta. It can be detected in maternal serum from 3 to 4 weeks post-LMP. The levels increase in the serum exponentially during the first trimester, reaching a plateau in the second and third trimesters. HCG is a glycoprotein, with a molecular weight of 40 000 Daltons, comprising two subunits alpha and beta, and can be assayed either as total HCG or just the free beta subunit. Levels of HCG in maternal serum have been found to be elevated up to twice the normal expected values in the second trimester in pregnancies affected with Down syndrome.

Multiples of the median

Since the levels of AFP, HCG and oestriol all vary with gestation, the values are expressed as **multiples of the median** (MOM). A MOM of 1 means that the values are as expected for the gestation, a value of 2 MOM is raised by twice the expected level, a value of 0.7 is 70% of expected level. In pregnancies affected with Down syndrome the level of HCG is raised (MOM $\geq$ 2.05) and AFP is reduced (MOM of $<$ 0.73).

Oestriol

Oestriol is a steroid hormone, produced by the syncytiotrophoblast from fetal precursors. It is assayed in the unconjugated form. The values found in the maternal serum for AFP and HCG (the Double Test), or with oestriol as well (the Triple Test), are combined with maternal age using a complex formula on a computer program (algorithm) to give each mother an individual risk. The results of the blood tests are used together with the mother's exact age, weight and gestation to calculate the risk. This is based on the age-related risk (prior odds) and the likelihood ratio. The values of the cut-off determine the detection rate and the false positive rate. The lower the cut-off the lower the false positive rate but also the lower the detection rates. The most widely used cut-off is a risk of 1 in 150 at birth.

- Women with a risk of *greater than* 1 in 150 at birth have a 'high risk' or 'screen positive' result and will be offered a diagnostic test (see below).
- Women with a risk of *less than* 1 in 150 at birth have a 'low risk' or 'screen negative' result. A risk of 1 in 150 means that for every 150 births, 149 would be expected to be normal and one to have Down syndrome. Remember that a ratio of 1 in 200 represents a greater risk than a ratio of 1 in 500.

This does not mean there is no risk of the baby having Down syndrome. There is still a small risk because some babies with Down syndrome are not detected by screening tests.

There is ongoing research investigating alternatives to increase the sensitivity of the maternal serum biochemistry using other biochemical analytes, such as pregnancy-associated plasma protein A (PAPPA) and inhibin A, to enable screening to be brought forward to the first trimester.

Pregnancy-associated plasma protein A

PAPPA is the largest of the pregnancy-associated proteins produced by both the embryo and placenta during pregnancy. It is a zinc-binding glycoprotein which enzymatically cleaves insulin-like growth factor-binding proteins. The protein is thought to have several functions, including preventing recognition of the fetus by the maternal immune system. Levels of PAPPA rise from first detection in the first trimester until term. Detection of this protein is used as a first and second trimester test for Down syndrome screening. Maternal serum concentrations are related to subsequent fetal growth and this relationship has suggested that it can be used as an indicator of adverse pregnancy outcomes such as intrauterine growth restriction, premature birth, pre-eclampsia and stillbirth.

Second trimester fetal anomaly scans

A fetal anomaly is defined as 'a marked deviation from the normal standard, especially as a result of congenital defects'. The objective of a fetal anomaly scan is to identify any possible structural abnormalities in the baby, and if this is not the case to identify anomalies, particularly those:

- not compatible with life
- associated with high morbidity and long term disability
- fetal conditions with the potential for treatment before birth
- fetal conditions that will require post-natal investigations or treatments

An ultrasound scan is usually offered to all pregnant women at between 18 and 20 weeks to screen for fetal anomalies. At this stage it is possible to detect abnormalities affecting the central nervous system, limbs, heart, kidney and gastrointestinal tract. However, not all abnormalities will be detectable. Some that are detected may be only very minor and not require any action. Others may indicate the presence of an underlying chromosome abnormality (see *Box 9.11*) which can be investigated by amniocentesis, chorionic villus or fetal blood sampling (see later), or allow for preparation for appropriate care after birth.

Box 9.11 Ultrasound findings that suggest a chromosome abnormality

Finding	Chromosome abnormality
Cardiac defect	trisomy 13, 18, 21 or deletion of chromosome 22
Choroid plexus cysts	trisomy 18, 21
Cystic hygroma	trisomy 13, 18, 21 or Turner syndrome
Exomphalos	trisomy 13, 18
Rocker bottom feet, overlapping fingers	trisomy 18
Growth retardation	triploidy

First trimester scan

At 11 weeks' gestation the fetus is approximately 7 cm in size and the fetal heart can be seen beating (*Figure 9.2*). Even this early some abnormalities can be seen. One example is an abnormal collection of fluid behind the fetal neck (nuchal translucency) which can be seen between 10 and 14 weeks and which has been found to be associated with an increased risk of chromosome abnormality (see *Box 9.12*). The risk is related both to the thickness of the nuchal translucency and the maternal age. Measurement of the nuchal area is now being increasingly used as a method of screening for Down syndrome. The advantage is that it can be done earlier than the second trimester maternal serum biochemistry and is also reported to have a higher detection rate of up to 90%. This earlier test helps to reduce parental anxiety and allows termination to be performed much earlier in pregnancy if appropriate.

Box 9.12 Increased risk of trisomy 13, 18 and 21 from nuchal translucency scan measurements

Size	Increased maternal age risk
3 mm	× 4
4 mm	× 21
5 mm	× 26
>6 mm	× 41

The finding of a normal karyotype in fetuses with a raised nuchal translucency thickness does not exclude other problems. There is an increasing awareness that these fetuses are at increased risk of other problems including an increased risk of intra-uterine death.

Combining screening tests

Combining biochemical and ultrasound screening allows for improved detection rates with a reduced need for invasive procedures and hence loss of normal pregnancies. Several combinations are available as outlined below. The methods used will depend on when in pregnancy the woman attends for screening and availability of local services.

First trimester combined

This is the preferred method to aid in early diagnosis and is undertaken before 13 weeks and 6 days of pregnancy; therefore the risk can be identified before 14 weeks of pregnancy. It also allows screening to be completed in one stage. The test uses ultrasound nuchal translucency measurement, plus serum biochemistry testing to measure free beta HCG and PAPPA.

Integrated testing

This requires the woman to attend at least twice for screening, once before 13 weeks + 6 days for the nuchal translucency scan and first trimester PAPPA, and then again between 15 weeks + 0 days and 20 weeks + 0 days for second trimester biochemistry testing to measure all types of HCG, unconjugated oestriol and AFP. The results of both samples have to be obtained to give the risk.

Serum integrated testing

This also involves two attendances, one in the first trimester for biochemical testing of PAPPA and one in the second trimester testing all types of HCG, unconjugated oestriol and AFP. It does not include the nuchal translucency scan.

Second trimester testing

This is required for the approximately 15% of women who attend later in pregnancy. It involves serum measurements of up to four markers, i.e. AFT, HCG, oestriol and PAPPA. *Table 9.2* shows the modelled outcome in 100 000 women screened for Down syndrome pregnancies with the various screening tests currently available. The integrated tests perform best with the lowest number of losses of unaffected fetuses.

Table 9.2 Modelled outcome in 100 000 women screened for Down syndrome pregnancies

Test	Unaffected women referred for CVS or amniocentesis	Down syndrome fetuses diagnosed (no.)	Unaffected fetuses lost* (no.)	Down syndrome fetuses diagnosed per unaffected fetuses lost (no.)
Double	6500	152	47	3.2
Triple	4200	152	30	5.1
Quadruple	2500	152	18	8.5
Combined	2300	152	17	9.0
Serum integrated	800	152	6	25.4
Integrated	300	152	2	76.3

Data derived from Table 20 in the SURUSS report (Wald *et al.*, 2003).

*Assuming 80% acceptance rate of amniocentesis or chorionic villus sampling and 0.9% loss rate.

Factors affecting maternal serum screening

There are a number of factors that can affect maternal serum screening and which need to be taken into account. These may be maternally related, pregnancy related or programme related.

Maternal factors include:

- maternal age – risk of trisomy increases with age
- weight – there is an inverse relationship between biochemical marker levels and maternal weight in that marker levels seem higher in lighter women and lower in heavier women. For each 20 kg above the average weight of 63.4 kg there is a reduction of 17% in AFP levels, a 16% reduction in HCG levels, and a 7% reduction in unconjugated oestriol levels.
- ethnic group – there are slight differences in the average levels of biochemical markers between different ethnic origins, for example, Afro-Caribbean women have higher levels of AFP and HCG compared to Caucasian women
- previous Down syndrome pregnancy – results in a higher chance in a subsequent pregnancy and it is important to note if it was inherited or not

Pregnancy related factors include:

- gestational age – levels vary with the stage of pregnancy
- assisted conception – with pregnancies from egg donation it is the age of the donor at the time of harvesting that is important, not the age of the pregnant woman. Also, pregnancies from assisted conceptions may have higher levels of HCG.
- multiple pregnancies – biochemical markers are higher in twin pregnancies compared to singleton pregnancies. It is not known with confidence what happens to markers in multiple pregnancies where one or more fetuses may be affected and serum screening cannot be used, but nuchal translucency screening can be. If the pregnancy started as a twin pregnancy and there has been a fetal demise this may affect levels.
- bleeding in pregnancy – recent bleeding will increase AFP levels

Programme related factors include:

- timing of screening – there are optimum times for screening depending on the method used
- accuracy of information – includes gestation and other relevant information

9.4 METHODS OF PRENATAL DIAGNOSIS

Once a pregnancy has been identified as at risk for any of the reasons identified earlier, then the next step is to undertake a diagnostic test to determine whether or not the pregnancy is affected. It should be remembered that a normal result does not guarantee a normal outcome as other problems may be present which cannot be detected. The choice of test will depend on various factors including the stage in pregnancy when the risk is identified. Ideally the earlier the test the better as this allows the couples more time to make their decisions. If this is in the first trimester the pregnancy can still be 'private' to the couple without family and friends being aware.

There are several methods available for prenatal diagnosis and they involve the removal of fetal tissue for analysis. These invasive tests carry a risk of losing the pregnancy from spontaneous miscarriage related to the procedure (see *Box 9.13*). The methods currently available are **amniocentesis, chorionic villus sampling** (CVS) and **fetal blood sampling** (FBS).

Box 9.13 Comparison of prenatal diagnosis techniques

Amniocentesis	Procedure risk 0.5%
	Performed in second trimester
	Widely available
	Karyotype, some biochemical disorders, DNA (rarely)
CVS	Procedure risk 1–2%
	Performed from first trimester
	Specialized centres only
	Rapid karyotype, DNA, biochemical disorders
FBS	Procedure risk 2–3%
	Performed from second trimester
	Specialized centres only
	Rapid karyotype

Amniocentesis. This involves the removal of amniotic fluid by the insertion of a needle, under ultrasound guidance, through the mother's abdomen into the amniotic cavity (*Figure 9.5*). It is usually undertaken at 14–16 weeks' gestation and 10–20 ml of fluid is removed. It is the most commonly used method of obtaining cells for fetal karyotyping, with about 40 000 procedures being performed in the UK each year; it can also be used to measure levels of certain chemicals for prenatal testing of some Mendelian disorders. The cells can be grown as a source of fetal DNA, although it does takes time to culture sufficient cells, which delays the result.

The composition of amniotic fluid varies with gestational age. The fetal cells, many of which are dead, are shed from the skin, lungs and kidneys. Biochemically, amniotic fluid resembles fetal urine. The volume increases up to 300 ml by the end of pregnancy.

The cells have to be cultured for an average of 6–14 days to obtain a fetal karyotype. Occasionally the cells fail to grow and the procedure needs to be repeated.

Chorionic villus sampling. CVS involves the removal of a small amount of the placenta. It can be carried out from 11 weeks' gestation, which makes it suitable for chromosome analysis where a high risk of chromosome abnormality is suspected (as determined by nuchal translucency scans), or for

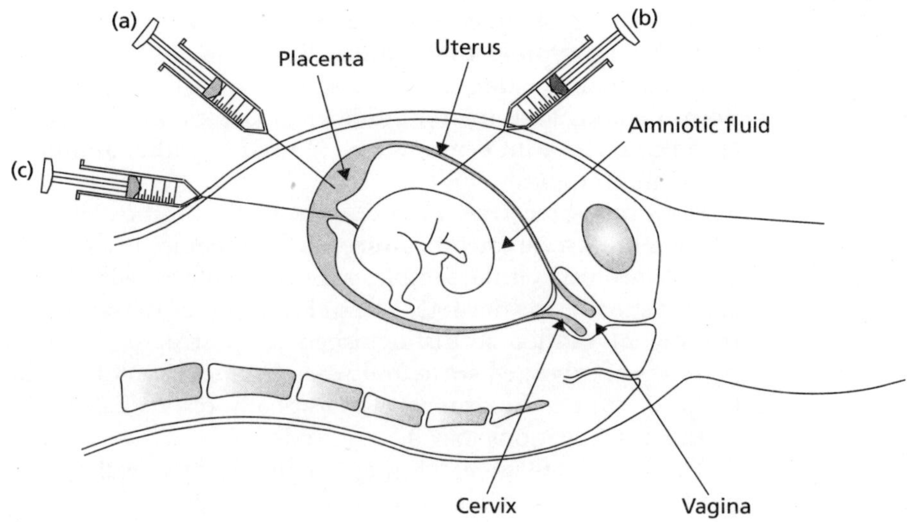

Figure 9.5
The various methods of prenatal diagnosis. (a) Chorionic villus sampling; (b) amniocentesis; (c) fetal blood sampling.

known carriers of chromosome abnormalities. The placenta, unlike amniotic fluid, contains spontaneously dividing cells.

Two methods can be used to obtain the fetal karyotype.

- Direct preparations take advantage of the spontaneously dividing cells in the outer syncytiotrophoblast layer. Using this technique it is possible to obtain a fetal karyotype within 24 hours. However, because of the poor quality of the chromosome preparations produced, analysis is limited to the detection of numerical or large structural abnormalities. The tissue can also be cultured to provide better quality chromosome preparations and a result is available in 8–14 days.
- The villus tissue is also a rich source of DNA and can be used for molecular diagnosis of some disorders such as cystic fibrosis or Duchenne muscular dystrophy. Some biochemical tests including some enzyme assays can also be done.

Fetal blood sampling. FBS, or **cordocentesis**, is the removal of a small amount of fetal blood, usually by inserting a needle into the site where the cord joins the placenta. This is the site where fetal movement is likely to be least. It is not usually performed before 18 weeks' gestation. The blood can be cultured for fetal karyotype analysis and a result obtained in 48 hours. The blood can also analysed for fetal blood group, infections, etc.

Methods for the rapid detection of Down syndrome

Chromosome analysis of amniotic fluid or chorionic villus samples takes between 1 and 3 weeks to obtain a result, and this is a time of great anxiety for parents and has implications on the options available for termination if the parents choose to do this. It is now possible to undertake rapid detection of Down syndrome using either FISH or QF–PCR (see *Chapter 6*) with results available within 24 hours of receipt of the sample by the laboratory. Both methods look for the number of copies of DNA sequences from the chromosomes of interest and can be used on either amniotic fluid or chorionic villus samples.

However, these tests do not detect all chromosome abnormalities for which a full karyotype on cultured cells is required. The other disadvantages include problems if the samples are contaminated with maternal blood, they are not suitable for the detection of low levels of mosaicism, and the fact that trisomy 21 cannot be distinguished from structural rearrangements. The presence of inherited structural rearrangements will have different implications for future pregnancies. Occasionally results can be inconclusive and further investigations may be required.

The big advantage is the speed with which a result can be obtained and most women with pregnancies identified at high risk of Down syndrome are now offered this rapid test.

Problems associated with prenatal diagnosis

Although the most common indication for fetal karyotyping is to detect Down syndrome in pregnancies identified as being at high risk, it is inevitable that other chromosome abnormalities will be detected in some patients by chance (see *Chapter 5*). The finding of unexpected chromosome abnormalities (see *Box 9.14*) at prenatal diagnosis can cause great anxiety for the couple and it is not always possible to give detailed and clear information as to the effect the chromosome abnormality may have, if any, on the child.

Some abnormalities, such as trisomy 13 and 18 are known to result in severe abnormality. Others, such as sex chromosome abnormalities (Turner syndrome 45,X and Klinefelter syndrome 47,XXY), can have less severe phenotypes and can provide difficult decisions for the parents.

Structural rearrangements may also be found. These may be balanced or unbalanced. In these cases it is essential to examine the parental chromosomes. Unbalanced rearrangements would be expected to result in fetal abnormality. However, identification of the balanced form of the rearrangement in one parent would be helpful in providing information as to the

Box 9.14 Examples of unexpected results that may be found during prenatal diagnosis

Sex chromosome abnormalities
De novo balanced rearrangements
Extra structurally abnormal chromosomes
Mosaicism
Maternal cell contamination
Uniparental disomy

chromosomes involved. It would also indicate a risk of unbalanced offspring in future pregnancies in the couple, or any other family member subsequently found to be a carrier. If the rearrangement were balanced, parental karyotyping would reveal if the rearrangement had arisen *de novo* or was inherited. The presence of a balanced inherited rearrangement in a fetus would be expected to be compatible with normal development. *De novo* rearrangements carry a 5–10% risk of abnormality since it is possible that a small amount of chromosome is missing but not detectable. High resolution techniques such as array CGH (see *Chapter 6*) are increasingly being used and can identify minute imbalances.

It is also possible that a break(s) has occurred within a gene or that a break(s) has altered the position of the gene leading to altered expression (see *Chapter 3*) and this can result in abnormality.

Extra structurally abnormal chromosomes (ESACs) are additional chromosomes whose identity is unknown. These can also be inherited without effect. If these arise *de novo,* the risk of fetal abnormality can be from 5 to 30% depending on the size and the material involved. Molecular cytogenetic techniques may be helpful in identifying the chromosome of origin, but even if the chromosome is identified it is not possible to know which genes are involved and therefore an accurate prediction of the expected phenotype cannot be given.

Other problems of interpretation of prenatal chromosome results

Fetal karyotyping following amniocentesis, CVS and FBS is considered to be reliable. Occasionally, however, the interpretation of results can be difficult because of contamination of the sample with maternal cells or by the presence of mosaicism.

Maternal cell contamination

Maternal cell contamination is usually identified in pregnancies where both male and female cells are found. This is thought to be due to the presence of a male fetus with the female cells being maternal in origin; if the fetus is female, it will only be noted when the fetus or the mother has a chromosome abnormality and two different cell lines are discovered. The estimated frequency is 0.5% of amniocentesis and CVS cultures. In FBS testing, the presence of fetal haemoglobin can identify the origin of the blood and hence the presence of any contamination. Maternal tissue contaminating a CVS will not have spontaneous metaphases and so will not affect the results from direct preparations.

Maternal cell contamination can also be a problem when undertaking DNA analysis. This may be a particular concern when using the highly sensitive PCR technique. However, direct visual differentiation between maternal and villus tissue in the CVS, and by confirming the result on DNA extracted from whole tissue or cultures, reduces the chance of error. Alternatively, comparison of fetal and maternal DNA polymorphisms may be helpful.

Mosaicism

Mosaicism is the presence of two or more cell lines with different karyotypes. Mosaicism detected at prenatal diagnosis may have several explanations.

- It may be a genuine reflection of the fetal karyotype (for example, 2–3% of Down syndrome and about 40% of Turner syndrome individuals have mosaic karyotypes).
- It may be due to an error in cell division in culture. The finding of abnormal cells in only one cell culture with a second culture having only normal cells indicates **pseudomosaicism** and is of no clinical significance. This is observed in 1–2% of cell cultures.
- It may reflect mosaicism confined to the placenta, with the fetus having a normal karyotype. This **confined placental mosaicism** (CPM) is detected in 1–2% of CVS cultures.

One particular concern arising from CPM is the possibility that the fetus was originally trisomic and the diploid cells have arisen by loss of one of the extra chromosomes (trisomic rescue). This can result in **uniparental disomy** (UPD) which is the presence of both chromosome homologues from the same parent. UPD is known to be one of the causes of some abnormal phenotypes such as Prader–Willi syndrome and Angelman syndrome (*Chapter 2*).

The discovery of mosaicism may require further tests to help to interpret the significance (if any) of the findings. The finding of mosaicism in other tissues confirms the original finding. However, the finding of only normal cells does not exclude the presence of an abnormal cell line.

Other screening before or during pregnancy

Although many genetic diseases are well defined, they are too rare to merit a whole population screening programme. Exceptions to this occur due to the relative high frequency of certain disorders in particular ethnic groups (see *Table 9.3*). In addition to screening for Down syndrome, screening tests for sickle cell anaemia and thalassaemia are also available in early pregnancy. Other disorders for which screening may be used includes Tay–Sachs disease and cystic fibrosis.

Table 9.3 Autosomal recessive disorders for which population screening is suitable

Disorder	Group
α-thalassaemia	China, south-eastern Asia and Mediterranean countries
β-thalassaemia	Indian sub-continent and Mediterranean countries
Sickle cell disease	Afro-Caribbean
Tay–Sachs disease	Ashkenazi Jews
Cystic fibrosis	Western European Caucasians

Finding out early in pregnancy gives the chance to talk to a counsellor and find out more about the disorders and care available. If the test shows the woman to be a carrier of an autosomal recessive condition, the baby's father will also be offered the test. If he is also a carrier the baby has a 1 in 4 chance of inheriting the disorder. A diagnostic test can then be offered. Other family members of carriers of autosomal recessive disorders may also be carriers and may request a test, especially if they are planning a pregnancy. This is called cascade screening (see *Chapter 10*).

Screening for haemoglobin disorders

The haemoglobin disorders, thalassaemia and sickle cell anaemia, are the most common of the severe inherited disorders in human. More than 5% of the human race are carriers without symptoms. Worldwide, 1 in 500 children have a haemoglobin disorder and the disorders are so common because carriers are protected against death from malaria. Carrier prevalence ranges from about 0.1% in northern Europe to over 30% in sub-Saharan Africa. Global migrations have disseminated the haemoglobin disorders widely and they are now common in the multi-ethnic populations of many formerly non-endemic areas, including north-west Europe.

It has been possible to detect carriers by a blood test for many years and prenatal diagnosis is available. Education and information are vital and can be targeted to the groups most at risk. Ideally the best time to have a test is before the tenth week of pregnancy. All pregnant women are offered a blood test for thalassaemia, but only women from certain ethnic backgrounds will be offered a test for sickle cell disease.

Sickle cell disease and thalassaemia major are serious inherited blood disorders that affect haemoglobin and people who have these conditions will need specialist care throughout their lives. There are also other less common haemoglobin disorders, many of which are not as serious.

People with sickle cell disease can have attacks of very severe pain, can get serious life-threatening infections, are usually anaemic, and need medicines and injections when they are children and throughout the rest of their lives to prevent infections. Carriers of sickle cell disease are healthy but may experience problems in rare situations such as when having an anaesthetic or during deep sea diving where their bodies may not get enough oxygen. Knowing that they are a carrier can help them to manage these situations.

People with thalassaemia major are very anaemic and need blood transfusions every 4–6 weeks. Carriers are healthy and do not have the disorders.

In very rare cases the screening test may show that the person has a haemoglobin disorder without knowing it. The test is between 95 and 99% accurate, although in a small number of cases the result may be unclear and further tests may be needed.

Newborn blood spot screening

As well as screening during pregnancy, there are screening tests available which can be carried out shortly after birth. These are carried out on a spot of blood taken from the baby, usually by means of a simple heel prick. This newborn blood spot screening is offered to everyone in the first week after birth. The purpose of the programme is to identify babies who may have rare but serious conditions which will benefit from early treatment which can improve health and prevent severe disability or even death. Whilst screening will identify babies more likely to have these conditions, as with all screening tests they are not 100% accurate.

The American College of Medical Genetics recommends that newborn babies be screened for 29 different conditions, but in the UK, screening is currently only available for:

- phenylketonuria
- congenital hypothyroidism
- sickle cell disease
- cystic fibrosis
- medium chain acyl-CoA dehydrogenase deficiency

Phenylketonuria

Phenylketonuria (PKU) is an example of an inborn error of metabolism, a genetically inherited metabolic defect which results in deficient production or synthesis of an abnormal enzyme. PKU is an autosomal recessive disorder leading to a deficiency of the enzyme phenylalanine hydroxylase, which normally converts the amino acid phenylalanine to tyrosine. PKU affects about 1 in 12 000 babies born in the UK. Affected individuals are unable to process phenylalanine in their food and if untreated they will develop serious irreversible mental disability. Early removal of phenylalanine from the diet provides effective treatment and will prevent severe disability and allow children to lead a normal life. If not detected early and children are later found to have PKU it may be too late for the special diet to make a difference.

Congenital hypothyroidism

Congenital hypothyroidism (CHT) affects about 1 in 4000 babies born in the UK. The thyroid gland usually starts working in the unborn baby at about 20 weeks' gestation, but CHT occurs when the baby's thyroid gland does not develop or function properly.

Affected individuals do not have enough of the hormone thyroxine, and if untreated they do not grow properly and can develop serious permanent physical and mental disability. Early treatment allows normal development, but if delayed it may be too late to prevent serious disability. CHT cannot be cured but it can be treated simply and effectively by taking tablets of thyroxine for life and this replaces the thyroxine the body cannot produce. The long-term outcome for children with CHT who are treated daily with thyroxine is very good and most will lead completely normal, full and active lives.

In most babies with CHT the thyroid gland is either very small, in the wrong place (called an ectopic thyroid), or is missing completely (agenesis). Currently there is no reliable way to detect CHT before birth by prenatal diagnosis and no way to prevent babies being born with CHT.

Most babies with CHT do not show any symptoms when born. The newborn blood spot screening test measures the levels of TSH in the baby's blood. TSH stimulates the thyroid gland to make more thyroid hormone and a raised TSH level at screening suggests that the baby's thyroid is not functioning properly and further tests need to be carried out. Raised TSH and low free T4 confirm the diagnosis. These cases of CHT are not usually inherited but seem to happen by chance to babies with no history of CHT in their family. These cases are **sporadic** and the cause is unknown. The chance of parents having another baby who is affected is very low.

It is also known that the thyroid gland may be of normal size or enlarged but still does not produce enough thyroid hormone – this is called dyshormonogenesis. This type of thyroid problem is usually inherited and affects between 10 and 20% of babies with CHT. Most inherited cases are autosomal recessive with a 1 in 4 risk of recurrence.

Sickle cell disease

This disorder affects 1 in 2500 babies born in the UK. It is an inherited disease that affects the red blood cells. Sickle cell disease (SCD) affects the normal oxygen-carrying capacity of red blood cells due to a change in the structure of haemoglobin. When deoxygenated, the red blood cells containing sickle cell haemoglobin are unable to pass freely through the capillaries and instead form clusters which block blood vessels. This blockage prevents oxygenation of the tissues in the affected areas, resulting in tissue hypoxia and consequent intense pain (known as sickle cell crisis). Other symptoms include severe anaemia, damage to major organs and infections.

Early detection in newborns and appropriate management can improve quality of life, especially when parents/guardians and patients learn to recognize and avoid the risk factors that can trigger painful 'crisis' attacks. Prophylactic antibiotics can lessen the number of crises by preventing infections.

Cystic fibrosis

About 1 in 2500 babies born in the UK have cystic fibrosis (CF). CF affects the lungs and can affect digestion and, as a consequence, babies may not gain weight and have frequent chest infections. Screening allows treatment with a high energy diet, medicines and physiotherapy, allowing them to live longer healthier lives.

CF screening initially involves measuring levels of immunoreactive trypsin (IRT) in the blood spot. Increased levels on two separate occasions are indicative of CF, although there may be other reasons for high levels of IRT (such as ethnicity (North Africans), renal insufficiency, congenital heart disease, congenital viral infection, spina bifida, and trisomies 13 and 18). Molecular testing will then be undertaken and this usually tests for the four most common CFTR mutations in the local population, in order to identify affected children. The wider family can then be investigated using cascade screening. This is important as the carrier frequency of CF is high, at 1 in 22 Caucasians.

It should be noted that several hundred different mutations have been identified in patients with CF and many of these mutations are very rare indeed. The frequency of some mutations varies in different populations and knowledge of the ethnic background can be important. It is also important that the couples understand this limitation of screening in that there is a possibility that, in the absence of the common mutations, there may still be a rare mutation present that is not detected by the test.

Medium chain acyl-CoA dehydrogenase deficiency

Medium chain acyl-CoA dehydrogenase deficiency (MCADD) affects about 1 in 10 000 babies born in the UK. It is the most common of the fatty acid oxidation disorders. Those affected have problems breaking down fats to make energy for the body and this can lead to serious illness or death. During long periods between eating, the body breaks down its own fat stores to produce energy. People with MCADD lack one of the enzymes needed to do this. They can break down the stored fat partly but not completely. There is a hold-up at the medium chain fat step, where the enzyme needed to complete the breakdown is not working properly. This causes the medium chain fats to build up and make toxic substances that may lead to serious symptoms. Once diagnosed, MCADD is usually quite straightforward to manage and children with this condition usually lead healthy normal lives. MCADD is treated by diet and avoiding long periods without food. MCADD is autosomal recessive and two copies of the most common mutation (935A>G) are found in 85–90% of those with symptoms.

9.5 ALTERNATIVES TO PRENATAL DIAGNOSIS

Assisted reproduction

Nearly 2000 children are born every year in the UK using donated eggs, sperm or embryos, usually as a treatment for infertility.

Some couples cannot contemplate the termination of a pregnancy and therefore prenatal diagnosis is not appropriate. Others wish to know the result to take time to prepare for the birth of an affected child. For these couples, alternatives may be more acceptable. The conception of an affected child can be avoided by donation of gametes, either sperm or eggs, or by the identification of a pregnancy as affected prior to implantation. These techniques can be used when one partner is known to carry a genetic disease such as cystic fibrosis or Huntington disease.

Artificial insemination by donor

This involves the collection of sperm from an unrelated male donor. The sperm is placed in the woman's vagina, cervix or womb at the time of ovulation with the aim of achieving natural fertilization and pregnancy. The success rate is about 10% per treatment, but this varies with the age of the woman from 13.55% under 35 years to 1.2% at over 43 years.

Egg donation

Egg donation is less widely used than artificial insemination by donor because of the risks to the donor and the difficulties of the technique. The donated eggs are collected by laparoscopy following hormone stimulation of a female donor with the aim of producing several eggs. These can be fertilized outside the body (*in vitro* fertilization – IVF), or mixed with the sperm and placed together in the Fallopian tubes (gamete intra-Fallopian transfer – GIFT) to achieve a pregnancy. The success rate is about 20% per cycle.

The HFEA register

The Human Fertilisation and Embryology Authority (HFEA) regulates fertility treatment in the UK. It has a duty to keep a record of all registrations, treatments and outcomes resulting from assisted reproduction techniques. The collection of records is known as 'the Register'. The aim is to give people conceived through egg or sperm donation an opportunity to learn about their genetic background and to help avoid biological relatives inadvertently marrying or having children. The Human Fertilisation and Embryology Act came into effect on 1 August 1991 and allowed people donating sperm, eggs or embryos to remain anonymous. They were asked to provide some non-identifying information which could be given to people choosing a donor for treatment, and to any person conceived using their donation (when they reach the age of 18). With a growing awareness of how important it would be for some donor-conceived people to find out more about their genetic origins, the Government lifted anonymity for donors in April 2005. People donating after this date are no longer able to remain anonymous.

Over the years, the volume of non-identifying information collected has grown. This is because of an increasing awareness that some donor-conceived people would find even the smallest pieces of information about their donors of great value. The non-identifying information about donors which can be made available to a donor-conceived person when they reach the age of 18 includes:

- physical description of donor (height, weight, eye, hair and skin colour), year and country of the donor's birth
- donor's ethnic group
- whether the donor had any children
- any other details the donor may have chosen to provide, such as their occupation, religion, interests and skills

From April 2005, a donor-conceived person can ask for the following identifying information as well as the non-identifying information:

- the donor's name (and their name at birth, if different)
- date and place of donor's birth
- last known address (or their address recorded at the time of registration)

After the Act came into force in 1991, there was a limit of 10 'live birth events', which meant that 10 couples or individuals could have children from the same donor. They might have had a single child, twins or triplets at the birth event, and could subsequently apply to use the same donor to have further children, who would be full genetic siblings to the first child(ren). This was allowed even if the limit of 10 live birth events had been reached. In 2005, the Act was simplified to allow 10 families to be created from any one donor's gametes.

Pre-implantation genetic diagnosis

The availability of *in vitro* fertilization has allowed techniques to detect genetic disorders to be used in the embryos prior to replacement into the mother and hence avoid affected pregnancies. Each approach involves the removal of 1 or 2 cells at the 3-day blastocyst stage, and these cells are then analysed using FISH or molecular techniques such as PCR.

Pre-implantation genetic diagnosis (PIGD) has now been used successfully in a number of areas, with over 60 genetic conditions now licensed for analysis by the HFEA, including single gene disorders such as cystic fibrosis, Huntington disease, and Marfan syndrome, and some chromosome abnormalities in known carriers. It can be used to screen out those embryos carrying chromosome aneuploidy, a high cause of pregnancy loss (see *Chapter 5*), and hence potentially increase the success rate by only returning chromosomally normal embryos to the woman.

Potentially the whole genome could be investigated using techniques such as array CGH (see *Chapter 6*). However, the high cost (several thousands of pounds), low success rate (the chance of a live baby is only about 25%, even in the best scenarios), and complexity of the techniques mean that pre-implantation genetic diagnosis is likely to be limited to couples with high risk of fetal abnormality. In 2006, only 151 patients in the UK underwent pre-implantation genetic diagnosis at HFEA-licensed centres, less than 0.5% of all IVF cycles.

Cell-free nucleic acid for non-invasive prenatal diagnosis

One major concern with methods of prenatal diagnosis is the 1% risk of loss of normal pregnancies associated with the invasive procedures used. More than a decade ago, it was found that the blood of pregnant women contains cell-free DNA from the fetus. Of the total DNA in maternal plasma 3–6% is fetal in origin. It is detectable as early as day 18 of gestation, the amount increases during gestation and it is derived predominantly from the placenta.

Cell-free fetal mRNA has also been detected in the mother's blood, derived from genes that are expressed in the placenta. Since a small subset of genes are actively expressed only during fetal development, this offers an attractive target for non-invasive prenatal diagnosis and for investigating the development process.

Fetal DNA is cleared rapidly from the maternal circulation, with a mean half-life of 16 minutes. As it is comprehensively cleared from the maternal circulation within an hour of birth, the presence in one pregnancy will not affect any subsequent pregnancies.

The discovery of this cell-free fetal DNA has led to the development of non-invasive prenatal diagnosis (NIPD), where genetic characteristics of the fetus can be analysed a mere few weeks into pregnancy by studying a sample of the mother's blood. This is safer and more convenient than invasive procedures such as amniocentesis, which carry a risk of miscarriage. Also, unlike other non-invasive prenatal tests such as ultrasound and serum screening, NIPD can offer definitive diagnoses.

There are broadly four potential applications for which cell-free fetal nucleic acid (DNA or RNA) can be used for the purpose of NIPD.

- The first application is the determination of fetal sex in families at risk of a sex-linked disease by detecting genes present on the Y chromosome. This is primarily used for the purpose of reproductive choice in families with rare X-linked disorders, such as Duchenne muscular dystrophy, where only males are affected. NIPD can be used to determine the sex of the fetus from as early as 7 weeks and only pregnancies with a male fetus will then require an invasive diagnostic test to determine whether the specific disease-causing mutation has been inherited. This test is currently available in a small number of places, is 98% accurate and results in a 45% reduction in the number of invasive tests for X-linked disorders.
- The second application is for detecting specific single gene disorders in families with a high risk of inherited disorders such as cystic fibrosis. As techniques develop, this application is likely to increase to a wider range of genetic conditions.
- A third application is the determination of fetal Rhesus D blood group status in Rhesus D-negative women by detecting the paternally inherited RHD gene. This can be offered to women at high risk of potentially life-threatening haemolytic disease of the newborn. Rhesus D-negative women are currently given immunoprophylaxis (anti-D therapy) at 28 and 34 weeks to prevent possible sensitization to a rhesus-positive fetus. Early confirmation of fetal RHD status would allow those women carrying a D-negative fetus to avoid unnecessary anti-D administration with its concomitant exposure to human blood products.
- A fourth application is the detection of fetal aneuploidy, particularly Down syndrome. Depending upon the accuracy of the test, it could potentially complement or even replace either the many thousand current biochemical screening tests per year and/or the invasive diagnostic tests.

This improved safety, earlier detection and relative ease of NIPD technology raises specific ethical and social implications, as the volume of diagnostic testing will increase if NIPD becomes universally available (see *Chapter 10*). There are concerns being expressed over the use of NIPD for conditions for which prenatal diagnosis is not currently offered due to the low severity of the disorder compared to the risk of the invasive procedure. Also, the current multistep process of diagnosing Down syndrome could be moved to a single blood test which may have implications for informed choice and the 'routinization' of antenatal testing. There are also a number of potential non-clinical applications including sex determination (and selection) for social reasons, and paternity testing. A number of companies are already offering NIPD for fetal sex determination on a direct to the consumer basis. Public access to this technology via the internet, possibly without the involvement of a health professional, highlights a current gap in regulation that needs to be considered.

9.6 THE FUTURE

In the UK around 750 000 women become pregnant each year and every one of these will be offered some form of screening test such as an ultrasound scan and/or a serum test. As a result, more than 35 000 women will be told that there is a risk that their unborn baby may have a serious abnormality. This causes a great deal of anxiety and uncertainty for the parents and their families. Most parents will ultimately be reassured but, unfortunately, some will receive the devastating news that their baby has a serious or sometimes fatal disorder or a condition where the outlook is uncertain.

Prenatal diagnosis has improved in recent years with the development of more sensitive screening for Down syndrome, improved ultrasound scanning, more tests available for genetic disorders, and faster methods for obtaining results. Non-invasive tests for fetal sex and some blood groups are now a reality, with the potential for the detection of other pathogenic disorders (see RAPID–www.rapid.nhs.uk).

The ultimate goal, however, has to be to develop more treatments for genetic disorders including gene therapy (see *Chapter 10*), and to identify ways of reducing the incidence of abnormalities in the first place as has been achieved with folic acid, where taking this supplement prior to conception has significantly reduced the incidence of neural tube defects.

SUGGESTED FURTHER READING

Abramsky, L. and Chapple, J. (2003) *Prenatal Diagnosis – The Human Side.* Cheltenham: Nelson-Thornes.

Gardner, R.J.M. and Sutherland, G.R. (2000) *Chromosome Abnormalities and Genetic Counselling,* 3rd edition. Oxford: Oxford University Press.

Sadler, T.W. (2006) *Langman's Medical Embryology.* Philadelphia: Lippincott, Williams and Wilkins.

Wald, N.J., Rodeck, C., Hackshaw, A.K., *et al.* (2003) First and second trimester antenatal screening for Down's Syndrome: the results of the Serum, Urine and Ultrasound Screening Study (SURUSS). *Health Technology Assessment,* 7(11).

The UK National Screening Committee – policy and education on all aspects of screening NSC antenatal and newborn screening programmes – www.screening.nhs.uk.

RAPID (Reliable Accurate Prenatal non-Invasive Diagnosis) is an integrated project (started in July 2009) to refine and implement safer antenatal testing by evaluating early non-invasive prenatal diagnosis (NIPD) based on cell-free DNA and RNA extracted from maternal plasma – www.rapid.nhs.uk

SELF-ASSESSMENT QUESTIONS

1. There are a number of screening methods for Down syndrome in pregnancy. How many Down syndrome fetuses will be detected per unaffected fetus lost using the following methods:
   (a) maternal serum double test?
   (b) maternal serum triple test?
   (c) the integrated test?
   Why is this important?
2. What is the risk of a child being born with Down syndrome if the maternal age is
   (a) 20 years?
   (b) 40 years?
3. What factors are combined in the triple test?
4. Explain the following terms:
   (a) MOMs
   (b) specificity
   (c) sensitivity
   (d) detection rate
5. Match the disorder to the most common method of prenatal diagnosis:

(a) Down syndrome	(i) DNA analysis on CVS
(b) cystic fibrosis	(ii) maternal serum AFP and ultrasound
(c) anencephaly	(iii) biochemical assay of amniotic fluid
(d) Duchenne muscular dystrophy	(iv) fetal sex and DNA analysis of males on CVS
(e) Tay–Sachs disease	(v) karyotyping of amniotic fluid

6. Give some examples of alternatives to prenatal diagnosis for families at risk of a genetic disorder.

CHAPTER 10

Genetic counselling and genetics and society

Learning objectives

After studying this chapter you should confidently be able to:

- **Outline the nature of genetic counselling**
  Genetic counselling is the communication of information and support of families at risk of a genetic disorder. It includes establishing an accurate diagnosis and calculation of risks of recurrence or occurrence in a family member.
- **Explain the process of genetic counselling**
  The process of genetic counselling aims to establish an accurate diagnosis on which to base the counselling, to provide information about prognosis and follow-up, to provide risks of developing or transmitting the disorder, to discuss ways in which the disease can be prevented or ameliorated, and to support the family in adjusting to the implications of the genetic disease and the consequent decisions that have to be made.
- **Understand the three types of genetic tests**
  Genetic tests may be diagnostic tests which are performed when a person already has some kind of medical problem and is seeking an explanation. Carrier tests are used to look for gene carriers that are not going to do the person any harm but may impact on any children they may have (i.e. may have reproductive implications). Predictive tests provide information as to whether an individual may get a particular genetic condition or not.
- **Understand issues relating to privacy of genetic information**
  Opinions vary about whether genetic information should be seen as different to other medical information. A fundamental difference is that genetic information has direct relevance for a chain of related individuals who may not agree about what should be done with it. One person's right to confidentiality could conflict with the rights of others to access knowledge that would enable them to make informed choices. The aim of medical ethics is to balance such conflicting rights.
- **Understand the principles of consent**
  Before a healthcare professional examines or treats a person or shares their information with others, they must seek consent or permission. For consent to be valid, the person must be able to give their consent, they must be given enough information to make a decision, and they must be acting under their own free will.

- **Be aware of some of the complex personal, social and ethical issues in human genetics**
  New genetic research is providing a better understanding of heredity and genetically inherited disease and bringing opportunities for prevention, management and cure. The current explosion of human genetic information also has the potential for abuse, for damage to an individual's rights, privacy and fair treatment. Genetics raises many issues for individuals and for society, including genetic testing of children, predictive testing, over-the-counter testing, non-paternity testing, genetic identity cards, and the concerns regarding the impact of genetic information on insurance and protection of private genetic information.

10.1 INTRODUCTION

The preceding chapters have shown the many ways in which chromosome abnormalities and mutations in DNA can result in genetic disease. The ways in which these can be detected using various molecular genetic and cytogenetic techniques have also been examined.

The Human Genome Project is an international scientific research project with a primary goal to determine the sequence of the DNA base pairs and so to identify the many thousands of genes of the human genome, from both a physical and functional viewpoint. The project began in 1991; a working draft was published in 2000 and a complete sequence in 2003, with further analysis still being published today. This information, along with the rapid development of new technologies, has revolutionized the study of human biology and laid the foundation for many practical applications in relation to human genetic disorders.

The genome of any given individual (except for identical twins and cloned organisms) is unique and we still have much to learn about how most of the 35 000 known genes work in relation to our health and disease. There can be few individuals who have not encountered genetic issues in some form, perhaps through genetic testing and genetic conditions themselves personally, or in their families, friends or colleagues. Hardly a day goes by without some item of news involving genetics. As more and more environmental diseases are successfully controlled, those that are wholly or partly genetically determined are becoming more important.

Genetic disorders place a considerable health, emotional and economic burden on affected people, their families and on the community (see *Box 10.1*). Increasing public awareness of genetic issues and the development of new technologies are leading to ever greater demands on genetic services. Genetic disorders affect not just individuals but also their families, and more and more people at different life stages now need access to these high quality genetics services, which provide information and support to individuals and to their families in whom genetics is known to play an important role.

The detection of some genetic diseases by prenatal diagnosis and screening programmes is now widespread (see *Chapter 9*), and offers parents a way of avoiding the birth of an affected child, or of being aware in advance that a

Box 10.1 Impact of genetic disorders

- A chromosome abnormality is present in at least 50% of all recognized first trimester miscarriages.
- Of all neonates, 2–3% have at least one major congenital abnormality, which has often been caused by genetic factors.
- 2% of all neonates have a chromosome abnormality or a single gene disorder.
- Genetic disorders and congenital malformations together account for 30% of all childhood hospital admissions and 40–50% of all childhood deaths.
- Approximately 1% of all adult malignancy is directly due to genetic factors; 10% of common cancers such as breast, colon and ovary have a strong genetic component.
- By 25 years of age, 5% of the population will have a disorder in which genetic factors play a part.

pregnancy is affected with some disorders. Many new ethical issues need to be considered by individuals and by society, including confidentiality, the sharing of information, the insurance industry, the fear of producing 'designer' babies on demand, and eugenics.

Genetics services incorporate a range of clinical, laboratory and screening techniques in order to provide an effective co-ordinated service to patients and their families. The aim of the genetics service is to help those affected by, or at risk of, a genetic disorder to live and reproduce as normally as possible.

As has been described in earlier chapters, genetic disorders can affect all body systems and all age groups:

- chromosomal abnormalities that cause birth defects, learning difficulties, and reproductive problems (see *Chapter 5*)
- genomic disorders that cause diseases or developmental syndromes because of structural alterations in the genome, e.g. microdeletion or microduplication syndromes (see *Chapter 5*)
- single gene disorders, for example cystic fibrosis, muscular dystrophy and inherited biochemical disorders (see *Chapters 6* and *9*)
- genetic subsets of complex disease, including cancer (see *Chapter 8*) and cancer-prone syndromes, heart disease and diabetes
- disorders which may have a genetic component such as neural tube defect, cleft lip and palate, and learning disabilities

10.2 GENETIC COUNSELLING

It is in the specialist field of genetic counselling that medically qualified and other healthcare professionals explain to patients and their families the complex issues involved in their particular genetic disorder. Risks and patterns of inheritance are described and appropriate laboratory and other diagnostic tests are also organized.

The role of health professionals in genetics is to provide information in a supportive environment and to respect the fact that patients may choose different courses of action. This respect for patient autonomy means that they should not be pressurized to make a particular choice.

Genetic counselling is the communication of information and support of families at risk of a genetic disorder. It includes establishing an accurate diagnosis and calculation of risks of recurrence or occurrence in a family member. The person seeking such advice is called the **consultand**, but this is not always the person actually affected. The **proband** is the affected individual through whom a family with a genetic disorder is ascertained.

Important ethical issues are raised by genetic testing including confidentiality, the right not to be tested, and testing of children. This chapter starts with a brief discussion of the aims of genetic counselling in helping families with genetic disease, and also raises some of the ethical issues that need to be considered. It is not possible to cover such a wide area in too much detail but some of the current areas of concern are outlined. There are no easy answers and no one right way.

The general principles of any aspect of patient contact including genetic counselling are:

- informed consent – a patient is entitled to an honest and full explanation before any procedure is undertaken, including details of risks, limitations and implications of each procedure
- informed choice – a patient is entitled to full information about all options available without duress, including the potential implications of the results to themselves and their families
- autonomy – the patient is in charge and at any stage can decide to proceed no further with any investigations
- confidentiality – which can only be breached in extreme circumstances

The counselling should be:

- non-directive – no attempt should be made to direct the consultand along a particular course of action
- non-judgmental – even if a decision is reached which seems ill-advised, or is contrary to the counsellor's own beliefs, it is the consultand who has to live with the decision
- supportive – taking into account the complex psychological and emotional factors that go with the process
- confidential – this can raise difficult issues and will be discussed later

Genetic counselling may be sought for various reasons, for example, following prenatal testing, diagnosis of abnormality in a child, or following miscarriage or death, and to test for carrier status (see *Box 10.2*). An appointment at a genetics clinic is similar to other medical appointments, but with each one being tailored to the patient who sets the agenda by deciding what they want to discuss and what they do not. The role of genetic counselling is to:

- establish an accurate diagnosis on which to base the counselling
- provide information about prognosis and follow-up
- provide risks of developing or transmitting the disorder

Box 10.2 Reasons for referral to genetic services

- Common conditions which may be genetic, e.g. breast and colon cancer.
- Unexplained developmental delay or disability.
- Diagnosis of a genetic disorder after miscarriage or at birth.
- Testing carrier status of family members for Mendelian disorders.
- Genetic management of high-risk pregnancies.
- Interpretation of abnormal prenatal results.

- discuss ways in which the disease can be prevented or ameliorated
- support the family in adjusting to the implications of the genetic disease and the consequent decisions that have to be made

Establishing the diagnosis

A crucial step in any genetic consultation is that of establishing the diagnosis. If this is incorrect then inappropriate or misleading information could be given, with potentially tragic consequences. Establishing the diagnosis involves taking a family history which is then displayed on a pedigree as shown in *Chapter 4*. The pedigree should include any abortions, stillbirths, infant deaths, multiple marriages and consanguinity, as these may be of relevance and give clues as to the nature of the condition. A full examination, which may include further investigations such as molecular genetics and chromosome studies, will be required, as may referral to specialists in other fields such as neurology, cardiology and ophthalmology.

Genetic tests may divided into three types, as follows:

- **diagnostic tests** which are performed when a person already has some kind of medical problem and is seeking an explanation for this
- **carrier tests** which are performed when looking for gene carriers that are not going to do the person any harm but may impact on any children they may have, i.e. there may be reproductive implications
- **predictive tests** which are performed to establish whether an individual may get a particular genetic condition or not

When people are offered a test they should be informed in advance of the purpose of the test, what it involves and what it will indicate. The explanation needs to consider the following points:

- whether it will give a definite diagnosis or only provide information about the risk
- whether other tests will need to be done
- what it will not show, such as other possible conditions or mutations not specifically tested for
- implications for insurance or employment, especially in relation to late-onset conditions (see later)

Informed consent must always be sought before a test is carried out.

Problems with establishing the diagnosis

Many disorders are recognized as showing genetic heterogeneity, that is, they can be caused by more than one genetic mechanism (see *Chapter 4* and *Box 10.3*). This can have important implications when trying to establish the mode of inheritance and hence the risks in other family members. Fortunately, molecular genetic techniques are increasingly able to identify specific mutations involved, thereby providing solutions to some of these problems.

It is essential that the geneticist keeps up to date with new developments, in particular the identification of new mutations which may help to provide a diagnosis where it was not previously possible.

Box 10.3 Examples of hereditary disorders that can show different patterns of inheritance

Disorder	Inheritance patterns
Cerebellar ataxia	AD, AR
Ichthyosis	AD, AR, XR
Polycystic kidney disease	AD, AR

AD – autosomal dominant; AR – autosomal recessive; XR – X-linked recessive.

Calculating the risk

The calculation of the risk of a genetic disorder requires the knowledge of its genetic basis. For example, does it have a known Mendelian pattern of inheritance or is it known that genetics plays a part in its aetiology? The calculation of risks has been discussed in detail in *Chapter 4*.

Discussing the options

Once a diagnosis has been made and the risks of recurrence or occurrence discussed, the counsellor will aim to provide the consultand with the information they need, including details of choices open to them, to arrive at their own informed decisions. This may include discussion of:

- reproductive options, including prenatal testing and the termination or continuation of an affected pregnancy (see *Chapter 9*), and the psychological and social implications for the individual and their family
- reproductive alternatives such as contraception, assisted reproduction, pre-implantation genetic diagnosis, gamete donation, adoption, or not having a family at all
- the options and implications of testing for carrier status
- the options and implications of testing for late-onset disorders

Predictive testing

There are a wide variety of late-onset genetic disorders where the actual clinical features of the condition may not be detectable for many years. The use of molecular genetic techniques enables predictive or presymptomatic testing to be offered before the individual actually becomes ill for disorders such as familial adenomatous polyposis coli and Huntington disease.

This raises some difficult ethical issues. Areas to be considered include implications of an abnormal result for future employment and insurance, family and reproductive issues, the genetic implications for relatives, and coping with the diagnosis. One advantage may be that in some cases testing allows earlier screening for the disorder and so treatment or other preventative measures can also be instigated earlier. Dietary management in carriers of familial hypercholesterolaemia can have a beneficial effect by reducing the risk of heart problems. However, when no treatment is available and the testing reveals an individual to be a carrier it can be very traumatic for the family.

A positive test can also have major implications for close relatives who themselves do not wish to be informed of their disease status. 'Diagnosis by proxy' can result from testing for these autosomal dominant disorders. One scenario may be, for example, when an individual requests testing for such a disorder because his paternal grandfather is known to be affected, but his own father does not wish to know. A negative result in the consultand does not change his father's 1 in 2 risk. However, a positive result indicates that his father must also be a carrier.

A variable proportion of at-risk individuals will decline the option of testing when fully informed of the benefits and consequences, choosing instead to live with the risk of having inherited a mutation, as opposed to the certainty they would gain via molecular genetic testing.

Genetic testing of children

It is valid to undertake a diagnostic test in a child suspected of having a genetic condition. Parents benefit from having an answer to a child's problems, and a diagnosis can lead to appropriate treatment and support from various agencies, including education, and is therefore in the child's best interest.

Screening of all babies is now routinely carried out at birth for a small number of conditions. Newborn screening (see *Chapter 9*) on blood spots allows early treatment and the reduction or even prevention of health problems.

Information may also be obtained during prenatal diagnosis. The fetus may be found to be a carrier when testing for a recessive disease such as cystic fibrosis, or may be found to be a carrier of a balanced chromosome rearrangement (when having a karyotype established for Down syndrome, for example) which may impact on their reproductive health.

The decision of parents as to whether or not to have their children tested for a genetic condition that occurs in adulthood and is known to be in their family, and at what age to test the child, can be a very difficult one. There can be no simple or single formula that will be appropriate for every family. It may be possible, for example, to undertake predictive testing to allow the advantages of early testing in conditions where screening can be carried out and so help to reduce the onset or severity of the condition.

Giving children unnecessary genetic labels very early in life can be deleterious to the child. It can affect family relationships, and it can affect how that child is treated within that family and also the child's expectations of him- or herself.

Genetic testing is a very important procedure that requires **consent** (see below) and a person can only give consent if they have knowledge and understanding. With a very small child there is no way they can have this knowledge and understanding and hence they cannot give informed consent to a test. It is not an age-related test, but a functional test about understanding. Some quite young children can have a very good understanding, particularly where they have chronic life-long diseases such as cystic fibrosis. It should be remembered that the child also has the right **not to know**, as the result may well impact on their whole life. The principal ethical concerns about predictive testing of children are the loss of the child's future autonomy as an adult to make their own decisions about testing, and the loss of the confidentiality to which an adult would be entitled.

Genetic test information is the property of the person on whom the test has been performed and it is entirely up to this person or to the parent acting on their behalf to decide who knows and what they know. In the genetics community it is usually preferable to wait until the child is old enough to take part in the conversation and decision-making about predictive testing.

When a child is diagnosed as having a genetic condition or chromosome abnormality, their future reproductive risk should be recorded in their medical notes or in a genetic register (see *Section 10.4*) to facilitate follow-up for counselling at an appropriate time.

Screening for genetic carrier status

We all carry some recessive genes. When two carriers of the same disorder have children they have a 1 in 4 chance of having a child affected by the disease in every pregnancy. Female carriers of X-linked disorders such as Duchenne muscular dystrophy have a 50% chance of an affected son and 50% of daughters will also be carriers. Often people only become aware that they are carriers on the birth of an affected child.

Genetic screening is a search in a population for people harbouring susceptibility-causing or disease-causing alleles or genetic markers that may lead to disease in the individual or in that person's offspring. In some groups screening has been offered before marriage and may influence the choice of partner (see *Chapter 9*). Carriers might use the knowledge of their carrier status in their reproductive choices; for example, how many children they have, if any, and whether or not to make use of prenatal diagnosis if available; alternatively they might not use the information at all.

The decreasing popularity of marriage in many societies often rules out premarital screening. Screening is now often offered during pregnancy and, if the woman is identified as a carrier, screening is extended to the partner. This limits the options available to the couple to prenatal testing of the current pregnancy or to taking the risk.

Once identified as a carrier, 'cascade screening' of other relatives should be considered. This can cause great anxiety for those involved. The exclusion of carrier status in a family makes an enormous difference to family members, as for them the burden is lifted.

Carriers may also be identified during newborn screening or prenatal testing. Consideration needs to be given as to when best to inform the child as discussed above.

A family in which a child has suffered from and perhaps died of a debilitating genetic disorder may see carrier testing of that child's siblings as hugely advantageous. In general carrier testing, by definition, will have no implications for the health of the child being tested, but the results may have an enormous effect on the adult life of that child, in particular on the reproductive decisions that such an individual feels able to make. There may be different viewpoints between parents and professionals. The issues relating to autonomy and confidentiality described above apply. The advantages afforded by 'advance notice' must be balanced against the potential disadvantages of being labelled as a 'carrier' many years before such information can be used to make the relevant, usually reproductive, choices.

Consanguinity

It has been estimated that each individual carries about 12 mutations associated with recessive conditions and there are about 3500 recessive conditions, most of them very rare, in the general population. In children of unrelated couples the risk of congenital and genetic problems (including recessive conditions) is usually estimated to be about 2–3%. However, if relatives have offspring together, the chance of the same recessive gene being present is increased.

We all share a proportion of our genes with our relatives (see *Box 10.4*). **Consanguinity** means 'related by blood'. In genetics, the term is usually used to describe a mating or marriage between close relatives. Consanguineous marriages between relatives, such as first cousins, increase the risk in children of recessive conditions but not of general birth defects and genetic problems of all kinds. First cousins share 12.5% of their genetic material and may have inherited the same mutation from a common ancestor. Therefore the risk in offspring of first cousins is usually estimated at approximately double the risk of congenital and genetic problems in children of unrelated couples, that is, about 4–6%. This is a special problem in genetic counselling because of the increased risk of autosomal recessive disorders. Premarital or preconception counselling can be particularly useful.

Risks for recessive conditions also vary between communities (i.e. risk is not uniform across the world) and so knowledge of the population group is

Box 10.4 Shared genes in relatives

Degree of genetic relationship	Proportion of shared genes
First degree	
Parent–child	1/2
Siblings	
Second degree	
Uncle–niece	1/4
Half-siblings	
Third degree	
First cousins	1/8
Fourth degree	
First cousins once removed	1/16
Fifth degree	
Second cousins	1/32

important in assessing the risk. Also bear in mind that in some ethnic groups cousin marriages are part of their culture. This can impact on public health needs. Recessive risk presents different challenges for genetic counselling because it includes risk for many different, mostly rare, conditions, some of which are treatable. Carrier and prenatal tests are now possible for an increasing number of recessive conditions, but screening for them all is not currently possible or realistic. It is often only after the birth of an affected child that clinicians know which condition is relevant within any particular family.

Questions have been raised by some as to whether first cousin marriages should be banned based on this elevated risk. However, most in the UK would find it inappropriate to castigate cousin marriages as this would also stigmatize late childbearing on the grounds of elevated risk of chromosomal abnormalities with advanced maternal age (see *Chapter 9*).

Marriage between first degree (that is parent and child or brother and sister) and second degree relatives (such as uncle and niece) is almost universally illegal. Offspring of such incestuous relationships are at high risk of severe abnormality, mental retardation and childhood death. Only about half of children born to first degree relatives are normal and this has important implications for counselling should a pregnancy occur.

Unexpected information when taking the family tree

It is not uncommon to discover non-paternity coincidentally during DNA testing of a family to investigate a Mendelian disorder. This information must remain strictly confidential, but it may substantially alter the risks to certain family members and is of great importance in subsequent genetic counselling. Deliberate testing to determine paternity is discussed later in this chapter.

In our modern society, families also often contain offspring from different partnerships which needs to be clearly recorded to assess appropriate genetic advice.

Role of genetic testing in adoption

The issues from adoption in relation to genetics may be important and arise for several reasons. These include:

- adoption as an option open to couples at risk of transmitting a genetic disorder
- adoption as an option open to couples at risk of developing a genetic disorder themselves, such as Huntington disease
- an adopted person wishing to know their own family history to complete their own pedigree and risk assessment

Since 2005, offspring from donor gametes have the right to some information about the donor as explained in *Chapter 9*.

Many years ago, adoption used to be mainly about finding places for babies from young unmarried women. More recently it has become more about finding appropriate families from a wide range of backgrounds for needy children in care. One complex area is that of a child being placed for adoption who has a family history of a genetic disorder and the adoption agency wishes to know how great the risk is before finalizing the placement. This raises several questions relating to genetic testing in children.

- When should genetic testing of children be done?
- Is it in the child's interest to be tested for a condition earlier than it would have been if the child was not being placed for adoption?
- Does a child have a right not to know?
- Who makes the decision for vulnerable children in care?
- Whose best interest would genetic testing serve – the baby, their prospective parents or the local authority?

It could be argued that this knowledge would enable a child to be placed with a family who are able to cope with, for instance, frequent hospital visits, but it also raises the question of whether adoptive parents should have access to more information than a natural parent. There could be concern about a child being valuable in their own right and that their value doesn't increase or diminish according to any disability, disease or other problem.

Investigating paternity

Advances in genetics and DNA testing are making it easier for people to find out not only about inherited illness but also about suspected non-paternity. Deliberate testing is carried out most commonly to determine whether a man is the biological father of a child, or to establish if two or more adults are related, either for private or for legal reasons, e.g. civil or criminal court cases involving disputed paternity.

Before thinking about going ahead with genetic testing to determine paternity, it is really important that consideration be given to the consequences if the result is not the one expected or wanted. The consequences can be devastating for the entire family, and particularly for the child involved. Consideration should be given to counselling first, to sort out whether paternity testing is the right thing to do and to help sort out how you will talk to the child about what is happening. It is important to remember that just because one child looks different or has different health problems to their brothers and sisters, this should not raise suspicion about who the father is. These differences can be quite normal.

Sometimes conventional blood testing will provide enough evidence to determine that a man cannot be the child's true father, or that there is a high probability that he is the father of the child. If conventional blood tests do not provide an answer, a DNA test can prove whether or not there is a genetic relationship. With the current tests available, and assuming a mother, child and possible father are all tested, then the results produced by an accredited company will either show he is or is not the father with a confidence level of 99.99% (the confidence level is actually usually greater than 99.999%).

With modern technologies it is possible to produce accurate results from the small number of cells that can be collected by scraping the inside of the mouth. This method is often less invasive than taking a blood sample, but the use of mouth swabs means that testing can be performed on a child of any age after birth. This has made it possible for people to conduct paternity tests without any professional medical involvement, for example, by buying sampling kits over the counter or via the internet and returning the sample by post to a genetic testing company. The result is that this market for private paternity testing services has grown rapidly. Advertising these services, however, has to comply with the Advertising Standards Codes to prevent false claims.

Paternity testing during pregnancy has also begun to be offered by several companies and the market has the potential to grow. The result may influence the decision by women as to whether or not to continue with the pregnancy. This test may involve the collection of fetal tissue using an invasive procedure which involves the risk of miscarriage (see *Chapter 9*).

A voluntary code of good practice has been devised by the Department of Health for organizations that provide genetic paternity testing services direct to the public. A key principle of this *Good Practice Guide* is that the best interests of the child should be a primary concern when commissioning all genetic paternity tests, be they for legal or for private reasons. The same requirements for consent under ordinary principles of medical law and the Human Tissue Act (see below) apply even if the test is being undertaken by private individuals. This includes consent to take the material, consent to the material being used and/or stored, and consent to any DNA in the material being analysed and the results used other than for the expected purpose. This technology is now being used for a number of other purposes (see *Box 10.5*).

Box 10.5 Commercial uses of DNA fingerprint technology

- To determine paternity for the assessment and collection of child maintenance via the Child Support Agency.
- To assist individuals who need to provide evidence of biological relationships for immigration purposes.
- To determine twin zygosity to establish if twins are identical or fraternal. Identical twins will have the same DNA profiles, as they developed from the same zygote during gestation. Fraternal twins will have similar, but not identical, DNA profiles, having developed in two separate zygotes during gestation. Fraternal twins' DNA profiles will be as similar as two non-twin siblings'.
- Genealogy ancestry by DNA testing services to discover anthropological roots.
- Dog owners interested in the ancestral heritage (a 'canine heritage test') of their mixed-breed dog to establish their true pedigrees.

10.3 STORAGE OF GENETIC MATERIAL – PLANNING FOR THE FUTURE

The rapid development of new genetic tests means that more families are able to be helped. In order to do this access to material for testing is essential from both living and dead relatives. This can be in the form of DNA, post-mortem tissue, or organs or live cells.

The organ retention scandals of the 1990s highlighted issues of consent, and the use and storage of organs and tissues was placed under the spotlight. Various enquiries found that the storage and use of organs and tissues without proper consent after people had died were commonplace at that time.

It became apparent that existing laws needed to be updated and the result was the Human Tissue Act (2004).

Human Tissue Act 2004 (Human Tissue Act Scotland 2006)

This Act became fully implemented on 1 September 2006 and sets out the law with respect to the removal, storage and use of human tissue for public display (in museums, for example) and a number of health-related purposes (including research), collectively referred to as scheduled purposes (see *Box 10.6*). The Human Tissue Act (2004) in England provides a legislative framework that applies to:

- the storage and use of dead bodies
- the removal, storage and use of 'relevant material' from a dead body
- the storage and use of 'relevant material' from a living person

'Relevant material' is defined as material derived from a human body that consists of, or includes, cells including waste products. It does not relate to human gametes or embryos, body hair or nails, and material created outside the human body (e.g. cell lines). It is now illegal under the Act to retain human tissue, including gametes, hair and nails, with the intention of

Box 10.6 The Human Tissue Act 2004 – what does it do?

The Act:

- makes consent the most important principle for the lawful retention and use of human tissue
- establishes the Human Tissue Authority to:
  - advise on and ensure compliance with the Act
  - develop national operational procedures and guidelines
  - licence activities using human tissue
- makes it an offence to be in possession of bodily material with the intention of analysing the DNA without the consent of the individual from whom it was obtained (or someone close to them if they have died)
- also tidies up consent for organ donation, and makes it illegal for DNA samples to be stolen – as they might be, perhaps, for paternity testing

analysing its DNA without the consent of the person from whom it came, or of people close to them if they have died. This section of the Act makes 'DNA theft' illegal.

10.4 PROTECTING GENETIC INFORMATION – CONSENT AND CONFIDENTIALITY IN CLINICAL PRACTICE

Genetic information can be viewed differently from other types of health information because:

- it is shared within families
- it may predict future health
- it may reveal unexpected details, e.g. non-paternity

The doctor–patient relationship encompasses a set of boundaries with accepted rights and responsibilities regarding information obtained in this setting. In clinical genetics the situation often arises whereby information is held about individuals who are not the direct patients. To tell or not to tell is a question that has to be asked frequently. On the one hand, it may be important to communicate genetic information to individuals who are not the direct patients because, as holders of this information, clinicians have some responsibilities. On the other hand, there is the duty to respect the privacy and autonomy of those who have not sought advice from the clinician. In addition, the situation may arise whereby patients need to be provided with medical information about their relatives, but where these same relatives may have specifically requested not to know such information. This difficult area includes discussions about identical twins, obligate carriers and the wider family.

Consent

The seeking of consent is a **process** to ensure that a person understands the nature and purpose of giving a sample or undergoing an intervention.

Appropriate information and the opportunity to discuss issues should be offered as part of the process. Generally, in medico-legal terms, 'valid consent' is ensured if anything that represents a **significant risk** and that would affect the judgment of a reasonable person has been explained. This recognizes that the consent process cannot always be totally comprehensive: all eventualities cannot be covered because of unexpected clinical findings or test results, and there is no need to explain every last possibility. General guidelines for obtaining consent for examination and treatment have been issued by the Department of Health and underpin all areas of clinical practice including genetics.

Special guidance is also available for obtaining consent from children and in adults who lack capacity. Young people aged between 16 and 17 years are presumed to have the competence to give consent for themselves. Younger children who understand fully what is involved in the proposed procedure can also give consent, although their parents will ideally be involved. If a competent child consents to treatment a parent cannot override that consent.

Other than in exceptional circumstances, consent should be obtained prior to a clinical or laboratory test with genetic implications being undertaken, and consent should have been obtained before medical genetic information is disclosed.

There are a number of issues associated with the clinical practice of genetics in particular. The first step when discussing a possible genetic condition with a family is often to take a detailed family history – the pedigree. This family history is usually obtained from one family member and is given in good faith, but it is likely to contain information about other family members who may not be aware that it is being given and recorded. However, a family member is unlikely to know confidential information other than what has been given by the person concerned. Most of the information given to construct a pedigree during a genetic consultation is likely to be known to a wide circle of people. Nevertheless the family information is held in confidence.

When a person asks for genetic testing it may be because of knowledge that a family member, usually the proband, has had a genetic test performed. The request for testing may be to determine carrier status, to assist with reproductive decisions, or to determine whether someone with a genetic predisposition (e.g. for breast or bowel cancer) needs to have surveillance or treatment.

Most individuals express a desire for pedigree information and genetic test results to be made available to other family members, even if that individual does not want to contact family members personally but would prefer a health professional to take on this role. For example, identification of a carrier of a condition, such as a chromosome translocation or an inherited disorder, may have implications for other family members and lead to the offer of tests for the extended family. This can raise questions of confidentiality. A carrier would be urged to alert close relatives to the possibility that they too could be carriers and so their children may be at risk and genetic counselling should be arranged. If a patient, for whatever reason, refuses to allow this information to be disseminated, despite the consequences being explained, then this wish would usually be respected.

Professionals may decide to breach confidentiality when the potential harm to the family member of not being informed, or the potential benefit of being informed, outweighs the potential harm to the individual whose confidentiality is being broken.

Genetic information should never be disclosed to third parties such as employers or insurance companies without an individual's written consent.

Genetics issues to be discussed during the consent process may include (as appropriate):

- the use and sharing of information (pedigree, diagnosis, affected/carrier status test results) with other family members for their benefit, either by the individual or by a health professional
- the nature of the testing to be undertaken and its implications
- the possible prolonged nature of the testing process – genetic testing can be complex and may take several months
- the possibility that testing may reveal unexpected results, e.g. finding a translocation in prenatal chromosome analysis for Down syndrome
- storing of samples for continuing and further investigations, e.g. as new tests for the disorder or more information become available; DNA is routinely stored permanently
- the fact that samples may be used for quality assurance, education and training
- the fact that information may be shared with health professionals, including those from other areas where family members may be living, e.g. details of the exact mutation to ensure the appropriate test is undertaken

Data Protection Act 1998

Most people will be aware, from the news or personal experience, about the need to protect personal data (covered by the Data Protection Act 1998), about the issues surrounding storage and use of tissue by laboratories (covered by the Human Tissue Act 2004), and the need for consent in clinical practice.

The Data Protection Act 1998 applies to all personal information processed by organizations. Data about an individual's DNA are personal data, and so organizations have a responsibility to ensure that any databases containing information related to an individual, whether processed and stored on a computer or in a manual filing system, are kept and processed in accordance with the data protection principles as set out in the Act. This applies whether the database is structured by reference to individuals or by reference to criteria related to individual specimens. Records should not be retained for longer than is necessary for the purpose for which they were originally collected.

10.5 GENETIC SUSCEPTIBILITY

The mapping of the human genome has opened up the possibility that individuals will be able to access information about their genetic susceptibility to a wide range of diseases. It is expected that such genetic profiling will be used to advise people about their individual susceptibility to common chronic illnesses such as coronary heart disease, diabetes and Alzheimer disease. However, these illnesses are known to have multifactorial aetiologies of which genetic susceptibility is just one part. These illnesses develop because of complicated interactions between many different genes, and behavioural and environmental factors. Developing accurate and useful tests is therefore not altogether straightforward.

Predictive testing for diseases with a multifactorial aetiology will involve taking samples from asymptomatic individuals and analysing the samples to provide information about the potential for future disease. One example of this is the testing for hypercholesterolaemia (high levels of serum cholesterol) as a measure of the risk of heart disease. Hypercholesterolaemia is a known risk factor for coronary heart disease, but there is no certainty that the discovery of high serum cholesterol in an individual will mean death from heart disease.

There may also be serious consequences from informing an individual that they have a genetic susceptibility to coronary heart disease when this is not the case (false positive), or informing them they do not have genetic susceptibility when in fact they do (false negative). Hypercholesterolaemia is generally symptomless, so understanding this diagnosis and the accompanying high risk status may be difficult for some people; they will find it hard to accept the diagnosis and the ensuing recommendations for behavioural change which will reduce their risk of coronary heart disease when they do not currently feel unwell. Those who do accept it may question their identity as 'healthy people'. Problems may also arise because predictive genetic testing can indicate susceptibility to heart disease much earlier than current markers of risk, such as high serum cholesterol.

Familial cancer

Cancer is a common disease and almost everyone in the UK has a close relative who has had cancer – it is estimated that in the UK about 1 in 3 people will get cancer during their lives. Everyone has a certain risk of developing cancer whether they have relatives with cancer or not. The cause of most cancers is unknown, but what is known is that some risk factors increase the chance of developing cancer. However, having a particular risk factor for cancer does not mean cancer will definitely occur, just as not having it does not mean you will not get ill.

These risk factors play an important role in the development of cancer. The main risk factor for cancer is age. Cancer is relatively rare in young people but relatively common in the old, with approximately 65% of cancers occurring in people over 65 years of age.

Smoking is the biggest cause of cancer that can be avoided. One in every three cases of cancer is caused by smoking and it is responsible for about 90% of lung cancer cases. The risk of cancer is reduced in those who take regular exercise, maintain a healthy weight, eat a balanced diet, and do not exceed the recommended levels of alcohol intake.

Many people worry about getting cancer, particularly about having a higher risk of developing cancer when there has been cancer in their family. Many people think that because they have one or two relatives with cancer this means that a 'cancer gene' is running in their family, but this is not normally the case. It is only likely that a cancer gene is present in a family if:

- there are two or more close blood relatives on the same side of the family with the same type of cancer
- cancers are occurring at a young age (i.e. before the age of 60 years)
- a close relative has had two different types of cancer (not one cancer that has spread)
- relatives have particular types of cancers that are known to run together, for example, breast and ovarian, or bowel and womb cancer

Inheriting a cancer gene results in a predisposition or susceptibility to cancer compared to other people in the population, but this does not usually mean that the carrier will definitely get cancer. Offspring of a person with a cancer predisposition gene have a 50:50 chance of inheriting the gene. The most common cancers that may, in some cases, be due to an inherited mutation are ovarian, bowel and womb cancers. Only between 5 and 10% of cancers are thought to be due to inherited genetic mutation.

There are a number of risk factors known to be associated with breast cancer (see *Box 10.7*). Genetic testing at present is only available for breast, ovarian and bowel cancer (see *Chapter 8*). Some people with a strong family history of breast or bowel cancer may be invited to have more regular screening than those who do not seem to be at increased risk, to allow earlier detection and therefore treatment.

Some rare genetic disorders can increase a person's risk of getting several different types of cancer at a young age (see *Box 10.8*). If there is thought to be the risk of a family predisposition to cancer, patients will normally be referred to a genetics centre or family cancer clinic. The centre or clinic will investigate the family history, including obtaining consent to access the health records of relatives on genetic registers, to find the specific type of cancer and whether or not there is an increased risk in family members.

Box 10.7 Risk factors for breast cancer

- An early first menstrual period (under 12 years of age).
- A late menopause (after 50 years of age).
- No, or late (over 30 years of age), child bearing.
- No breast feeding or breast feeding for less than 12 months in total.
- Contraceptive pill – risk decreases after stopping.
- Combination hormone replacement therapy (HRT) – decreases when stopped.

Box 10.8 Genetic disorders with increased cancer risk

- Li–Fraumeni syndrome.
- Multiple endocrine neoplasia type 1 (MEN1).
- Von Hippel–Lindau disease.
- Neurofibromatosis.
- Retinoblastoma.

Genetic registers

A genetic register contains information obtained from genetic counselling and the production of a family pedigree and provides a valuable tool for the genetics services.

The aim of such a register is primarily to establish, as completely as possible, all those people at risk of developing or transmitting a particular disorder, so that appropriate counselling can be offered at different life stages. It also permits the long-term follow-up of family members. This is important for at-risk children who may not need any investigations or counselling for many years, for informing families of new information and research, and for continued follow-up and support of people with long-term or late-onset conditions. The genetic register is held on computer and is subject to the Data Protection Act 1998; no one is included without having given written consent.

10.6 GENETIC INFORMATION AND THE INSURANCE INDUSTRY

Not all genetic tests are relevant from an insurance perspective. Life insurers are keen to know predictive healthcare information about illnesses that are likely to affect policy holders during the term of their insurance policy. Most people only take out life insurance to cover the term of their mortgage for their home, and so the most relevant genetic diseases for the life insurer are those that affect people in their mid-twenties and mid-sixties. There are only a small number of diseases which affect people in this time frame and for which reliable genetic tests are available.

It is now possible for individuals with a family history of some late-onset conditions, such as Huntington disease, to be tested to determine whether or not they are likely to develop the disorder. With increasing scientific understanding of the contribution made by genetics to a wide range of late-onset conditions (such as some cancers and diabetes), it may become possible to identify a proportion of the population as being at a greater risk of developing one of these conditions later in life than is the case for the general population.

There is increasing concern that insurance companies would use genetic information to discriminate against people at risk of developing a genetic disorder. Currently, there is a voluntary code of practice provided by the

Association of British Insurers. The applicant for insurance always has the choice whether or not to take genetic tests. The code does not allow insurers to insist that someone takes a genetic test as a condition of offering them insurance. However, when applying for insurance, any existing genetic test result must be given to the insurer unless the insurer has said that such information is not required. Any future genetic tests do not need to be given to the insurer after the policy has been issued.

Positive test results for predisposition to certain diseases do not automatically mean refusal or an increase in premiums. The information provided will always be kept confidential and the insurers will take professional advice on the information received.

Until accurate genetic tests become more widespread for common multifactorial diseases, the impact on insurance is likely to be minimal.

10.7 GENE THERAPY

Gene therapy has been in the headlines promising great advances in the treatment of genetic disease. There are two broad types of gene therapy, **somatic** and **germ-line**.

Somatic gene therapy can be defined simply as the delivery of functional genes to somatic tissue for the treatment of disease. In somatic gene therapy, a therapeutic gene is administered to the individual in order to make changes to the somatic cells in the body, but not to the germ cells that are involved in reproduction. The therapy therefore affects only the person to whom it is given.

Germ-line therapy is aimed at genetic alteration of the germ-line cells for the treatment of disease in future generations. In germ-line gene therapy, the therapeutic gene is inherited by the progeny of the treated individual and becomes a stable part of their genetic make-up. For ethical reasons, germ-line therapy is not being developed in humans at present.

There are two discrete ways in which somatic gene therapy can be applied:

- *ex vivo* gene therapy involves modifying the individual's cells outside the body; cells are removed from the body, genetically altered in culture and returned to the individual, for example, by blood transfusion or bone marrow transplantation
- *in vivo* gene therapy involves genetically altering the cells by direct administration of the therapy to the individual; administration is usually by injection or in an aerosol spray to the lungs

The early growth of the industry was based on high hopes that gene therapy would prove to be an effective treatment for many diseases for which existing therapy is poor or non-existent. Since then there has been a chequered history, including the high profile death of a patient in 1999. Progress in improving the efficiency of gene transfer vectors (see *Chapter 6*) has remained slow.

In the longer term there are considerable grounds for optimism that gene therapy will provide a series of important new cures in the coming decades.

Over the last 20–30 years incremental advances in vector development, together with increased understanding of the underlying genetic defect, have led to the implementation of gene therapy for some disorders, including, for example, some specific severe combined immunodeficiency (SCID) diseases. These are monogenic defects that arise from abnormalities of lymphocyte development and function leading to severe recurrent opportunistic infections in the first year of life. Without therapy these conditions are uniformly fatal. Retroviral-mediated correction of autologous haematopoietic stem cells has shown significant correction of function in patients.

10.8 OVER-THE-COUNTER GENETIC TESTS

All aspects of healthcare are in general moving towards being more easily accessible for patients. This trend includes a growing availability of over-the-counter products, including the consumer-led demand for testing to improve health; this is testing the so-called 'worried well'.

There are broadly two kinds of genetic test which are available to buy over the counter or from the internet:

- non-related tests, e.g. paternity or genealogy tests
- health-related tests

Testing can be done from very small samples, such as mouth scrapes, which can then be sent for testing without any input from healthcare professionals. However, individuals need to consider carefully a range of questions before proceeding (see *Box 10.9*).

Box 10.9 Questions for the consumer of genetics tests

The Human Genetics Commission advises that before buying an over-the-counter test it is important that consumers think carefully and consider questions such as:

- why am I considering taking this test?
- what do I hope the test will tell me?
- how important is it to know what the test will tell me?
- will the test be able to answer my question?
- what do I hope to be able to do after getting the test result that I cannot do now?
- what if I get a result that I am not expecting?
- would it be a good idea to get professional advice (e.g. from my doctor) before buying the test and if so how can I get it?
- will the information the test provides have implications for my relatives?
  - am I going to tell them and, if so, how?
  - will they want to know?
  - should I discuss this with them beforehand?
- do I need to take the test now?
- is there another way of finding this information out?
- would the result of the test have any impact for my work or my insurance prospects now or in the future?
- is the company a reputable one?

Non-invasive prenatal diagnosis (see *Chapter 9*) offers tremendous benefit to patients and medical practitioners alike, and holds considerable future promise, but concerns over the technique have also been raised. Questions include whether or not it will encourage sex selection for non-medical reasons and what of the broader ethics of selective termination? Also, how can these tests be regulated, with direct availability via mail order and the internet, and possibly from unreliable or unscrupulous providers?

Much work is needed to realize the clinical relevance of findings from genome-wide association studies which claim to predict susceptibility to disease. Despite this, there is a rapidly emerging market of commercial genetic testing companies who are marketing tests based on these associations direct to the consumer, to provide health information or lifestyle choices. Commentators have raised concerns about the scientific validity of these tests, the quality of the testing services, and the appropriate regulatory and policy response.

The UK Human Genetics Commission, in its 2003 report *Genes Direct* and the 2008 follow-up report *More Genes Direct,* has made a number of recommendations in relation to overseeing genetic tests provided direct to the public. The Commission recommended stricter controls on genetic testing but did not believe that there should be statutory prohibition of some or all direct genetic tests. The recommendations are clustered around three areas:

- pre-market review
- quality assurance and advice
- advertising

If the promise of genomics to improve health is to be realized without causing harm or loss of public trust, developments in the evaluation and control of the supply of these tests may be as important as the science itself.

10.9 BABIES BY DESIGN – THE ETHICS OF GENETIC CHOICE

The use of embryo selection to avoid inherited disease by pre-implantation genetic diagnosis (PIGD) is already here, even for late-onset diseases and carrier status. It is possible that, in the not too distant future, there will be the first attempts at gene alteration through precise gene targeting techniques, and eventually there could be a move beyond immediate *prevention* to *enhancement,* with the goal of making better than well.

Naturally, concerns have been raised that this will change having a baby from a gift to a consumer product, with man playing God.

10.10 PHARMACOGENETICS AND PERSONALIZED MEDICINE

The aim of drug therapy is to administer the appropriate drug in the correct dose to produce the desired effect, with minimum toxicity and side-effects. The current approach uses trial and error to guide the choice of drug and its dose, and hence there is wide variability in efficacy and side-effects between individuals.

Many diverse factors affect an individual's response to drugs. These include the age, weight, sex and ethnicity of the patient, the nature of the disease, and the patient's diet. Environmental effects, concurrent drug treatments and the extent to which the patient adheres to their treatment may be equally or even more important.

Recent important advances in genomic research have opened the way to new strategies for public health. The emerging science of **pharmacogenetics** seeks to determine how people's genetic make-up affects their response to medicines. The core hypothesis underlying pharmacogenetics is that genetic factors play a significant role in the well-recognized differences between individuals in response to medication and susceptibility to adverse effects. It is hoped that this information will help to design safer and more efficient drugs, and that pharmacogenetics can be applied as a clinical tool for the prediction of treatment outcome.

Most current examples of drugs linked to genetic tests are anti-cancer therapies where the functional variation is acquired during the lifetime. The success story of Herceptin® in the treatment of breast cancer and imatinib (Glivec®) in chronic myeloid leukaemia is likely to be only the beginning of what can be achieved (see *Chapter 8*).

Other examples of the benefits of pharmacogenetics involve studying how we metabolize drugs. The **cytochromes** are the major enzymes involved in drug metabolism, accounting for approximately 75% of total metabolism. Cytochrome P450 (abbreviated as CYP, P450, or CYP450) is the most important element of drug metabolism, that is, the chemical modification or degradation of drugs. Cytochromes are found predominantly in the liver but can also be found in the intestines, lungs and other organs.

Many drugs may increase or decrease the activity of various CYP isozymes in a phenomenon known as enzyme induction and inhibition. This is a major source of adverse drug interactions, since changes in the CYP enzyme activity may affect the metabolism and clearance of various drugs. For example, if one drug inhibits the CYP-mediated metabolism of another drug, the second drug may accumulate within the body to toxic levels, possibly causing an overdose. Hence, these drug interactions may necessitate dosage adjustments or the selection of drugs which do not interact with the CYP system.

Humans have 57 genes and more than 59 pseudogenes divided between 18 families of cytochrome P450 genes and 43 subfamilies. It is hoped that genetic tests for isoform expression will become available to help anticipate drug interactions in patients.

Warfarin is an effective and routinely used oral anticoagulant, given to prevent blood clot formation in people who have coronary heart disease or venous thrombosis. Warfarin is metabolized in the liver by a member of the cytochrome P450 family, CYP2C9, but a variant gene alters the rate of its metabolism. People with the variant gene break the drug down more slowly than usual, and so require lower doses to achieve the same anticoagulant effect. Having too much anticoagulant can lead to potentially dangerous bleeding and increased susceptibility to some drug interactions. Patients are routinely started on a low dose of warfarin, their blood clotting is monitored, and the dose gradually adjusted until the appropriate level is reached; this can be seen as a sort of biological assay of drug effectiveness in the individual patient being treated.

Pharmacogenetics is unlikely to revolutionize or personalize medical practice in the immediate future. However, it is likely to become increasingly important in drug discovery and development as knowledge of the relative importance of genetic factors helps to identify optimal populations for a particular medicine.

10.11 GENETIC 'IDENTITY CARDS' AND HOROSCOPES

Genome sequencing has developed at a rapid pace. In 1985 it took 3 years to decode a single gene; by 2003 we could sequence the entire genome in just a few years, and today we can sequence genes in a matter of minutes. Within the next 2 decades it is expected that it will be technically possible to sequence the genome of each new baby – providing them with a run-down of each and every one of their genes and their associated risk of developing certain diseases. This genetic 'identity card' will enable them to see into their future health issues and this 'horoscope' will, in theory, enable them to seek preventative measures and adopt healthier lifestyles.

However, concerns have been raised that this could lead to 'genetic apartheid' with people discriminated against because of DNA defects. In addition, if it were only available privately to those who could afford it, there would be a risk of creating a genetic underclass and increasing the nation's health inequalities. This genetic screening could lead to a national database of everyone's DNA, with the potential for erosion of privacy and civil liberties.

The promotion of widespread genetic testing by vested interests distorts the health research agenda, diverts resources from more valuable approaches and risks a loss of public trust in medical research. It should be remembered that genes are actually poor predictors of complex disease in most people; the impact of smoking, poor diet, poverty and pollution are not limited to individuals with bad genes.

10.12 DISABILITY AND GENETICS

Advances in genetic knowledge and the development of prenatal screening programmes promise improved health on the basis of selective termination of pregnancies affected by impairment. This has occurred at the same time as progress in recognizing the rights of disabled people and the development of anti-discriminatory legislation and the challenge by disabled people of the negative attitude to impairment.

Disability rights and genetics at their extremes represent polarized paradigms. We all carry deleterious genes and, as knowledge and therapy improve, questions on normality and difference will become salient for everyone. The deliberate testing of children to establish whether they have the same disability as their deaf parent, or the intentional creation of 'saviour siblings' to help treat their seriously ill sibling has sparked fierce debate.

10.13 THE FUTURE

Throughout history countless genetic diseases have blighted the lives of families, causing chronic ill health, progressive disability, and often resulting in premature death. Today we are on the brink of being able to lift the threat of genetic disease from millions of families. The challenge will be to use the knowledge that results from our endeavours for the benefits of all who need and hope for treatment and cure. Despite all of this progress, the actual personal decisions that stem from DNA analysis remain very difficult for those involved.

The 'genetics genie' is already leaking out of the bottle. Genetic information raises concerns that preoccupy many of us, partly as a function of it being new and everybody still feeling their way. We all need to look closely at what happens in this field.

SUGGESTED FURTHER READING

Clarke, A. and Ticehurst, F. (2006) *Living with the Genome, Ethical and Social Aspects of Human Genetics*. Basingstoke: Palgrave Macmillan.

Gardner, R.J.M. and Sutherland, G.R. (2000) *Chromosome Abnormalities and Genetic Counselling*, 3rd Edition. Oxford: Oxford University Press.

Genetics Interest Group (1999) *Genetics – What has it got to do with me?* London: Genetics Interest Group.

Harper, P.S. (2004) *Practical Genetic Counselling*, 6th Edition. London: Arnold.

Harper, P.S. and Clarke, A.J. (1997) *Genetics, Society and Clinical Practice.* Oxford: BIOS Scientific Publishers.

Joint Committee on Medical Genetics (2006) *Consent and Confidentiality in Genetic Practice. Guidance on Genetic Testing and Sharing Genetic Information*. Available at www.bshg.org.uk.

Read, A.P. and Donnai, D. (2007) *New Clinical Genetics.* Bloxham: Scion Publishing.

Skirton, H. and Patch, C. (2009) *Genetics for the Health Sciences.* Bloxham: Scion Publishing.

Understanding cancer genetics – how cancer sometimes runs in families – www.cancerbackup.org.uk.

SELF-ASSESSMENT QUESTIONS

1. What is genetic counselling?
2. What is the role of genetic counselling?

3. What is presymptomatic testing?
4. What would you do if:
   (a) your father/child were at risk of carrying a genetic disorder that might affect you?
   (b) you suspected non-paternity?
   (c) you were offered a test to see if you had susceptibility to coronary heart disease ?
   (d) you had a family history of cancer?
   (e) you were pregnant with a child known to have Down syndrome?

Answers to self-assessment questions

Chapter 1

1. The nucleus and the mitochondria.
2. DNA comprises an antiparallel right-handed double helix with a sugar–phosphate backbone. Phosphate groups are attached to the pentose sugar via the 3′ and 5′ carbons. The bases attach to the deoxyribose at carbon atom position 1. They comprise two purines (adenine and guanine) and two pyrimidines (cytosine and thymine), which pair A with T (using two hydrogen bonds) and C with G (using three hydrogen bonds). These lie stacked flat at right angles to the helix.
3. Complementarity is the state in which the double helix of the DNA molecule exists, due to the pairing of bases A with T and C with G. Complementarity enables the copying of one strand of the double helix. If the sequence of bases on one strand is known, the other strand will have a complementary sequence of bases such that the sequence
   `AGGTTCGGAT` should have
   `TCCAAGCCTA` on the opposing antiparallel strand.
4. Exons are the conserved coding regions of genes. Introns (or intervening sequences) lie between the exons of genes, and are less conserved. A mutation may have clinical consequences in an exon, whereas mutations in introns generally result in polymorphisms, which are harmless differences in DNA sequences between individuals. As these are non-coding sequences there are no clinical manifestations.
5. Alphoid (α) repetitive sequences are found at the centromeres of chromosomes. Telomeres are found at the ends of chromosomes. Triplet repeats are found throughout the genome, but may have a clinical effect if they exceed a certain number and are located near or within genes.
6. G_1 (variable), S (8 hours), G_2 (4 hours), M (1 hour).
7. Mitosis takes place in somatic cells. There is one division, resulting in 46 chromosomes (diploid: $2n$). Meiosis takes place in the germ cells. There are two divisions, resulting in 23 chromosomes (haploid: n).

Chapter 2

1. Gene A is a structural gene; the protein contributes to the structure of the red blood cell. Gene B is a controlling gene; its protein (which may be a transcription factor) interacts with the promoter of gene C and helps to switch it on.
2. During the process of transcription, the whole of one DNA strand is copied (including the introns), producing primary transcript RNA. The introns then form loops which are spliced out, thus leaving a much shorter mRNA strand which is a copy of the exons of the DNA.
3. The gene for steroid sulphatase lies in an area which is not totally inactivated like the rest of the inactive X (Xp22.3). As a female has two X chromosomes she will produce approximately twice the amount of enzyme as a male with one active X.
4. The *HOX* or homeobox genes are important in pattern formation. They ensure that the correct type and number of appendages are present with respect to their symmetrical orientation along the spinal column.
5. The sex-determining region on the Y; *SRY*.
6. Female, as this is the 'default' state of the embryo in the absence of the Y chromosome. The second X in normal females is not totally inactivated; the remaining active genes must therefore be required to provide a dosage necessary for

normal female development, so in their absence a clinical phenotype will result.

7. Somatic recombination is the mechanism responsible for rearranging the genetic subunits which produce immunoglobulins (antibodies). Without somatic recombination, many separate genes would be required to code for the hundreds of thousands of antibodies needed in the immune system.

Chapter 3

1. A mis-sense mutation results in a base change in the DNA which may have a clinical effect if the resulting amino acid is different (or differently charged) to the normal amino acid. A nonsense mutation produces a stop codon in the DNA sequence, resulting in a prematurely terminated (i.e. shorter) protein.
2. A triplet repeat is a repeating run of three bases. Some repeats cause clinical syndromes if more than a critical number are present in an individual. In Huntington disease, more than 37 copies of a CAG repeat produces an affected phenotype. The repeat lies within the gene and is transcribed, so it must interfere with the function of the normal gene.
3. In fragile X, if the CGG repeat is present in more than 200 copies in a male, the FRAXA gene which lies near to the repeats becomes methylated, leading to cessation of transcription and hence translation.
4. Either a position effect, as the gene for Hunter syndrome has been separated from a controlling element, or skewed X-inactivation.
5. Maternal uniparental disomy of chromosome 14. This indicates that imprinting is occurring on the maternal chromosome 14, and that the active gene(s) on the father's 14 are required for normal development.
6. Normally active genes may be silenced by removal to a new position close to a centromere. This is a position effect due to the inactive heterochromatin found in centromeres.
7. *HOX* (homeobox) genes, as they are involved in pattern formation.

Chapter 4

1. Because the recombination fraction is less than 0.5, the genes must be linked. However the linkage is not very tight (closely linked genes usually have a θ of less than 0.05).
2. Autosomal dominant (myotonic dystrophy, Huntington disease), autosomal recessive (cystic fibrosis, phenylketonuria), X-linked recessive (haemophilia, colour blindness).
3. Two-thirds.
4. One in three.
5. This is essentially a Y-linked disorder, as the deletion is on the father's Y chromosome (which he passes on to his son). All his male children would inherit the deletion and would themselves be infertile due to azoospermia. All his daughters would inherit his normal X and would not be at risk.
6. Applying Bayes' theorem to the father:

	Affected	Not affected
Prior risk	1/2	1/2
Conditional risk	15/100	1
Joint risk	15/200	100/200

15/115 = 1/7.67

Therefore his risk of developing HD is about 1/8. The patient's chance of inheriting HD (should his father carry the gene) is 1/2. The final risk is therefore $1/2 \times 1/8 = 1/16$.

Chapter 5

1. Nutrients: vitamins, salts, growth factors; temperature; pH.
2. (a) Free trisomy 21.
   (b) An unbalanced Robertsonian 14;21 translocation.
3. Edward syndrome will have 47 chromosomes, with three copies of chromosome 18; Cri du chat syndrome will 46 chromosomes, with a deletion of part of the short arm of one chromosome 5; Turner syndrome will most commonly have 45 chromosomes with a single X sex chromosome, although other karyotypes are also possible.
4. The resulting chromosomes will be ABCDEBA with duplication of A and deletion of F and G, or GFEDCFG with duplication of F and G and deletion of A and B.

5. (a) A microdeletion is a deletion too small to detect reliably by G-banding analysis and usually needs *in situ* hybridization or molecular analysis to be visualized.
(b) A terminal deletion appears to have a single breakpoint and involves the loss of the end of the chromosome while an interstitial deletion involves two breaks resulting in loss of a segment within a chromosome arm.
6. Autosomal trisomies, of which trisomy 16 is the most common, 45,X Turner syndrome and triploidy. Approximately 50% of spontaneous abortions that occur within the first three months of pregnancy will have a chromosome abnormality.
7. A reciprocal translocation involves exchange between two arms of different chromosomes, while a Robertsonian translocation fuses two acrocentric chromosomes into a single metacentric chromosome, reducing the diploid number to 45.
8. The ISCN is 46,XX,inv(3)(p21;q26). This is a pericentric inversion with the breakpoints in each arm.
9. Risk can be assessed by taking account of the method of ascertainment (live-born abnormal offspring or miscarriages only), size of imbalance, sex of carrier, and prediction of viability of unbalanced combinations.

Chapter 6

1. A probe is a double-stranded piece of DNA complementary to the region of interest.
2. (i) Detection of microdeletions such as the 22q11deletion.
(ii) Detection of numerical changes such extra copies of the *HER2* gene in breast cancer.
(iii) Painting of structural abnormalities to identify extra material on a chromosome or extra structurally abnormal chromosomes.
3. (i) Prader–Willi syndrome, chromosome 15q11-q13, features include neonatal hypotonia, childhood obesity, hypogonadism, small hands and feet.
(ii) Wolf–Hirschhorn syndrome, 4p16.3, features include developmental delay, 'Greek warrior helmet' facies, short philtrum, cleft lip.
(iii) Williams syndrome, chromosome 7q11.23, features include elfin facies, attention deficit disorder with cocktail party manner, congenital heart defects.
4. The correct stringency is a balance of temperature and salt concentration such that all loosely matched probe DNA is washed off, leaving only the tightly bound probe. Stringency can be increased by increasing the temperature or decreasing the salt concentration. This will wash off the less well hybridized DNA, which if left will result in higher background.
5. FISH can have a resolution of 10 kb to 1 Mb. This is much higher than the 2–5 Mb resolution of conventional cytogenetics.
6. Filters are used which permit or prevent the transmission of light emitted from the fluorescent dyes used to label the FISH probes. Dual pass filters allow the detection of two different wavelengths and triple pass filters allow the detection of three different wavelengths of light.
7. Steps include comparison with parental samples – the presence in a normal individual suggests it is a polymorphism (but does not totally exclude the possibility that it is pathogenic) – and comparison with similar sequences already reported as polymorphisms in databases such as DECIPHER .

Chapter 7

1. Whole blood, CVS, cultured cells. The cells must be nucleated, as the DNA is derived from the chromosomes in the nucleus.
2. A low percentage gel. These are medium-sized fragments of DNA so a 0.8% gel would be adequate. Low percentage gels allow most sizes of fragments to travel but do not resolve small sizes well. High (2–4%) percentage gels are more suitable for small fragments as they can migrate easily and produce a crisp band.
3. Three for the normal sequence, two for the mutant sequence, as the restriction site for *Taq*1 is T*CGA.
No, as it is part of an intron.
Polymorphisms.
4. | | |
|---|---|
| AGCCACGAATTCAT | 14 bp = 1 (runs least far) |
| CGAATTCAT | 9 bp = 2 |
| AGCCT | 5 bp = 3 |
| CGAG | 4 bp = 4 (runs furthest) |
5. Southern blotting. Fragile X or myotonic dystrophy.

6. A probe which hybridizes to a locus lying close to the location of a disease gene of interest. The disadvantage is that there is a risk of recombination between the probe and the polymorphism which tracks with the disorder, such that an incorrect prediction is made. The advantage is that as long as the probe is tightly linked to a disease locus, recombination is minimal, and the exact location of the gene need not be known. Gene tracking is then possible.
7. The advantage of PCR is that very little DNA is required – usually nanograms compared with several micrograms for Southern blots. If the sequence of the exon is known, PCR primers can be designed to flank the exon, which can then be amplified in a PCR reaction.
8. Large deletions can most easily be detected by MLPA or Southern blotting. Sequencing would be too costly as a primary screen.

Chapter 8

1. Sporadic. Viruses, chemicals and radiation.
2. Growth factors are normally produced in order to initiate appropriate cell growth in a particular cell type or at a specific time in development. An oncogenic mutation may result in a permanent switching on of the growth factor gene, or other inappropriate gene expression – so there is overgrowth and uncontrolled cell proliferation.
3. The *MYC* proto-oncogene on chromosome 8 is translocated next to an active domain on chromosome 14 (the gene for immunoglobulin heavy chain production); *MYC* is therefore activated.
4. The gene for aniridia must lie near to the Wilms' tumour gene locus to produce the WAGR complex. WT is known to result in a deletion of 11p13. If a constitutional del(11p13) is found, the baby has an increased chance of a second somatic hit on the other chromosome 11 at an early age. A tumour suppressor gene is involved.
5. If MDM2 did not degrade p53, the excess p53 would cause a decrease in the amount of BCL2 and an increase in BAX, leading to apoptosis.
6. Both the *RB1* and the *APC* genes can be inherited constitutionally (i.e. they are familial). There is almost 100% certainty (especially with the polyps of FAP) that there will be a second sporadic hit, therefore these genes appear to act in a dominant manner.
7. Mismatch repair genes proof-read DNA such that normally any potential oncogenic changes are repaired. If these genes themselves are mutated, there will be an increased rate of mutation in other genes, thus increasing the risk of other mutagenic changes.

Chapter 9

1. (a) 3.2.
   (b) 5.1.
   (c) 76.3.
   The screening test carries no risk but the invasive tests (amniocentesis and chorionic villus sampling) carry a risk of losing the pregnancy of about 1%. Therefore it is important to use a method that detects most affected pregnancies (sensitivity) with the lowest number of invasive procedures, due to false positive results from the screening test. This will in turn have the lowest number of normal babies lost due to the invasive procedure.
2. (a) 1 in 1540.
   (b) 1 in 100.
   A Down syndrome child can be born to a mother at any age but the risk increases with increasing maternal age.
3. Maternal serum AFP, hCG, oestriol and maternal age.
4. (a) Multiples of the median (MOMs). A MOM value of 1 means that the values are as expected for the gestation, a value of 2 MOM is raised by twice the expected level and a value of 0.7 is 70% of the expected level. MOMs are used as a simple way to express results when the values measured vary with the gestation of pregnancy.
   (b) Specificity is the extent to which a test detects only affected individuals. False positives are unaffected individuals detected by the test. The aim of a screening test is to have high specificity with low numbers of false positives.
   (c) Sensitivity is the proportion of abnormal cases detected. False negatives are the affected cases that are missed by the screening test. The aim of a screening test is to have high sensitivity with low numbers of false negatives.
   (d) Detection rate is the number of abnormals detected. No screening programme has a 100% detection rate but methods are improving all of the time.

5. (a) matches with (v).
   (b) matches with (i).
   (c) matches with (ii).
   (d) matches with (iv).
   (e) matches with (iii).
6. - Sperm of egg donation if one parent is a carrier of a genetic disorder so it cannot be passed on to offspring.
   - Pre-implantation genetics diagnosis to screen the embryos and only replace those found not to carry the disorder.
   - Non-invasive prenatal diagnosis using cell free nucleic acid – this has limited use at present but is expected to grow in the next few years.
   - Refrain from a pregnancy or adoption – some couples cannot consider prenatal diagnosis and termination of an affected pregnancy.

Chapter 10

1. Genetic counselling is the non-directive and non-judgmental communication of information and advice about conditions where genetics is known or believed to be a part.
2. The role of genetic counselling is to:
   - establish a diagnosis
   - provide information
   - estimate risks
   - discuss methods of prevention
   - provide support to the family/individual
3. Presymptomatic or predictive testing is the use of a genetic test in an asymptomatic person to predict the likely onset of disease in the future.
4. Only you can answer – they are all individual decisions. There is lots of information available in books and on the internet. However, there is no substitute for advice from a trained healthcare professional.

Glossary of disorders

Modes of inheritance

AD = autosomal dominant; AR = autosomal recessive; XL = X-linked recessive; XD = X-linked dominant.

Methods of analysis

C = cytogenetics; F = FISH; M = molecular genetics; Q = QF–PCR; L = MLPA.

The OMIM (**O**nline **M**endelian **I**nheritance in **M**an) database is accessible at www.ncbi.nlm.nih.gov/Omim. Each Mendelian character is given a MIM number (in parentheses below each 'disorder') which can be used to access details of each genetic disease.

Disorder	Mode of inheritance	Approximate frequency (if known)	Chromosome locus/karyotype	Key clinical features	Method of analysis	Mechanism
Achondroplasia (100800)	AD	1/12 000	4p16.3	Large head, shortening of limb bones leading to dwarfism	M	Mutations in *FGFR3*
Adult polycystic kidney disease (APKD 1) (173900)	AD	1/1000	16p13	Gradual formation of renal cysts leading to loss of glomerular filtration	M	Gene mutation and conversion
Angelman syndrome (AS) (234400)		1/25 000–1/40 000	15q11–q13	Absent speech, severe retardation, inappropriate laughter, jerky movements, seizures	C, F, M	Deletion, UPD, imprinting error, mutation in *UBE3A* gene
Ataxia telangectasia (AT) (208900)	AR	1/100 000–1/300 000	11q22.3	Ataxia, immunodeficiency and cancer. Increase in chromatid damage when exposed to ionizing radiation in G_2	C, M	Mutations in the *ATM* gene result in a kinase deficiency disrupting cell cycle and signalling
Barth syndrome (302060)	XL	Rare: 1/100 000–1/300 000?	Xq28	Dilated cardiomyopathy, neutropenia, skeletal myopathy, abnormal mitochondria, elevated levels of 3-methylglutaconic acid	M	Mutations in the TAZ/G4.5 gene (300394)
Basal cell naevus syndrome (109400)	AD	1/57 000 40% new mutations	9q22–q31	Skin cancer, macrocephaly, short metacarpals, mild MR	M	Mutations in the *PTCH* TS gene
Beckwith–Wiedemann syndrome (130650)	AD	1/15 000	11p15.5	Overgrowth and large tongue, prone to Wilms' tumour	C, M	Duplication, genomic imprinting

Disorder	Mode of inheritance	Approximate frequency (if known)	Chromosome locus/karyotype	Key clinical features	Method of analysis	Mechanism
Bloom syndrome (210900)	AR	100 cases since 1950		Immunodeficiency. Acute leukaemias. Spontaneous increase in SCEs	C	Deficiency of helicase in DNA replication
Breast cancer (*HER2*) (164870)			17q21.1	Amplification of *HER2* occurs in 30% of breast cancer	F, M	Over-expression of tyrosine kinase receptor (*ERBB2*)
Burkitt lymphoma (113970)	Acquired		8q24	Cancer of the jaw in African children	C, M	Activation of *MYC* oncogene
Campomelic dysplasia (114290)	AD		17q24.3–q25.1	Bowing of long bones, large head, males have ambiguous genitalia	M	Mutations in *SOX9* or positional effect
CCHS (congenital central hypoventilation syndrome) (209880)	AD in 5–10%, rest sporadic	1/50 000 (SW England)	4p12	Idiopathic abnormal control of respiration during sleep	M	Expansion at *PHOX2B* gene (603851)
Chronic myeloid leukaemia (CML)	Acquired		t(9;22)(q34;q11)	Raised levels of myeloid cells in adults, weight loss, tired, night sweats, splenomegaly	C, F	Fusion of *ABL* oncogene on 9 with *BCR* on 22
Congenital adrenal hyperplasia (CA21H) (201910)	AR	1/5000–1/12 000	6p21.3	With 21-hydroxylase deficiency, girls have virilized genitalia due to a deficiency of cortisol and an increase in adrenocortical hormone	M	Deletions and gene conversion of *CYP21*
Cri-du-chat syndrome		1/20 000	5p15	High pitched cat-like cry, mental retardation, microcephaly, round face	C	Deletions

Disorder	Mode of inheritance	Approximate frequency (if known)	Chromosome locus/karyotype	Key clinical features	Method of analysis	Mechanism
Crouzon syndrome CFD1(123500)	AD	16.5/10^6	10q26	Premature fusion of the skull bones leading to beak-shaped nose, small jaw, hypertelorism, protruding eyes	M	Mutations in *FGFR2*
Cystic fibrosis (219700)	AR	1/2000	7q31	Defect in chloride channel leading to sticky mucus in lungs, malabsorption of fats, secretion of excessive salt in sweat	M	Mutations in the *CFTR* gene
DiGeorge syndrome (188400)		1/20 000	22q11	Cardiac defects, abnormal facies, thymic hypoplasia, cleft palate, hypocalcaemia	F	Deletion, ?position effect, haploinsufficiency
Down syndrome		1/700	Trisomy 21 (critical region 21q22.3)	Neonatal hypotonia, epicanthic folds, low set ears, single palmar crease, cardiac anomalies, low IQ	C Q	?Dosage effect
Duchenne muscular dystrophy (310200)	XLR	1/3500 males	Xp21.2	Progressive muscle wasting leading to wheelchair by early teens and death by 20 years of age	M, L	Deletions, some duplications
Edwards syndrome		1/3000	Trisomy 18	Mental retardation, round head, rockerbottom feet, clinodactyly, cardiac abnormalities	C Q	?Dosage effect
Familial adenomatous polyposis coli (FAP) (175100)	AD	1/10 000	5q21–q22	Polyps of the colon, becoming malignant	M	Tumour supressor mutations or deletions in the *APC* gene

Disorder	Mode of inheritance	Approximate frequency (if known)	Chromosome locus/karyotype	Key clinical features	Method of analysis	Mechanism
Fanconi's anaemia (FANCA-227650)	AR	1/100 000–1/300 000	Various complementation groups	Pancytopaenia. Absent radius and thumbs. Acute myeloid leukaemia. Sensitivity to alkylating agents – spontaneous increase in chromatid damage	C	Molecular basis unknown. Deficiency of excision repair?
Facioscapulo-humeral dystrophy (FSHD) (158900)	AD	1/20 000	4q35	Muscle weakness of the face, scapula, shoulders and upper arms	M	?Position effect resulting in gene silencing
Fragile X (309550)	Unusual X-linked	1/2500–1/4000	Xq27.3	Mental retardation, long face, large ears, macro-orchidism	C, M	Expansion and methylation of triplet repeat CGG
Galactosaemia (230400) (606999)	AR	1/44 000	9p13	MR, hepatomegaly, cataracts	M	Mutations in *GALT* gene galactose-1-phosphate uridyltransferase
Haemochromato-sis (235200)	AR	10% white population carriers	6p21.3	Elevated iron levels. Cirrhosis of liver, diabetes, hypermelanotic pigmentation of skin, heart failure	M	Mutation in *HFE* gene
Hereditary non-polyposis coli (HNPCC) (120435-6)	AD	5–10% of all colorectal cancer	MSH2: 2p21–p22 MLH1: 3p21.3	Colon cancer	M	Mutation or deletion of mismatch repair gene
Holoprosen-cephaly (HPE3) (142945)	AD	1/16 000–1/53 000	7q36	Developmental abnormality of the face leading to hypertelorism, midline clefting, single nostril, cyclopia	M	Mutation in *SHH*

Disorder	Mode of inheritance	Approximate frequency (if known)	Chromosome locus/karyotype	Key clinical features	Method of analysis	Mechanism
Hereditary Sensory Motor Neuropathy/ Charcot–Marie–Tooth (CMT 1A) (118220)	AD	70% of all HSMN (1/3000)	17p11.2	Motor and sensory neuropathy	M, L	Duplication of *PMP22* in 70%, increased gene dosage
Huntington disease (143100)	AD	1/10 000–1/18 000	4p16.3	Defect in the caudate nucleus of the brain leading to severe motor disturbance. Late onset	M	Expansion of triplet repeat CAG. Gain of function
Incontinentia pigmenti (IP) (308300) (308310)	XLD		?Xp11 (sporadic) Xq28 (familial)	Disturbances in skin pigmentation, abnormalities of the eye	M	Gene *NEMO* at Xq28
Klinefelter syndrome		1/1000	47,XXY	Mild learning difficulties, tall, some have gynaecomastia, infertile	C Q	?Dosage due to extra X
Marfan syndrome (154700)	AD	1/5000	15q21.1	Arachnodactyly, cardiac and eye abnormalities	M	Mutations in fibrillin gene
Multiple endocrine neoplasia type 2A (MEN 2A) (171400)	AD		10q11.2	Medullary thyroid carcinoma, pheochromocytoma, parathyroid adenomas	M	Mutation in the *RET* oncogene
Miller–Dieker lissencephaly syndrome (247200)			17p13.3	Mental retardation, smooth brain (lissencephaly), mid-face hyperplasia, small mandible, depressed nasal bridge	C, F	Deletion, haploinsufficiency

Disorder	Mode of inheritance	Approximate frequency (if known)	Chromosome locus/karyotype	Key clinical features	Method of analysis	Mechanism
Myotonic dystrophy (160900)	AD	1/8000	19q13.2–13.3	Loss of muscle tone leading to drooping eyelids and downturned mouth. Abnormal grip	M	Expansion of triplet repeat CTG
Neuroblastoma (256700)		1/10 000	2q35 (MYCN) 1p36 (LOH)	Embryonal tumour of the thoracic cavity	C, F, M	Activation of the *MYCN* oncogene, with amplification
Neurofibromatosis type 1 (NF1) (162200)	AD		17q11	Café-au-lait patches, neurofibromata	M	Mutations in 20% of cases
Pallister–Killian syndrome		Rare	iso (12p)	Severe mental retardation, facial dysmorphism, hypertelorism, sparse eyebrows. Tissue specific; tetrasomy 12p mainly in skin fibroblasts, rare in lymphocytes	C, F	Tetrasomy 12p, hence ?dosage effect
Patau syndrome		1/5000	Trisomy 13	Mental retardation, holoprosencephaly, cleft lip, polydactyly, genital and heart defects	C, Q	?Dosage effect
Phenylketonuria (PKU) (261600)	AR	1/12 000	12q24.1	Mental retardation treatable by diet, light pigmentation, ezcema, epilepsy, odd posture	M	Mutations in the phenylalanine hydroxylase (*PAH*) gene
Prader–Willi syndrome (PWS) (176270)		1/20 000	15q11–q13	Neonatal hypotonia, childhood obesity, hypogonadism, small hands and feet	C, F, M	Deletion, UPD, imprinting error

Disorder	Mode of inheritance	Approximate frequency (if known)	Chromosome locus/karyotype	Key clinical features	Method of analysis	Mechanism
Retinoblastoma (180200)	AD		13q14.1–q14.2	Embryonal tumour of the eye	C, F, M	Loss of TS gene(s)
Rett syndrome (312750)	XLD	1/10 000	Xq28	Arrested neurodevelopment at 6–18 months, regression of acquired skills, mental retardation	M	Mutations in the *MECP2* gene encoding the methyl-CpG-binding protein
Spinal Muscular Atrophy Type 1 (253300)	AR	1/10 000	5q13	Symmetrical muscle weakness and atrophy	M, L	Mutations of *SMN1* and/or *SMN2*
Triploidy			69,XXX, 69,XXY, or 69,XYY	Severe interuterine growth retardation with small trunk. Syndactyly. Molar placenta depending on karyotype	C	Dosage effect, imprinting
Turner syndrome		1/2500 of female births	45,X and variants	Neck webbing due to fetal oedema, short, primary amenorrhoea, wide spaced nipples	C	Dosage effect, haploinsufficiency
WAGR (194070)			11p13	Wilms' tumour, aniridia, gonadoblastoma, mental retardation syndrome	C, M	Deletion and position effect on contiguous gene complex
Waardenburg syndrome type 1 (193500)	AD	1–2/100 000 (2–3/100 000 deaf population)	2q35	Deafness, abnormal iris pigmentation, white forelock	M	Mutation in *PAX3* gene. Haploinsufficiency, loss of function
Williams syndrome (194050)	AD	1/10 000	7q11.23	Elfin facies, attention deficit disorder with cocktail party manner, congenital heart defects	F	Deletion of elastin gene

Disorder	Mode of inheritance	Approximate frequency (if known)	Chromosome locus/karyotype	Key clinical features	Method of analysis	Mechanism
Wolf–Hirschhorn syndrome (194190)	AD		4p16.3	Developmental delay, 'Greek warrior helmet' facies, short philtrum, cleft lip	C, F	Deletion. ?*HOX7* involved in phenotype
Xeroderma pigmentosum (*XPA*-278700)	AR	1/250 000	Various complementation groups	Photosensitivity, skin cancers, neurological disorders	M	Deficiency of excision repair

Internet resources

There are numerous sources of information on the internet relating to genetics, some of variable quality. Inclusion in the list below does not indicate that the authors agree with its content, but the list gives some ideas of where to start looking for information. Each of these sites contains further links that can be explored.

Antenatal Results and Choices (ARC) – a registered charity that provides support to parents during their pregnancy including the antenatal screening process: www.arc-uk.org.

Association of British Insurers (ABI) – represents the collective interests of the UK insurance industry. The ABI speaks out on issues of common interest; helps to inform and participate in debates on public policy issues; and also acts as an advocate for high standards of customer service in the insurance industry. The site contains free downloads on genetics and insurance: www.abi.org.uk.

Association for Clinical Cytogenetics – the professional association for Clinical Cytogeneticists working in the UK: www.cytogenetics.org.uk.

The Association of Genetic Nurses and Counsellors – an organization representing Genetic Associates, Nurses, Counsellors and other non-medical staff working within clinical genetics: www.agnc.org.uk.

Atlas of Genetics and Cytogenetics in Oncology and Haematology – a peer-reviewed online journal and database (with free access) devoted to genes, cytogenetics, and clinical entities in cancer, and cancer-prone diseases: http://atlasgeneticsoncology.org.

Bioethicsweb – searchable catalogue of internet sites and resources covering biomedical ethics: www.bioethics.ac.uk.

Bionews – digest of news and developments in human genetics and assisted reproduction and a free weekly newsletter: www.bionews.org.uk.

Breast cancer information core – An open access on-line breast cancer mutation database: http://research.nhgri.nih.gov/bic/.

British Society for Human Genetics (BSHG) – founded in 1996, the BSHG provides a forum for professionals involved in genetics as a clinical service and research. Its membership is in excess of 1750 and includes a wide spectrum of clinical, laboratory and research disciplines: www.bshg.org.uk.

Cancerbackup (merged with Macmillan Cancer Support in April 2008) – provides high quality, expertly developed information for all people affected by cancer: www.cancerbacup.org.uk.

Chromosomal Variation in Man – database consisting of a systematic collection of important citations from the world's literature reporting on all common and rare chromosomal alterations, phenotypes, and abnormalities in humans. The database is organized by variations and anomalies, numerical anomalies, and chromosomal breakage syndromes. It contains over 24 000 entries updated continuously since 1974. Access to the database is open to the public and freely available: www.wiley.com/legacy/products/subject/life/borgaonkar/access.html.

CLIMB (National Information Centre for Metabolic diseases) – providing information and support for children and families living with inherited metabolic diseases: www.climb.org.uk/.

Clinical Genetics Society (CGS) – set up in 1970 to bring together doctors and other professionals involved in the care of individuals and families with genetic disorders, with the following aims: to advance and promote the science and practice of clinical genetics, to bring together workers who have a common interest in clinical genetics, to understand, prevent, cure and alleviate conditions with a genetic aetiology, and to publish and disseminate reports, statements, and research findings: www.clingensoc.org.

Clinical Molecular Genetics Society (CMGS) – aims to advance the science of clinical molecular genetics and to further public education. Has a strong interest in promoting the quality of molecular genetic (DNA) testing in the UK through training, education, research, data-collection and quality schemes: www.cmgs.org.

Clinical Trials Unit (CTU) – a centre of excellence for clinical trials, meta-analyses and epidemiological studies, which seeks to strengthen and expand the evidence base for healthcare nationally and internationally. Coordinates a number of high-quality trials in cancer and HIV. Also undertakes research in areas such as rheumatoid arthritis, respiratory disorders, infectious diseases, and haematological disease: www.ctu.mrc.ac.uk.

Contact-a-Family – UK-wide charity providing advice, information and support to the parents of all disabled children: www.cafamily.org.uk.

DECIPHER – DatabasE of Chromosomal Imbalance and Phenotype in Humans using Ensembl Resources: https://decipher.sanger.ac.uk/information/.

Department of Health Genetics Unit – information on genetics in the UK: www.doh.gov.uk/genetics.

DNA from the beginning – includes a series of introductory genetics tutorials plus videos: www.dnaftb.org.

ENSEMBL – the Ensembl project produces genome databases for vertebrates and other eukaryotic species, and makes this information freely available online: www.ensembl.org/index.html.

European Cytogeneticists Association Register of Unbalanced Chromosome Aberrations (ECARUCA) – a database which collects and provides cytogenetic and clinical information on rare chromosomal disorders, including microdeletions and microduplications. It works to improve patient care and collaboration between genetic centres in the field of clinical cytogenetics: http://agserver01.azn.nl:8080/ecaruca/ecaruca.jsp.

Fetal Anomaly Screening Programme – provides information on fetal anomaly ultrasound screening, specifically using ultrasound scanning as a screening tool and including biochemistry for Down Syndrome: www.fetalanomaly.nhs.uk.

Frequency of Inherited Disorders Database (FIDD) – a database of epidemiology references on genetic disorders established for use in a clinical context, in medical research, for epidemiological studies and in planning for genetic services: http://archive.uwcm.ac.uk/uwcm/mg/fidd/introduction.html.

Gene Therapy Advisory Committee (GTAC) – advises on the acceptability of proposals for gene therapy research on human subjects, on ethical grounds, taking account of the scientific merits of the proposals and the potential benefits and risks: www.advisorybodies.doh.gov.uk/genetics/gtac/.

GeneCards® – a searchable, integrated database of human genes that provides concise genomic, proteomic, transcriptomic, genetic and functional information on all known and predicted human genes. Information includes orthologies, disease relationships, mutations and SNPs, gene expression, gene function, pathways, protein–protein interactions, related drugs and compounds and direct links to cutting-edge research reagents and tools such as antibodies, recombinant proteins, clones, expression assays and RNAi reagents www.genecards.org/.

GeneTests – an American publicly funded medical genetics information resource developed for physicians, other healthcare providers, and researchers. Includes useful reviews and overview documents on various inherited diseases: www.genetests.org.

GeneWatch UK – a not-for-profit group that monitors developments in genetic technologies from a public interest, environmental protection and animal welfare perspective. GeneWatch believes people should have a voice in whether or how these technologies are used and campaigns for safeguards for people, animals and the environment; works on all aspects of genetic technologies, from GM crops and foods to genetic testing of humans: www.genewatch.org.

Genetic Interest Group (GIG) – a national alliance of organizations with a membership of over 130 charities which support children, families and individuals affected by genetic disorders. Its primary goal is to promote awareness and understanding of genetic disorders so that high quality services for people affected by genetic conditions are developed and made available to all who need them: www.gig.org.uk.

Genetics and Family Medicine – Australian website created to provide GPs with online access to the resource, *Genetics in Family Medicine: the Australian Handbook for General Practitioners.* Information on a range of genetic disorders: www.gpgenetics.edu.au/.

Genetics and Insurance Committee (GAIC) – a non-statutory advisory and non-departmental public body with a UK-wide remit, including to develop and publish criteria for the evaluation of specific genetic tests, their application to particular conditions and their reliability and relevance to particular types of insurance, and to provide independent wide-ranging oversight of how insurers are using genetic tests: www.advisorybodies.doh.gov.uk/genetics/gaic/index.htm.

Genome Database – an encyclopaedia of the current state of knowledge of the human genome: www.gdb.org.

The Human Fertilisation and Embryology Authority (HFEA) – the UK's independent regulator, overseeing the use of gametes and embryos in fertility treatment and research: www.hfea.gov.uk.

Human Genetics Commission – UK government advisory body on new developments in human genetics and how they impact on individual lives: www.hgc.gov.uk.

Human Genome Organization (HUGO) – an international organization of scientists involved in human genetic and genomic research. Established in 1989 by a collection of the world's leading human geneticists, the primary ethos of HUGO is to promote and sustain international collaboration in the field of human genetics: www.hugo-international.org.

HUMGEN – a comprehensive international database on the legal, social and ethical aspects of human genetics: www.humgen.umontreal.ca/int/.

Leukaemia Research Fund – the only charity in the UK dedicated exclusively to researching blood cancers and disorders including leukaemia, Hodgkin's and other lymphomas, and myeloma. Contains useful information for patients and families: www.lrf.org.uk/.

Mendelweb – general genetics information: www.mendelweb.org/.

Metabolic Biochemistry Network – a group of specialist laboratories providing tests for the diagnosis and management of patients with inherited metabolic disorders across the UK. The Network has an assay directory to source laboratory testing services in the UK for specialist metabolites and enzymes for inherited metabolic disorders. There is also an active training and education initiative and best practice guidelines aimed to help local non-specialist laboratories and clinical teams: www.metbio.net.

Mitelman Database of Chromosome Aberrations in Cancer – part of the cancer genome anatomy projects; the data are culled from the literature and relate chromosomal aberrations to specific tumour subclasses, with molecular biology and clinical associations: http://cgap.nci.nih.gov/Chromosomes/Mitelman.

National Centre for Biotechnology (NCBI) – established in 1988 as a national resource for molecular biology information, NCBI creates public databases, conducts research in computational biology, develops software tools for analysing genome data, and disseminates biomedical information – all for the better understanding of molecular processes affecting human health and disease. Excellent genetics educational resources pages: www.ncbi.nlm.nih.gov.

National Down Syndrome Cytogenetic Register (NDSCR) – the NDSCR has accumulated over 17 000 anonymous records since it started in January 1989 and it is now probably the largest single dataset on Down syndrome and provides an opportunity to search for possible causal factors. It also enables the study of the response of parents and clinical services to the new technologies of screening and diagnosis: www.wolfson.qmul.ac.uk/ndscr/.

National Genetics Education and Development Centre – a centre to facilitate the development, provision and evaluation of educational opportunities and resources in genetics for the NHS and to raise awareness of genetics in healthcare. The website gives health professional educators and trainers access to genetics education resources. Those teaching genetics or learning about the impact of genetics knowledge on practice have access to searchable databases of resources and courses specific to their practice: www.geneticseducation.nhs.uk/.

NHS National Library for Health – includes online genetics library for patients, carers, and the general public and a specialist screening library: www.library.nhs.uk.

National Screening Committee – advises ministers and the NHS in the UK about all aspects of screening policy and supports implementation. Using research evidence, pilot programmes and economic evaluation, it assesses the evidence for programmes against a set of internationally recognized criteria. Lots of information on all aspects of screening in the UK: www.nsc.nhs.uk.

National Institutes of Health (NIH) – US site with a huge resource for genetic conditions and scientific research: www.nih.gov.

Online Mendelian Inheritance in Man (OMIM) – an online catalogue of inherited disease for specialist and professional use with a wealth of links to many other resources: www.ncbi.nlm.nih.gov.omim.

Orphanet – a database of information on rare diseases and orphan drugs. Its aim is to contribute to the improvement of the diagnosis, care and treatment of patients with rare diseases. Orphanet includes a Professional Encyclopaedia, which is expert-authored and peer-reviewed, a Patient Encyclopaedia, and a Directory of Expert Services (which includes information on relevant clinics, clinical laboratories, research activities and patient organizations): www.orpha.net.

Professional Education for Genetic Assessment and Screening (PEGASUS) – an online resource for health professionals involved in antenatal and newborn screening: www.pegasus.nhs.uk/.

Public Health Genetics Foundation – an independent policy research organization that works to achieve the timely, responsible and effective translation of biomedical science (particularly genome-based knowledge and technologies) into improved health for individuals and populations. It includes a comprehensive site with links to a variety of policy and ethics related documents on genetics: www.phgfoundation.org.

Royal College of Pathologists – the College is a professional membership organization committed to setting and maintaining professional standards and to promoting excellence in the practice of pathology. The members work in hospital laboratories, universities and industry worldwide. The College aims to advance the science and practice of pathology, to provide public education, to promote research in pathology and to disseminate the results: www.rcpath.org/.

Sanger Institute – a genome research institute primarily funded by the Wellcome Trust, it provides large-scale sequencing, informatics and analysis of genetic variation to further our understanding of gene function in health and disease and to generate data and resources of lasting value to biomedical research: www.sanger.ac.uk.

Society for the Study of Inborn errors of Metabolism (SSIEM) – the Society fosters the study of inherited metabolic disorders and related topics and it exists to promote exchange of ideas between professional workers in different disciplines who are interested in inherited metabolic disease; it is an excellent source of information and links: www.ssiem.org/.

Special Non-Invasive Advances in Fetal and Neonatal Evaluation Network (SAFE) – an EC-funded Network of Excellence which aims to implement routine non-invasive prenatal diagnosis and cost-effective neonatal screening; Unique is a partner in SAFE: www.safenoe.org/.

UK Genetics Testing Network (UKGTN) – advises the NHS on genetic testing across the UK; lists all molecular genetics tests available in

NHS genetics laboratories in the UK: www.ukgtn.nhs.uk/gtn.

UK Newborn Screening Programme Centre – information for parents and professionals on the newborn screening programme: www.newbornbloodspot.screening.nhs.uk/.

Unique – a source of information and support to families and individuals affected by any rare chromosome disorder: www.rarechromo.co.uk.

Your Genome – a general website on genetics sponsored by the Wellcome Trust: www.yourgenome.org.

Index